Dedicated to my parents
John and Charlotte Walecka

Preface

In 1984, Felix Bloch's wife, Lore, asked my advice concerning the completion of his unfinished manuscript on the "Fundamentals of Statistical Mechanics". The manuscript was based on a course Felix had offered for many years at Stanford. In the end, I wound up completing the book, which was originally published by Stanford University Press in 1989 as *Fundamentals of Statistical Mechanics: Manuscript and Notes of Felix Bloch, prepared by John Dirk Walecka.* This book was later reprinted by World Scientific Publishing Company [Walecka (2000)]. Several instructors have told me how much they enjoyed using this book.

Bloch's work was based on Gibbs' analysis at the beginning of the last century (see [Gibbs (1960)]). Gibbs starts from classical dynamics and builds from it the statistical mechanics of many-particle assemblies. The validity of the statistical methods in Gibbs' approach depends on an appropriate dynamical behavior in phase space (it holds for "quasi-ergodic coarse-grain averaging"). Quantum statistical mechanics, which came later, was introduced by Bloch through an analogous phase-space averaging. Although many examples were used to illustrate the principles, Felix always maintained that it was not a course on applications. In completing the book, I stuck close to Bloch's unfinished manuscript, and also employed a detailed set of lecture notes graciously loaned to me by Brian Serot, who had taken the course from Felix in 1976.

I had, in fact, previously taught a one-quarter, first-year graduate statistical mechanics course myself a few times at Stanford. My starting point was a complementary set of statistical assumptions due to Boltzmann. Since the counting is simpler there, the course started with quantum mechanics and then went to the classical limit. Boltzmann's approach is very powerful, for a few basic statistical hypotheses allow one to successfully analyze a truly wide variety of applications. Both approaches end up in the same place, arriving at the canonical partition function for the canonical ensemble, from which the Helmholtz free energy follows immediately. From there, the extension to the grand partition function for the grand canonical ensemble, which yields the thermodynamic potential, takes a similar path. In further contrast to the Bloch course, mine was essentially all on applications.

The course was very satisfying, and because the course was sufficiently different both in starting approach and in the variety of applications, I decided to write up my original lecture notes as a text entitled *Introduction to Statistical Mechanics*. This book is meant to *complement* the Bloch Book, and is in no sense a replacement for it. The last few times I taught the one-semester graduate course on statistical mechanics at the College of William and Mary, I integrated the material and covered both approaches.

The power of statistical mechanics is illustrated by the wide variety of applications covered in the present volume, including

- molecular spectroscopy
- paramagnetic and dielectric assemblies
- chemical equilibria
- normal modes in solids and the Debye model
- virial and cluster expansions for imperfect gases
- law of corresponding states
- quantum Bose and Fermi gases
- black-body radiation
- Bose-Einstein condensation
- Pauli spin paramagnetism
- Landau diamagnetism
- regular solutions
- order-disorder transitions in solids
- spin lattices and the Ising model
- U(1) lattice gauge theory

Furthermore, coverage is extended through the problems to include

- white-dwarf stars
- Thomas-Fermi screening in metals
- Thomas-Fermi theory of atoms
- nuclear symmetry energy
- thermal current in metals

- keratin molecules in wool
- lattice gas
- numerical Monte Carlo methods
- density functional theory
- quark-gluon plasma, *etc.*

An appendix is devoted to non-equilibrium statistical mechanics through an analysis of the Boltzmann equation and its extension to the Vlasov and Nordheim-Uehling-Uhlenbeck equations. An application to heavy-ion reactions is given there.

Over 130 problems have been included, some after each chapter and the appendix. The reader is urged to attempt as many of them as possible in order to consolidate knowledge and hone working skills. For the most part, the problems are not difficult, and the steps are clearly laid out. Those problems that require somewhat more effort are so noted.

The book assumes a knowledge of quantum mechanics at the level of [Walecka (2008)], and of classical mechanics at the level of [Fetter and Walecka (2003a)]. A knowledge of complex variables at the level of appendix A of that latter reference is also assumed,[1] as well as familiarity with multi-variable calculus. The reader is assumed to have a basic knowledge of thermodynamics; however, the first chapter of this book provides an appropriate review of that subject. Given these prerequisites, the reader should find the present volume to be self-contained.

The goal of this text is to provide the reader with a clear working knowledge of the very useful and powerful methods of statistical mechanics, and to enhance the understanding and appreciation of more detailed and advanced texts, such as [Fowler and Guggenheim (1949); Mayer and Mayer (1977); Tolman (1979); Landau and Lifshitz (1980); Ma (1985); Huang (1987); Chandler (1987); Kubo (1988); Negele and Ormond (1988); Kadanoff (2000); Davidson (2003); Fetter and Walecka (2003)].

The author is grateful to Professor Carl Garland of M.I.T. for introducing him to this beautiful subject, and he would also like to acknowledge the excellent book [Rushbrooke (1949)], from which he first studied it. Other existing introductory texts, such as [Reif (1965); Ter Haar (1966); Kittel and Kroemer (1980); Wannier (1987)], also played an important role in the author's education.

I would like to thank Dr. K. K. Phua, Executive Chairman of World

[1]The complex analysis plays a central role in the method of steepest descent used to analyze the microcanonical ensemble.

Scientific Publishing Company, and my editor Ms. Lakshmi Narayanan, for their help and support on this project. I am greatly indebted to my colleague Paolo Amore for his reading of the manuscript.

Williamsburg, Virginia
April 5, 2011

John Dirk Walecka
Governor's Distinguished CEBAF
Professor of Physics, emeritus
College of William and Mary

Contents

Preface vii

1. Introduction 1

1.1 Review of Thermodynamics 2
1.1.1 First Law 2
1.1.2 Second Law 3
1.1.3 Free Energies 7
1.1.4 Equilibrium 9
1.1.5 Third Law 12
1.2 Basic Statistical Hypotheses 12
1.2.1 Some Definitions 12
1.2.2 Statistical Assumptions 12

2. The Microcanonical Ensemble 17

2.1 Independent Localized Systems 17
2.2 The Boltzmann Distribution 21
2.3 The Partition Function 25
2.3.1 Einstein's Theory of the Specific Heat 27
2.4 Method of Steepest Descent 29
2.5 Independent Non-Localized Systems 39
2.5.1 Perfect Gas of Structureless Particles 41
2.5.2 Validity 48
2.6 Transition to Classical Dynamics 50
2.6.1 Classical Mechanics 50
2.6.2 Quantum Mechanics 51
2.6.3 Compute $\Omega(E, V, N)$ 55

3. Applications of the Microcanonical Ensemble 63
3.1 Internal Partition Function 63
3.2 Molecular Spectroscopy 64
3.2.1 Diatomic Molecules 64
3.2.1.1 Born-Oppenheimer Approximation 66
3.2.1.2 Partition Function 70
3.2.1.3 Heat Capacity 72
3.2.1.4 Symmetry of the Wave Function 73
3.2.1.5 Ortho- and Para-Hydrogen H_2 79
3.2.1.6 Typical Spectrum 80
3.2.1.7 Selection Rules 81
3.2.2 Polyatomic Molecules 85
3.2.2.1 Symmetric Top 85
3.2.2.2 Partition Function 87
3.2.2.3 Hindered Rotation 90
3.2.3 Comparison of Spectroscopic and Calorimetric Entropies 92
3.2.3.1 Sources of $k_B \ln \Omega_0$ 93
3.3 Paramagnetic and Dielectric Assemblies 96
3.3.1 Classical Gas of Permanent Dipoles 97
3.3.2 Magnetic Moments in Quantum Mechanics 101
3.3.3 Polarization in a Dielectric Medium 104
3.3.4 Paramagnetic Susceptibility 109
3.3.5 Thermodynamics 111
3.4 Chemical Equilibria 112
3.4.1 Some Preliminaries 112
3.4.2 Chemical Reactions and the Law of Mass Action . . 114
3.4.3 Chemical Potentials 118
3.4.4 Solid in Equilibrium with Its Vapor 120
3.4.5 Surface Adsorption 123

4. The Canonical Ensemble 127
4.1 Constant-Temperature Partition Function 127
4.1.1 Independent Localized Systems 130
4.1.2 Independent Non-Localized Systems 132
4.2 Classical Limit 132
4.3 Energy Distribution 135
4.4 Summary of Results So Far 137

4.4.1 Microcanonical Ensemble 138
4.4.2 Canonical Ensemble 138

5. Applications of the Canonical Ensemble 141

5.1 Solids 141
5.1.1 Einstein Model 141
5.1.2 Normal Modes 142
5.1.3 Debye Model 145
5.1.3.1 Normal-Mode Spectrum 145
5.1.3.2 Thermodynamics 148
5.1.3.3 Discussion 149
5.1.4 Improved Normal-Mode Spectrum 151
5.1.4.1 Longitudinal Waves in a Rod 151
5.1.4.2 Lattice Model 154
5.2 Imperfect Gases 158
5.2.1 Configuration Integral 159
5.2.2 Second Virial Coefficient 160
5.2.3 General Analysis of Configuration Integral 165
5.2.3.1 Linked-Cluster Expansion 165
5.2.3.2 Summation of Series 169
5.2.3.3 Interpretation 171
5.2.3.4 Virial Expansion 171
5.2.4 Law of Corresponding States 173
5.2.4.1 Derivation 176

6. The Grand Canonical Ensemble 181

6.1 Grand Partition Function 182
6.2 Relation to Previous Results 186
6.2.1 Independent Non-Localized Systems 186
6.2.2 Independent Localized Systems 187
6.2.3 Imperfect Gases 188
6.3 Fluctuations 189
6.3.1 Distribution of Energies in the Canonical Ensemble . 189
6.3.2 Distribution of Particle Numbers in the Grand Canonical Ensemble 190

7. Applications of the Grand Canonical Ensemble 195

7.1 Boltzmann Statistics 195

7.2 Quantum Statistics 196
7.2.1 Grand Partition Function 197
7.2.2 Bose Statistics 197
7.2.3 Fermi Statistics 198
7.2.4 Distribution Numbers 198
7.2.5 Energy 199
7.3 Bosons 199
7.3.1 Electromagnetic Radiation 200
7.3.1.1 Normal Modes 200
7.3.1.2 Chemical Potential 201
7.3.1.3 Spectral Weight 202
7.3.1.4 Equation of State 203
7.3.1.5 Discussion 205
7.3.2 Bose Condensation 207
7.3.2.1 Non-Relativistic Equation of State 207
7.3.2.2 Transition Temperature 209
7.3.2.3 Discontinuity in Slope of C_V 214
7.3.2.4 Liquid ^{4}He 217
7.4 Fermions 218
7.4.1 General Considerations 219
7.4.1.1 Non-Relativistic Equation of State 219
7.4.1.2 Distribution Numbers 221
7.4.1.3 Zero Temperature 222
7.4.2 Low-Temperature C_V 224
7.4.3 Pauli Spin Paramagnetism 228
7.4.3.1 Grand Partition Function 229
7.4.3.2 Magnetization 230
7.4.4 Landau Diamagnetism 233
7.4.4.1 Charged Particle in a Magnetic Field 234
7.4.4.2 Counting of States 238
7.4.4.3 Grand Partition Function and Magnetization 240
7.4.4.4 High-Temperature Limit 240
7.4.4.5 Low-Temperature Limit 242

8. Special Topics 249

8.1 Solutions 249
8.1.1 Perfect Solutions 249
8.1.1.1 Canonical Partition Function 249
8.1.1.2 Helmholtz Free Energy 251

8.1.2 Regular Solutions 254
8.1.2.1 Improved Model of Localized Systems 255
8.1.2.2 Configuration Partition Function 257
8.1.2.3 Bragg-Williams Approximation 258
8.1.2.4 Quasi-Chemical Approximation 260
8.2 Order-Disorder Transitions in Crystals 261
8.2.1 λ-Point Transitions 261
8.2.2 Configuration Partition Function 263
8.2.2.1 Bragg-Williams Approximation 264
8.2.2.2 Ising Solution for Z=2 268
8.3 The Ising Model 270
8.3.1 Heisenberg Hamiltonian 270
8.3.2 One-Dimensional Ising Model 271
8.3.2.1 Canonical Partition Function 272
8.3.2.2 Matrix Solution 272
8.3.3 Two-Dimensional Ising Model (Z=4) 277
8.3.4 Mean Field Theory 278
8.3.5 Numerical Methods 282
8.4 Lattice Gauge Theory 283
8.4.1 The Standard Model 283
8.4.1.1 Quantum Electrodynamics (QED) 284
8.4.2 Partition Function in Field Theory 284
8.4.3 $U(1)$ Lattice Gauge Theory 285
8.4.3.1 Mean Field Theory (MFT) 289
8.4.3.2 Numerical Monte Carlo 291
8.4.3.3 Strong-Coupling Limit 291
8.4.3.4 Improved Analytic Approximations 292
8.4.4 Non-Abelian Theory $SU(n)$ 295

9. Problems 297

Appendix A Non-Equilibrium Statistical Mechanics 335

A.1 Boltzmann Equation 335
A.1.1 One-Body Dynamics 335
A.1.2 Boltzmann Collision Term 337
A.1.3 Vlasov and Boltzmann Equations 340
A.1.4 Equilibrium 340
A.1.5 Molecular Dynamics 342

A.2 Nordheim-Uehling-Uhlenbeck Equation 342
A.3 Example—Heavy-Ion Reactions 343

Bibliography 345

Index 349

Chapter 1

Introduction

The basic problem in statistical mechanics is to bridge the gap between the bulk thermodynamic properties of matter and a classical, or quantum mechanical, description of the microscopic systems of which it is composed. On the one hand we have the *thermodynamic properties of the macroscopic sample*, such as

- Temperature
- Pressure
- Volume
- Energy
- Entropy
- Heat capacity, *etc.*

On the other hand, we have the *mechanical description of the microscopic systems*, such as

- Coordinates $\{\mathbf{p}, \mathbf{q}\}$ for each system
- Complete set of quantum numbers (for example, $|nlm\rangle$) for each system, *etc.*

We here focus on the *equilibrium* properties of matter. A discussion of rate processes, and the approach to equilibrium, is relegated to Appendix A.

The science of *thermodynamics* develops general relations between the thermodynamic quantities that hold for any substance. While it is assumed that the reader has some familiarity with thermodynamics, we provide a review of the relevant material in the following section.

In contrast, a complete microscopic determination of the properties of a macroscopic sample is impossible for $\sim 10^{23}$ systems. As a consequence, one is *forced* to perform some sort of microscopic average in order to compare with an experimental measurement of a thermodynamic quantity. A restatement of the basic problem in statistical mechanics is then

> *Basic Problem: To define the thermodynamic properties of a macroscopic sample in terms of the properties of the microscopic systems of which it is composed.*

Once this has been accomplished, these thermodynamic properties can then be computed for any appropriate mechanical description of the microscopic systems.

In order to establish a common background, we first present the promised review of thermodynamics.[1]

1.1 Review of Thermodynamics

1.1.1 *First Law*

It is an experimental fact that in all closed cycles with any sample

$$\oint (\underset{\text{Heat in}}{đQ} - \underset{\text{Work out}}{đW}) = 0 \qquad ; \text{ All closed cycles} \tag{1.1}$$

This is a statement of the *conservation of energy.* The first term is the total heat supplied *to* a sample during a closed cycle, and the second term is the work performed *by* the sample, for example, by its expansion, during the cycle. The notation $(đQ, đW)$ indicates that these are not exact differentials depending only on the initial and changed equilibrium states of the sample, but they may depend on just how this change was carried out (they are "path-dependent"). Now any quantity satisfying the relation in Eq. (1.1) can be written as

$$đQ - đW = dE \qquad ; \text{ state function} \tag{1.2}$$

where dE is a change in a state function, here the total energy, depending only on the initial and changed states of the sample. This is readily established, since Eq. (1.1) implies that the following quantity is *path-independent*, and hence can be used to define a quantity that depends only

[1]See, for example, [Zemansky (1968); Gibbs (1993)].

on the initial and final states[2]

$$\int_1^2 (đQ - đW) \equiv E(2) - E(1) \qquad ; \text{ independent of path} \qquad (1.3)$$

It is important to note that the first law provides a relation between heat and work for *any* process.

1.1.2 *Second Law*

The second law, again based on experiment, is a statement of just *how* one can convert heat into work. There are two equivalent versions of the second law, due to Kelvin and Clausius, respectively

(1) *Kelvin*: It is impossible to construct an engine that, operating in a cycle, will produce no effect other that extraction of heat from a reservoir and performance of an equivalent amount of work.
(2) *Clausius*: It is impossible to construct a device that, operating in a cycle, will produce no effect other than the transfer of heat from a cooler to a hotter body.

Note that the second law does *not* say that it is impossible to convert heat into work.[3] The key word in both versions is "cycle", where the sample returns to its original state.

Consider the *Carnot cycle* for a perfect gas, illustrated in Fig. 1.1.

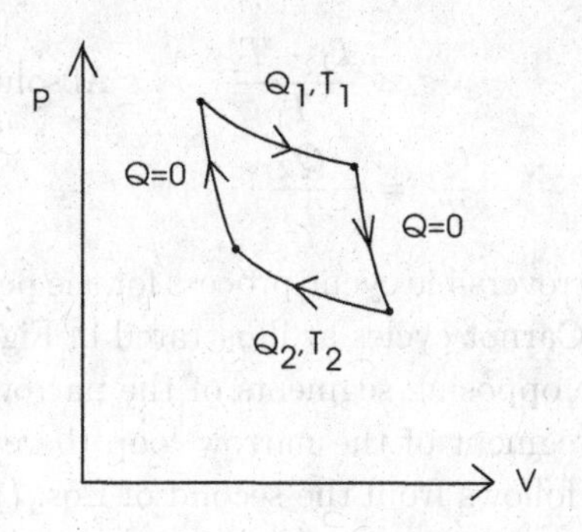

Fig. 1.1 Carnot cycle for a perfect gas.

[2]See Prob. 1.1.

[3]Take, for example, the isothermal expansion of a perfect gas which has $E(T)$; then $dE = 0$, and $đW = đQ$.

All the changes here are carried out as *reversible (quasistatic) processes.* There is an isothermal compression in contact with a heat bath at temperature T_2, with a corresponding heat flow Q_2, an adiabatic compression to the higher temperature T_1 of a second heat bath,[4] an isothermal expansion with heat transfer Q_1, and then an adiabatic expansion back to the original state at T_2. Through the use of the above statements of the second law, and consideration of engines in tandem where appropriate, one arrives at the following conclusions:

- Q_1 and Q_2 cannot both be positive or negative;
- Define the *efficiency* by (here Q_1 is positive)

$$\varepsilon \equiv \frac{W}{Q_1} = \frac{Q_1 + Q_2}{Q_1} \qquad ;\ \text{efficiency} \tag{1.4}$$

  All reversible engines operating between the same two heat reservoirs have the *same* efficiency;
- The efficiency of an irreversible engine is *less* than that of a reversible engine;[5]
- The efficiency of a reversible Carnot engine is a function only of the temperatures of the two heat baths

$$\varepsilon = f(T_1, T_2) \tag{1.5}$$

- An absolute (or Kelvin) temperature scale can then be defined according to

$$\begin{aligned} \varepsilon &\equiv \frac{T_1 - T_2}{T_1} \qquad ;\ \text{Absolute scale} \\ \Longrightarrow \qquad \frac{Q_1}{T_1} &= -\frac{Q_2}{T_2} \end{aligned} \tag{1.6}$$

- Now consider *any* reversible cyclic process for the perfect gas. Replace it by many narrow Carnot cycles as illustrated in Fig. 1.2. The adiabatic work done on the opposing segments of the narrow loops cancels. For each isothermal segment of the narrow loop there is a reversible heat flow dQ_R, and it follows from the second of Eqs. (1.6) that

$$\frac{dQ_1}{T_1} + \frac{dQ_2}{T_2} = 0 \qquad ;\ \text{each Carnot cycle} \\ \text{reversible heat flow} \tag{1.7}$$

[4] "Adiabatic" implies there is no heat flow.

[5] As an example of an *irreversible* process in thermodynamics, withdraw a thin partition confining a gas to one-half of a box, so that the gas now fills the whole box.

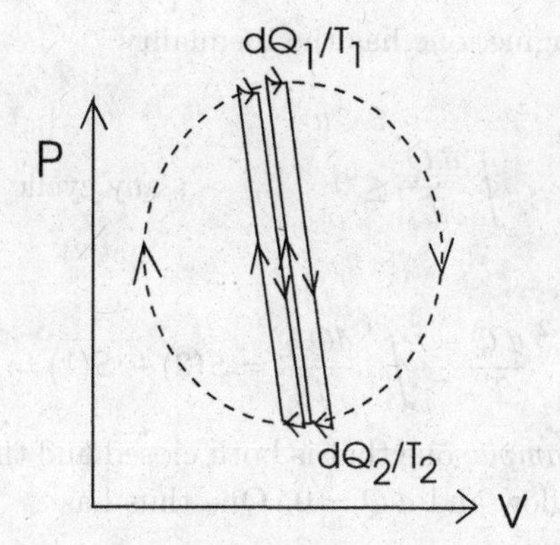

Fig. 1.2 Arbitrary reversible (quasistatic) cyclic process for a perfect gas replaced by many narrow Carnot cycles. Adiabatic work done on opposing segments of the narrow loops cancels. The reversible heat flows dQ_R during each Carnot cycle satisfy Eq. (1.7). [See Prob. 1.3].

For the entire cyclic process one therefore has

$$\oint \frac{dQ_R}{T} = 0 \tag{1.8}$$

- Exactly as in the preceding argument, one can use the expression in Eq. (1.8) to define a *state function*, the *entropy*

$$\int_1^2 \frac{dQ_R}{T} \equiv S(2) - S(1) \qquad ; \text{entropy}$$
$$\frac{dQ_R}{T} = dS \tag{1.9}$$

  The integral in the first line is *independent of the path*, and hence it can be used to define the difference in entropy for the two configurations. The integral can be evaluated by finding *any* reversible path (combination of adiabats and isotherms) between the two equilibrium configurations;
- Furthermore, since the efficiencies of all reversible engines are identical, Eq. (1.8) holds for all such engines, and the relation in Eq. (1.9) can be used to define the entropy difference for *any* substance;
- In general, since the efficiency of an irreversible engine is less than that

of a reversible engine, one has the inequality

$$\varepsilon \leq \varepsilon_R$$

$$\Longrightarrow \quad \oint \frac{đQ}{T} \leq 0 \qquad ; \text{ any cyclic process} \tag{1.10}$$

Similarly

$$\int_1^2 \frac{đQ}{T} \leq \int_1^2 \frac{dQ_R}{T} = S(2) - S(1) = \Delta S \tag{1.11}$$

- For an *isolated sample*, one that is both closed and thermally insulated, there is no heat flow and $đQ = 0$. One thus has

$$\Delta S \geq 0 \qquad ; \text{ isolated sample} \tag{1.12}$$

The entropy of an isolated sample can only increase.

As an application of these results, consider the isolated *combination* of a heat bath at temperature T and a sample in contact with this heat bath, where an amount of heat $đQ$ flows from the heat bath to the sample (Fig. 1.3).

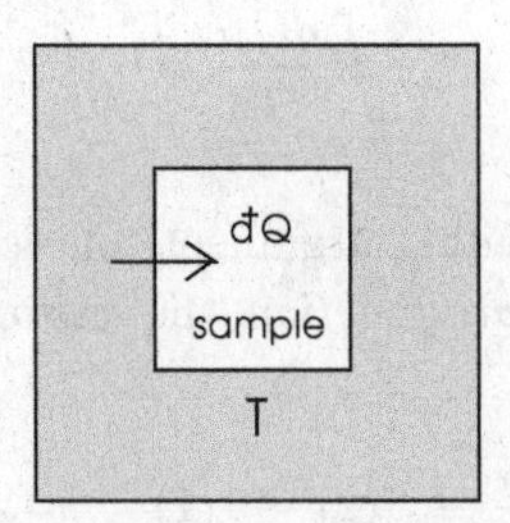

Fig. 1.3 Isolated combination of a heat bath at temperature T and a sample in contact with this heat bath, with an infinitesimal amount of heat flowing from the heat bath to the sample.

Equation (1.11) applied to the infinitesimal change for the *sample* gives

$$dS - \frac{đQ}{T} \geq 0 \qquad ; \text{ sample} \tag{1.13}$$

Now $-đQ/T$ is just the infinitesimal entropy change of the isothermal *bath*.[6] Hence Eq. (1.13) yields the following relation for the isolated combination

[6]For condensed matter with no pressure-volume work there is no path dependence, and thus for the bath $-đQ = dE = TdS$.

in Fig. 1.3

$$(dS)_{\text{sample}} + (dS)_{\text{bath}} \geq 0 \quad ; \text{ sample + bath} \tag{1.14}$$

The entropy of the combined, isolated entity can again *only increase*[7]

$$(dS)_{\text{total}} \geq 0 \tag{1.15}$$

1.1.3 *Free Energies*

Everything is now reversible (quasistatic), and for simplicity, we focus here on pressure-volume work. The work done by, and incoming heat flow to, a sample are then given by[8]

$$\begin{aligned} dW &= PdV && ; \text{ work done} \\ dQ &= TdS && ; \text{ incoming heat} \end{aligned} \tag{1.16}$$

A combination of the first and second laws then reads

$$dE = TdS - PdV \quad ; \text{ first and second laws} \tag{1.17}$$

The *Helmholtz free energy*, a state function useful at fixed volume, is defined by

$$A \equiv E - TS \quad ; \text{ Helmholtz free energy} \tag{1.18}$$

The differential of this quantity gives

$$\begin{aligned} dA &= dE - TdS - SdT \\ &= -PdV - SdT \end{aligned} \tag{1.19}$$

where Eq. (1.17) has been employed. If the quantity $A(T, V)$ has been determined by some means, then the pressure and entropy can be determined by partial differentiation

$$P = -\left(\frac{\partial A}{\partial V}\right)_T \quad ; \; S = -\left(\frac{\partial A}{\partial T}\right)_V \tag{1.20}$$

The *Gibbs free energy*, useful at fixed pressure, is defined by

$$G \equiv E + PV - TS \quad ; \text{ Gibbs free energy} \tag{1.21}$$

---

[7] Compare Prob. 1.4.
[8] See Prob. 1.6.

The differential of this quantity gives

$$\begin{aligned} dG &= dE + PdV + VdP - TdS - SdT \\ &= VdP - SdT \end{aligned} \tag{1.22}$$

If $G(T, P)$ is known, then the volume and entropy can be similarly determined from this relation

$$V = \left(\frac{\partial G}{\partial P}\right)_T \qquad ; \; S = -\left(\frac{\partial G}{\partial T}\right)_P \tag{1.23}$$

With *open samples*, and again with pressure-volume work and everything reversible, the change in energy dE is extended to include a contribution from the change in the number of constituents dn_i through their *chemical potentials* μ_i

$$dE = TdS - PdV + \mu_1 dn_1 + \cdots + \mu_m dn_m \qquad ; \text{ open samples} \tag{1.24}$$

Some comments:

- The sum can go over either *components* or *species*;
- Components are the underlying chemical constituents whose numbers can be varied independently;
- Species refers to *all* chemical constituents, whose numbers may be constrained;[9]
- In either case, the chemical potential follows as

$$\mu_i = \left(\frac{\partial E}{\partial n_i}\right)_{S,V,n_j \neq n_i} \qquad ; \text{ chemical potential} \tag{1.25}$$

The chemical potential is the *same*, whether calculated for a component or a species. We shall prove this result later when we talk about chemical reactions.

The changes in the Helmholtz and Gibbs free energies then take the following form for open samples

$$\begin{aligned} dA &= -SdT - PdV + \sum_i \mu_i dn_i \qquad ; \text{ open samples} \\ dG &= -SdT + VdP + \sum_i \mu_i dn_i \end{aligned} \tag{1.26}$$

---

[9]For example, suppose there is a chemical reaction $A + 2B \rightleftharpoons AB_2$ taking place. Then the components are the atoms of A and B, whose numbers $n_1 = n_A + n_{AB_2}$ and $n_2 = n_B + 2n_{AB_2}$ can be varied independently. The species are (A, B, AB_2), whose numbers are constrained by the previous relations.

The chemical potentials are similarly determined as

$$\mu_i = \left(\frac{\partial A}{\partial n_i}\right)_{T,V,n_j \neq n_i} = \left(\frac{\partial G}{\partial n_i}\right)_{T,P,n_j \neq n_i} \tag{1.27}$$

1.1.4 *Equilibrium*

We have seen from the second law that the entropy of an isolated sample can only increase. This observation can be re-formulated in the following fashion:

An isolated sample with given (E, V) will be in equilibrium if

$$(\delta S)_{E,V} \leq 0 \qquad ; \textit{equilibrium} \tag{1.28}$$

where δS now means any possible variation, not necessarily one leading to an equilibrium state.

One example of such a variation is illustrated in Fig. 1.4. Equation (1.28) implies that at fixed (E,V), an isolated sample in equilibrium will be one that *maximizes* its entropy.[10]

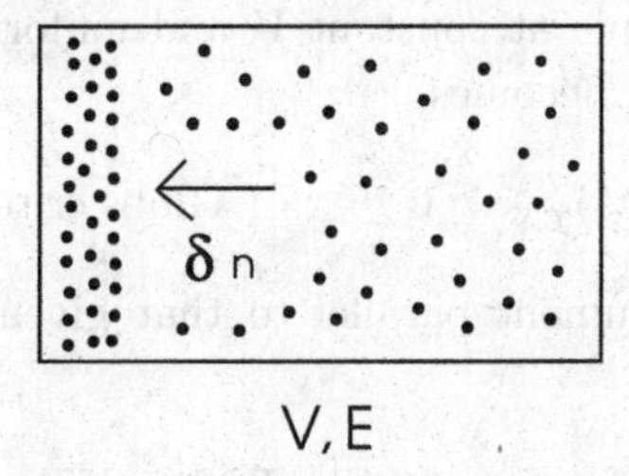

Fig. 1.4 Example of a variation $(\delta S)_{E,V}$ in an isolated sample, arising from a change in particle density, that does not lead to an equilibrium state.

Gibbs presented an alternative criterion for equilibrium

$$(\delta E)_{S,V} \geq 0 \qquad ; \text{Gibbs criterion} \tag{1.29}$$

[10]To get $(\delta S)_{E,V}$ for the sample, just subdivide it, compute dS for each piece, and then sum. The variation δS plays a role similar in spirit to the variation in classical mechanics. There, the trajectory is changed slightly from the actual dynamical path, and some integrated quantity, in that case the action, is minimized (or made stationary) to determine the actual path —this is Hamilton's principle. We thus employ three types of small elements in this work: df is the differential of a function; δf is an arbitrary variation in f; and Δf is an interval in f, which may also be small.

The equilibrium state of a sample at fixed (S,V) will be the one of *minimum* energy. This criterion can be obtained in the following fashion. The condition of the second law in Eq. (1.13) can be re-stated as[11]

$$\delta S - \frac{đQ}{T} \geq 0 \tag{1.30}$$

In a similar vein, the first law then becomes

$$\begin{aligned} \delta E &= đQ - P\delta V \\ &\leq T\delta S \qquad ; \text{ fixed } V \\ &\leq 0 \qquad ; \text{ fixed } S \end{aligned} \tag{1.31}$$

where we have used Eq. (1.30), and the fact that (S,V) are fixed. Thus

- For an *allowable* change, $(\delta E)_{S,V} \leq 0$, and the sample can lower its energy at fixed (S,V);
- At *equilibrium*, $(\delta E)_{S,V} \geq 0$, and any possible variation can only raise the energy at fixed (S,V), or leave it unchanged. The energy is *minimized* at fixed (S,V).[12]

Re-stated for a sample at constant V and uniform constant T, Gibbs criterion for equilibrium becomes

$$(\delta A)_{T,V} \geq 0 \qquad ; \text{ Gibbs criterion} \tag{1.32}$$

This follows by an argument parallel to that given above, since for an allowable transition

$$\begin{aligned} (\delta A)_{T,V} &= \delta E - T\delta S - S\delta T \\ &= đQ - T\delta S \leq 0 \end{aligned} \tag{1.33}$$

Gibbs criterion in Eq. (1.32) implies that at fixed (V,T), the equilibrium state of a sample will be one of *minimum Helmholtz free energy.*

A similar derivation gives

$$(\delta G)_{T,P} \geq 0 \qquad ; \text{ Gibbs criterion} \tag{1.34}$$

At fixed (P,T), the equilibrium state of a sample will be one of *minimum Gibbs free energy.*

---

[11] See Prob. 1.9; Eq. (1.30) and the first of Eqs. (1.31) provide the basis for our discussion of the various equilibria, which in this context implies *stable* equilbria.

[12] An illustration of these results can be found in Prob. 1.10.

As an example of the utility of these results, consider two phases $(1, 2)$ in equilibrium at fixed (T, P) as illustrated in Fig. 1.5.

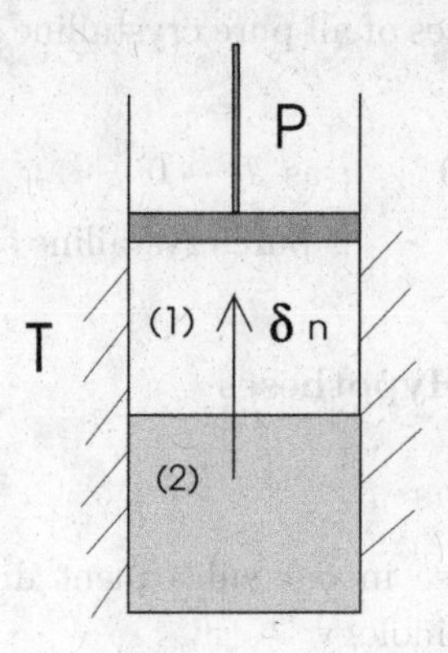

Fig. 1.5 A possible variation for two phases in equilibrium at fixed (T, P).

A possible variation is to make a change $(\delta n_1, \delta n_2)$ in the number of systems in each phase; however, since the total number of systems $n_1 + n_2$ is conserved, one has a constraint on these variations

$$\delta n_1 = -\delta n_2 \equiv \delta n \tag{1.35}$$

The Gibbs free energy is at a minimum for a sample in equilibrium, and therefore $(\delta G)_{T,P} = 0$. Hence

$$\begin{aligned}(\delta G)_{T,P} &= \mu_1 \delta n_1 + \mu_2 \delta n_2 \\ &= (\mu_1 - \mu_2)\delta n = 0 \end{aligned} \tag{1.36}$$

Thus, in equilibrium at fixed (T, P), the chemical potentials must be the same in the two phases

$$\mu_1 = \mu_2 \quad ; \text{ two phases in equilibrium} \tag{1.37}$$

If this were *not* true, then one could lower the Gibbs free energy by transferring systems from one phase to another.[13]

[13]That is, with an appropriate sign for δn, one could make $(\delta G)_{T,P} < 0$.

1.1.5 *Third Law*

The discussion so far has only dealt with *changes* in entropy. The entropy scale is set by the third law of thermodynamics, also known as the Nernst heat theorem. The entropies of all pure crystalline solids *vanish* at absolute zero

$$S \to 0 \qquad ; \text{ as } T \to 0 \qquad \text{pure crystalline solid} \tag{1.38}$$

1.2 Basic Statistical Hypotheses

1.2.1 *Some Definitions*

To be a little more precise in our subsequent discussion, we henceforth employ the following terminology:[14]

- The label *system* will now be reserved for the microscopic units, the *particles*;
- A *thermodynamic sample*, consisting of a collection of particles, will be referred to as an *assembly*. It is here characterized by (E, V, N).
- A collection of a given *distribution* of systems or assemblies will be said to form an *ensemble*.[15]

1.2.2 *Statistical Assumptions*

We must now face the question, "How do we carry out an average over the microscopic quantities to arrive at the thermodynamic variables?" There are two possible approaches:

First Way: Take the observed value of a dynamical quantity $f(t)$ to be its *time average over some time interval* τ

$$f_{\text{obs}} = \frac{1}{\tau}\int_t^{t+\tau} f(t)dt \tag{1.39}$$

In equilibrium, this property should not vary with time, and hence one can equally well take

[14] The terminology follows [Rushbrooke (1949)]; this book, when augmented with the method of steepest descent, provides the basis for our discussion of classical statistics.

[15] We will subsequently be interested in three of these — the microcanonical, canonical, and grand canonical ensembles.

$$f_{\text{obs}} = \text{Lim}_{\tau \to \infty} \frac{1}{\tau} \int_t^{t+\tau} f(t)dt \tag{1.40}$$

One needs to know the dynamical development $f(t)$ to evaluate this expression. We shall return to this shortly.

Second Way: Imagine taking a moving picture of an assembly. In each frame, we have the positions of the particles and their assigned momenta (Fig. 1.6).

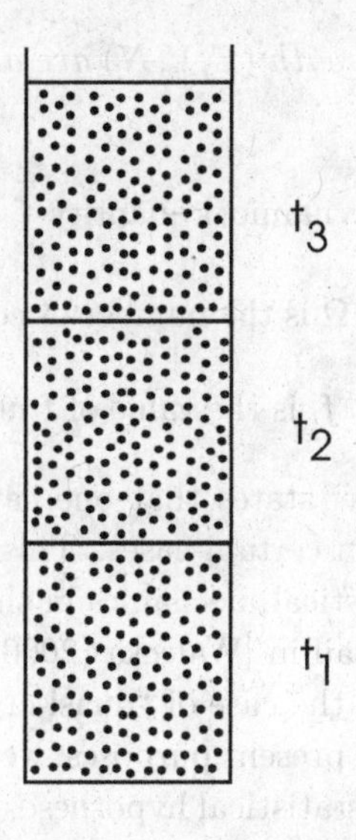

Fig. 1.6 Moving picture of an assembly, showing the position of the particles, where each particle is also assigned a momentum. Each frame constitutes a *complexion*.

Several comments:

- Over an infinite time, one would expect to see all possible configurations of particles and momenta consistent with a given (E, V, N);
- Call one of these pictures a *complexion*;
- Each distinct set of ways of describing an assembly is a *complexion* of the assembly; for example, as above, one distribution of momenta and coordinates $\{\mathbf{p}_i, \mathbf{q}_i\}$ within a given (E, V, N) forms one complexion;
- Within a quantum theory, a complexion can be a set of quantum numbers for each system, for example, a set of vibrational quantum numbers $\{n_i\}$ for a collection of simple harmonic oscillators (Fig. 1.7).

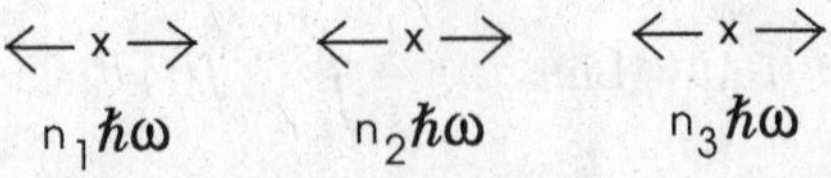

$\leftarrow \times \rightarrow \quad \leftarrow \times \rightarrow \quad \leftarrow \times \rightarrow$

$n_1\hbar\omega \quad n_2\hbar\omega \quad n_3\hbar\omega$

Fig. 1.7 A set of vibrational quantum numbers $\{n_i\}$ for a collection of simple harmonic oscillators.

We make *two basic statistical assumptions*:

Assumption I:

All complexions consistent with (E, V, N) *are a priori equally probable.*

$$(1.41)$$

The observed value of the dynamical quantity f is then given by

$$f_{\text{obs}} = \frac{1}{\Omega}\sum_i f_i \qquad ; \ \Omega \text{ is the number of complexions} \qquad (1.42)$$

$$; \ f_i \text{ is the value of f in the } ith \text{ complexion}$$

The "ergodic hypothesis" states that one can relate the time average to this average over states in certain cases. This is the basis of J. Willard Gibbs' development of statistical mechanics in the early 1900's (see [Gibbs (1960)]), as discussed in detail in [Walecka (2000)]. In the latter reference, the validity is established in the case of "quasi-ergodic coarse-grain averaging" in phase space. For our present purposes, we will take the assumptions in Eqs. (1.41) and (1.46) as statistical hypotheses fully verified by the many successful experimental applications.

Note that in applying any statistical approach, it is assumed that the assembly is not in a dynamical configuration that is *inconsistent* with other dynamical configurations or complexions, such as the one illustrated in Fig. 1.8, where non-interacting particles would simply continue to bounce

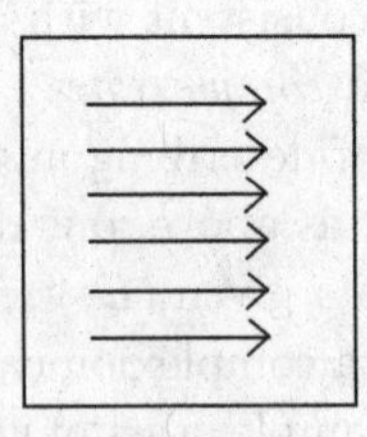

Fig. 1.8 A configuration of non-interacting particles in a box that is inconsistent with other dynamical configurations or complexions.

back and forth between the walls. It is evident that this is not a configuration that can be prepared experimentally, and that any slight perturbation would destroy it.

To make a connection with the thermodynamic functions, we go through the *entropy*. We note the following features of the entropy:

(1) It is a measure of *randomness*, increasing with increasing randomness as illustrated by the following examples:

- Expansion of gases (see Prob. 1.4);
- Mixing of gases;
- Evaporation of liquids;
- Heating of an assembly.

The first three examples illustrate an increasing randomness in the position of a system, while the fourth illustrates an increasing randomness in the energy of a system, as the energy distribution spreads with heating. Hence we observe that the entropy is an increasing function of the number of complexions

$$S \to + \qquad ; \text{ as } \Omega \to + \tag{1.43}$$

(2) The entropy is an *extensive* quantity, that is, the entropy of two assemblies is the *sum* of the entropies. In contrast, the number of complexions is the *product* of the number of complexions for each, since for each complexion in the second assembly, there is one in the first (Fig. 1.9). Hence

$$S = S_1 + S_2 \qquad ; \text{ while } \Omega = \Omega_1\Omega_2 \tag{1.44}$$

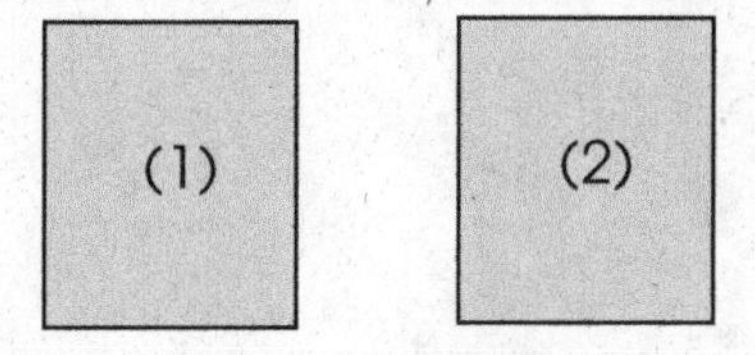

Fig. 1.9 The total entropy of two assemblies is the sum of the entropies, while the total number of complexions is the product of the number of complexions for each.

(3) At absolute zero, there is only one complexion, since everything is in the ground state of the assembly (assumed non-degenerate), while the

third law says the entropy vanishes there. Thus[16]

$$S = 0 \qquad ; \text{ when } \Omega = 1 \tag{1.45}$$

These observations suggest that $S = k \ln \Omega$, where k is a constant. Thus our second statistical assumption is

Assumption II:

$$S = k_{\rm B} \ln \Omega \qquad ; k_{\rm B} \text{ is Boltzmann's constant} \tag{1.46}$$

Here $k_{\rm B}$ is Boltzmann's constant.

Assumptions I,II are due to Ludwig Boltzmann in the late 1800's, in a series of papers in the journal *Wiener Berichte*.[17] Experiment has repeatedly established their validity, and we shall see just how powerful they are.

Note that Eq. (1.46) is a truly remarkable expression, relating a rather exotic thermodynamic variable, arrived at through consideration of the efficiency of reversible heat engines, to what is basically a counting problem, determining the number of complexions of a macroscopic assembly of microscopic systems at a given total (E, V, N)!

---

[16] This is actually a (valid) quantum argument—we shall return later to the issue of the absolute entropy in the classical case.

[17] See also [Boltzmann (2011)].

Chapter 2

The Microcanonical Ensemble

We proceed to evaluate the total number of complexions Ω, and the entropy S, in the simplest case of identical, independent, localized systems.

2.1 Independent Localized Systems

Make the following two assumptions (see Fig. 2.1):

- Assume that the N systems are *localized* in a lattice. Assume that they may oscillate about their mean positions, but that those mean positions are *distinguishable* and *fixed*. Label the positions by $\{1, 2, \cdots, N\}$;
- Assume that the systems are *independent*, so that the state of one of them does not depend on the others, except for the constraint that the overall energy E of the assembly is fixed.

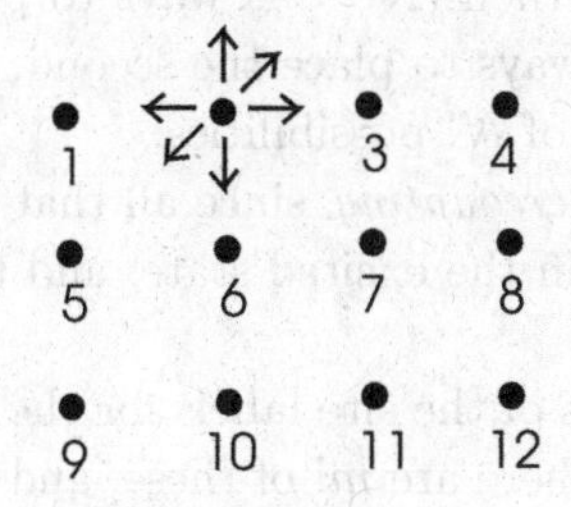

Fig. 2.1 An assembly of independent localized systems.

We start with *quantum mechanics*, since the counting and results are much simpler in this case. We shall return later to the transition to classical dynamics. There is then a set of energy levels $\{\varepsilon_1, \varepsilon_2, \cdots\}$ available to each independent, localized system.

To illustrate the idea, we start with an even simpler situation, where there are only *two* levels $\{\varepsilon_1, \varepsilon_2\}$ available to each system.[1] They are separated by $\varepsilon = \varepsilon_2 - \varepsilon_1$. Assume that $N\varepsilon \ll E$, where E is the total energy of the assembly. In this case, the excitation energy of any configuration is immaterial, and the subsidiary condition on the total energy can be ignored.[2]

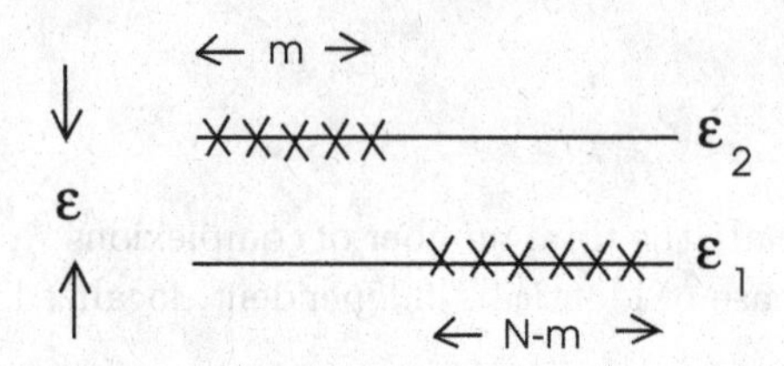

Fig. 2.2 Complexions with m excited systems and $N - m$ unexcited systems.

First, compute the number of complexions where m of the systems are excited and $N - m$ of the systems are in the ground state (Fig. 2.2). The basic counting problem is then to determine *the number of ways one can distribute m excitations on N sites.*

- Start laying down the configuration in Fig. 2.2 on the N labeled lattice sites in Fig. 2.1, starting from the left in Fig. 2.2. There are N ways to select the first site on which to place the first excitation, then $N-1$ ways to select the second, and $(N - m + 1)$ ways to place the last excitation. There are then $(N - m)$ ways to place the first unexcited system, $(N - m - 1)$ ways to place the second, and 1 way to place the last, for a grand total of $N!$ possibilities;
- However, this is an *overcounting*, since all that matters is that a given m of the systems are in the excited state, and the remaining $(N - m)$ are not;
- The $m!$ redistributions of the site labels for the excitations refer to the *same configuration.* There are $m!$ of these, and we must divide by this number;
- Similarly, we must divide by the number $(N-m)!$ of the site labels for

[1]This actually describes the physical situation of the contribution to the entropy of a nuclear spin 1/2 (compare Prob. 1.5).

[2]Let $\varepsilon \to 0$, for example.

the unexcited systems.

- Thus, the number of complexions where there are m excited systems on N sites is[3]

$$\Omega(N, m) = \frac{N!}{m!(N-m)!} \qquad ; \ m \text{ excitations on } N \text{ sites} \qquad (2.1)$$

The total number of complexions, with no constraint on the excitation energy ($N\varepsilon \ll E$), is then obtained by summing over all possible m

$$\Omega = \sum_{m=0}^{N} \frac{N!}{m!(N-m)!} \qquad ; \text{ number of complexions} \quad N\varepsilon \ll E \qquad (2.2)$$

Now observe that the sum in Eq. (2.2) can be *done exactly!* Recall the *binomial theorem*

$$(x_1 + x_2)^N = \sum_{m=0}^{N} \frac{N!}{m!(N-m)!} x_1^m x_2^{N-m} \qquad ; \text{ binomial theorem} \qquad (2.3)$$

Just set $x_1 = x_2 = 1$ in this expression, to obtain

$$\Omega = \sum_{m=0}^{N} \frac{N!}{m!(N-m)!} = 2^N \qquad (2.4)$$

The answer is obvious here, since each of the systems on each of the N sites can exist in one of two states. The entropy of this assembly then follows from Eq. (1.46) as

$$S = k_{\rm B} \ln \Omega = N k_{\rm B} \ln 2 \qquad (2.5)$$

We now note the following interesting (and quite unexpected!) feature of this result. Suppose we were to just pick out the *largest term* in the sum in Eq. (2.2), corresponding to the *most probable* distribution. To find this term, we want to maximize

$$t(m) = \frac{N!}{m!(N-m)!} \qquad (2.6)$$

Since $\ln t(m)$ is an increasing function of $t(m)$, we could just as well maximize $\ln t(m)$. This makes life a little easier, for then *Stirling's formula* can

[3]See Prob. 2.1. Readers should spend some time becoming familiar and comfortable with this counting argument, since it lies at the heart of much of what we shall do.

be employed for large N[4]

$$\ln N! = N \ln N - N + O(\ln N) \qquad ; \text{Stirling's formula} \tag{2.7}$$

Assume this result can also be used for $\ln m!$ and $\ln (N-m)!$ (the validity of this can be verified at the end). The problem then is to maximize

$$\begin{aligned}\ln t(m) &= \ln N! - \ln m! - \ln (N-m)! \\ &\approx N \ln N - N - m \ln m + m - (N-m) \ln (N-m) + (N-m) \\ &= N \ln N - m \ln m - (N-m) \ln (N-m)\end{aligned} \tag{2.8}$$

To find the maximum, set the derivative with respect to m equal to zero

$$\frac{\partial}{\partial m} \ln t(m) = -\ln m - 1 + \ln (N-m) + 1 = 0 \tag{2.9}$$

The solution to this equation gives the most probable value $m^\star$ as

$$m^\star = \frac{N}{2} \tag{2.10}$$

This answer is again obvious, since the most probable configuration here is to have half of the systems in the excited state. Insertion of this result in Eq. (2.6) yields

$$\begin{aligned}\ln t(m^\star) &= \ln \left[\frac{N!}{(N/2)!(N/2)!}\right] \approx N \ln N - N - 2\left(\frac{N}{2} \ln \frac{N}{2} - \frac{N}{2}\right) \\ &= N \ln 2\end{aligned} \tag{2.11}$$

The entropy arising from this largest term is then

$$S = k_{\rm B} \ln t(m^\star) = N k_{\rm B} \ln 2 \tag{2.12}$$

The largest term in the sum gives exactly the same answer!

How can this be? The answer is that we are interested in very large N, and we are dealing with logarithms. Since each term in the sum in Eq. (2.2) is positive, one has the evident inequalities

$$t(m^\star) \leq \Omega \leq (N+1) t(m^\star) \tag{2.13}$$

It follows that

$$\ln t(m^\star) \leq \ln \Omega \leq \ln t(m^\star) + \ln (N+1) \tag{2.14}$$

[4]We shall later derive Stirling's formula [see Eq. (4.41) and Prob. 4.1.].

Now $\ln t(m^\star)$ is given in Eq. (2.11), it is proportional to N, and we are interested in the regime where

$$\ln N \ll N \tag{2.15}$$

For example, with $N \sim 10^{23}$, one has $\ln N \sim 53$. Hence the $\ln(N+1)$ on the r.h.s. of Eq. (2.14) is *completely negligible.*[5] The reader is urged to spend some time pondering these numbers, which lie far outside the realm of everyday experience, and which form one of the keys to the success of statistical mechanics.

2.2 The Boltzmann Distribution

Consider next the problem where each system has the full set of energy levels $\{\varepsilon_1, \varepsilon_2, \cdots\}$ available to it. First focus on those complexions where the energy levels are occupied by $\{n_1, n_2, \cdots\}$ systems, respectively; these are known as the *distribution numbers* $\{n_i\}$ (see Fig. 2.3).

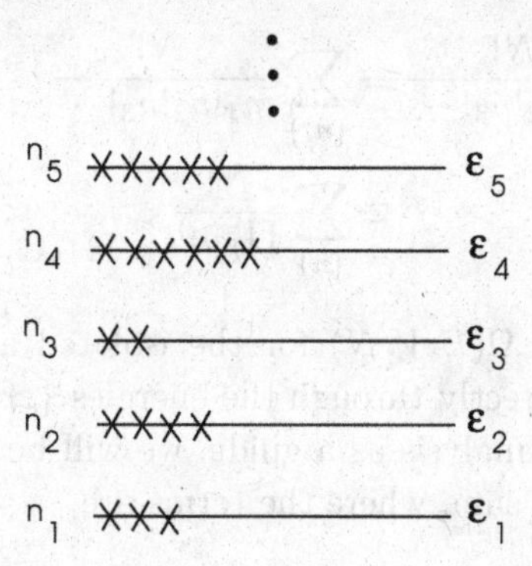

Fig. 2.3 One set of distribution numbers $\{n_1, n_2, n_3, \cdots\} \equiv \{n_i\}$.

The constraints that both the total number of systems N and total energy E of the assembly are fixed must now be explicitly taken into account

$$N = \sum_i n_i \quad ; \; E = \sum_i n_i \varepsilon_i \qquad ; \text{fixed} \tag{2.16}$$

[5]Note that $\ln(N+1) = \ln N + O(1/N)$. Note also that $\ln N$ is the same order as the terms neglected in the use of Stirling's formula.

It is a straightforward generalization of the previous counting argument to observe that the total number of complexions with this configuration is

$$\Omega(N; n_1, n_2, \cdots) = \frac{N!}{n_1!n_2!n_3!\cdots} \qquad ; \ \{n_i\} \text{ excitations on } N \text{ sites} \tag{2.17}$$

Since $0! = 1! = 1$, there is no difficulty with the unoccupied, or singly-occupied, levels in this expression.

The total number of complexions of the assembly at fixed (E, V) is then obtained by summing over all possible sets of distribution numbers $\{n_i\}$, subject to the constraints in Eq. (2.16)

$$\Omega = \sum_{\text{all sets of } \{n_i\}} \frac{N!}{n_1!n_2!n_3!\cdots}$$
$$\sum_i n_i = N \qquad ; \ \sum_i n_i\varepsilon_i = E \tag{2.18}$$

We will frequently employ the shorthand notation

$$\sum_{\text{all sets of } \{n_i\}} \frac{N!}{n_1!n_2!n_3!\cdots} \equiv \sum_{\{n_i\}} \frac{N!}{n_1!n_2!n_3!\cdots}$$
$$\equiv \sum_{\{n\}} \frac{N!}{\prod_i n_i!} \qquad ; \text{ shorthand} \tag{2.19}$$

Equations (2.18) yield $\Omega(E, V, N)$ for the isolated assembly, the volume here entering only indirectly through the energies $\{\varepsilon_i\}$.

With the previous analysis as a guide, we will be content with finding the largest term in the sum, where the terms are

$$t(n) = \frac{N!}{\prod_i n_i!} \tag{2.20}$$

The desired $\ln \Omega$ will then be evaluated as

$$\ln \Omega = \ln t(n^\star) \tag{2.21}$$

Here n stands for the set of distribution numbers $\{n_i\}$, and $n^\star$ for their most probable values, the ones that maximize $t(n)$.

Two observations:

- As before, we could just as well maximize $\ln t(n)$;

- Stirling's formula can then be employed to simplify the resulting expressions. We shall use Stirling's formula both for $\ln N!$ and for the $\ln n_i!$.[6]

The maximization of $\ln t(n)$, subject to the two constraints in Eq. (2.16), is a well-defined mathematical problem. To solve it, one employs *Lagrange's method of undetermined multipliers.* Let each distribution number vary by a small amount δn_i. Then

$$\begin{aligned}\delta \ln t(n) &= \frac{\partial \ln t(n)}{\partial n_1}\delta n_1 + \frac{\partial \ln t(n)}{\partial n_2}\delta n_2 + \frac{\partial \ln t(n)}{\partial n_3}\delta n_3 + \cdots \\ &\delta n_1 + \delta n_2 + \delta n_3 + \cdots = 0 \\ &\varepsilon_1 \delta n_1 + \varepsilon_2 \delta n_2 + \varepsilon_3 \delta n_3 + \cdots = 0\end{aligned} \tag{2.22}$$

Multiply the second equation by a constant α, the third by a constant β, and then add these to the first

$$\begin{aligned}\delta \ln t(n) = &\left(\frac{\partial \ln t(n)}{\partial n_1} + \alpha + \beta\varepsilon_1\right)\delta n_1 + \left(\frac{\partial \ln t(n)}{\partial n_2} + \alpha + \beta\varepsilon_2\right)\delta n_2 \\ &+ \left(\frac{\partial \ln t(n)}{\partial n_3} + \alpha + \beta\varepsilon_3\right)\delta n_3 + \cdots\end{aligned} \tag{2.23}$$

Now

- Choose α so that the coefficient of δn_1 vanishes;
- Then choose β so that the coefficient of δn_2 vanishes;
- The remaining δn_i, with $i \geq 3$, are now *independent, unconstrained* variations;
- At the maximum of $\ln t(n)$, $\delta \ln t(n)$ must vanish for all these independent variations. If not, one could make $\delta \ln t(n) > 0$, which violates the condition of a maximum.

Hence, the conditions for the maximization of the logarithm of the expression in Eq. (2.20), subject to the constraints in Eq. (2.16), are

$$\frac{\partial \ln t(n)}{\partial n_i} + \alpha + \beta\varepsilon_i = 0 \qquad ; \text{ all } i \tag{2.24}$$

Here (α, β) are constants, determined at the end so that the constraints are satisfied, and everything now stands on the same footing in Eq. (2.24).

[6] In the next section, we shall find a way to actually carry out the sum in Eqs. (2.18) for large N, which validates the present approach. It turns out here that $n_i^\star \propto N$ for any (β, ε_i), and that is why Stirling's approximation works for the $\ln n_i!$.

The partial derivative is now readily evaluated as

$$\begin{aligned}\frac{\partial \ln t}{\partial n_i} &= -\frac{\partial}{\partial n_i}\ln n_i! = -\frac{\partial}{\partial n_i}(n_i \ln n_i - n_i) \\ &= -\ln n_i \end{aligned} \tag{2.25}$$

Hence, when written in terms of the solution $n_i^\star$, Eq. (2.24) becomes

$$\begin{aligned} \ln n_i^\star &= \alpha + \beta\varepsilon_i \\ \Longrightarrow \qquad n_i^\star &= e^{\alpha}e^{\beta\varepsilon_i} \qquad ; \text{ all } i \end{aligned} \tag{2.26}$$

The constants (α, β) are to be chosen to satisfy

$$\begin{aligned} N &= \sum_i n_i^\star = e^{\alpha}\sum_i e^{\beta\varepsilon_i} \\ E &= \sum_i \varepsilon_i n_i^\star = e^{\alpha}\sum_i \varepsilon_i e^{\beta\varepsilon_i} \end{aligned} \tag{2.27}$$

Our basic statistical hypothesis in Eq. (1.46) now yields the *entropy* of the assembly as

$$\begin{aligned} S &= k_{\mathrm{B}} \ln \Omega = k_{\mathrm{B}} \ln t(n^\star) \\ &= k_{\mathrm{B}}\left[N \ln N - N - \sum_i n_i^\star \ln n_i^\star + \sum_i n_i^\star\right] \end{aligned} \tag{2.28}$$

where $\ln t(n^\star)$ follows from Eq. (2.20) through the use of Stirling's formula. Equations (2.26) and (2.27) can now be employed to give

$$\begin{aligned} S &= k_{\mathrm{B}}\left[N \ln N - \sum_i n_i^\star(\alpha + \beta\varepsilon_i)\right] \\ &= k_{\mathrm{B}}\left[N \ln N - \alpha N - \beta E\right] \end{aligned} \tag{2.29}$$

The logarithm of the first of Eqs. (2.27) yields

$$\alpha = \ln N - \ln\left(\sum_i e^{\beta\varepsilon_i}\right) \tag{2.30}$$

Substitution of this expression into Eq. (2.29) then gives

$$S = k_{\mathrm{B}}\left[N \ln\left(\sum_i e^{\beta\varepsilon_i}\right) - \beta E\right] \tag{2.31}$$

This is our basic result for the entropy of an assembly of identical, independent, localized systems. The most probable occupation numbers in Eq. (2.26) are said to form a *Boltzmann distribution.*

2.3 The Partition Function

In principle, the expression for the mean energy

$$\frac{E}{N} = \frac{\sum_i \varepsilon_i n_i^\star}{\sum_i n_i^\star} = \frac{\sum_i \varepsilon_i e^{\beta\varepsilon_i}}{\sum_i e^{\beta\varepsilon_i}} \tag{2.32}$$

determines $\beta(E, V, N)$, although the inversion is non-trivial.

At this point, *thermodynamic* arguments are of great assistance. The first and second laws for a closed assembly, summarized in Eq. (1.17), state that

$$dE = TdS - PdV \tag{2.33}$$

It follows from this relation that[7]

$$\left(\frac{\partial S}{\partial E}\right)_{V,N} = \frac{1}{T} \tag{2.34}$$

This derivative can be evaluated from Eq. (2.31) and thereby introduces the *absolute temperature* T into the analysis. Constant volume V implies that the ε_i are fixed, and if we note that there is an implicit dependence $\beta(E)$ arising from the subsidiary condition in Eq. (2.32), then

$$\left(\frac{\partial S}{\partial E}\right)_{V,N} = -k_B\beta + k_B\frac{\partial}{\partial\beta}\left[N\ln\left(\sum_i e^{\beta\varepsilon_i}\right) - \beta E\right]\frac{\partial\beta}{\partial E} \tag{2.35}$$

The partial derivative with respect to β of the term in square brackets *vanishes*, since

$$\begin{aligned}\frac{\partial}{\partial\beta}\left[N\ln\left(\sum_i e^{\beta\varepsilon_i}\right) - \beta E\right] &= N\frac{\partial/\partial\beta\left(\sum_i e^{\beta\varepsilon_i}\right)}{\sum_i e^{\beta\varepsilon_i}} - E \\ &= N\frac{\sum_i \varepsilon_i e^{\beta\varepsilon_i}}{\sum_i e^{\beta\varepsilon_i}} - E = 0\end{aligned} \tag{2.36}$$

[7]In analogy to Eqs. (1.20) and (1.23), if the equation of state is known in the form $E = E(S, V, N)$, then the temperature is determined by $T = (\partial E/\partial S)_{V,N}$.

where the last relation follows from Eq. (2.32). Hence Eqs. (2.34) and (2.35) reduce to the simple form

$$\left(\frac{\partial S}{\partial E}\right)_{V,N} = \frac{1}{T} = -k_{\rm B}\beta$$
$$\Longrightarrow \qquad \beta = -\frac{1}{k_{\rm B}T} \tag{2.37}$$

This relation provides the *statistical definition of temperature.*

If the result in Eq. (2.37) is substituted into Eq. (2.31), then the *Helmholtz free energy* is immediately determined as

$$A = E - TS = -Nk_{\rm B}T\ln\left(\sum_i e^{-\varepsilon_i/k_{\rm B}T}\right) \tag{2.38}$$

This result is rewritten as

$$A = -Nk_{\rm B}T\ln\text{(p.f.)} \qquad ;\ \text{Helmholtz free energy}$$
$$\text{(p.f.)} \equiv \sum_i e^{-\varepsilon_i/k_{\rm B}T} \qquad ;\ \text{partition function} \tag{2.39}$$

The second line defines the *partition function*, which plays a central role in the subsequent analysis. Note that the first line gives the Helmholtz free energy in the form $A(T, V, N)$,[8] which is exactly what is required in Eqs. (1.19) and (1.20)!

Several comments:

- Equations (2.39) form our principal result for identical, independent, localized systems;
- All thermodynamic properties now follow from the partition function;
- The sum in the partition function goes over *the energy levels of a single system*;
- If there is *degeneracy* in the single-particle spectrum, just let the ε_i become equal. Each ε_i will occur ω_i times, and

$$\text{(p.f.)} = \sum_i \omega_i e^{-\varepsilon_i/k_{\rm B}T} \qquad ;\ \text{with degeneracy } \omega_i \tag{2.40}$$

[8] Again, the volume dependence enters here through the $\{\varepsilon_i\}$.

- The quantities $(n_i^\star, N, E)$ are now given by Eqs. (2.26) and (2.27) as

$$
\begin{aligned}
n_i^\star &= e^{\alpha} e^{-\varepsilon_i/k_{\rm B}T} \\
N &= \sum_i n_i^\star = e^{\alpha} \sum_i e^{-\varepsilon_i/k_{\rm B}T} \\
E &= \sum_i \varepsilon_i n_i^\star = e^{\alpha} \sum_i \varepsilon_i \, e^{-\varepsilon_i/k_{\rm B}T}
\end{aligned}
\tag{2.41}
$$

- The fractional occupation of each level follows from Eqs. (2.41) as

$$
\frac{n_i^\star}{N} = \frac{e^{-\varepsilon_i/k_{\rm B}T}}{\sum_i e^{-\varepsilon_i/k_{\rm B}T}} \tag{2.42}
$$

  This is the reason for the name *partition function*;
- A collection of systems distributed over the energy levels in this fashion is said to form a *microcanonical ensemble.*
- The *ratio* of occupation numbers in this ensemble is

$$
\frac{n_i^\star}{n_j^\star} = \frac{e^{-\varepsilon_i/k_{\rm B}T}}{e^{-\varepsilon_j/k_{\rm B}T}} \tag{2.43}
$$

- The mean energy per particle in the ensemble is given by Eqs. (2.41) as

$$
\frac{E}{N} = \frac{\sum_i \varepsilon_i \, e^{-\varepsilon_i/k_{\rm B}T}}{\sum_i e^{-\varepsilon_i/k_{\rm B}T}} \tag{2.44}
$$

2.3.1 *Einstein's Theory of the Specific Heat*

As a first application, consider Einstein's theory of the specific heat of solids [Einstein (1907)]. Imagine that a crystal is made up of $3N$, one-dimensional, simple harmonic oscillators (Fig. 2.1). Assume further that there is a single oscillator frequency ν, with a corresponding energy spectrum for each oscillator of

$$
\varepsilon = h\nu(n + 1/2) \qquad ; \; n = 0, 1, 2, \cdots, \infty \tag{2.45}
$$

In this case, the (p.f.) can be evaluated analytically as

$$
\begin{aligned}
(\text{p.f.}) &= e^{-h\nu/2k_{\rm B}T}\sum_{n=0}^{\infty} e^{-nh\nu/k_{\rm B}T} \\
&= e^{-h\nu/2k_{\rm B}T}\left[1+\left(e^{-h\nu/k_{\rm B}T}\right)+\left(e^{-h\nu/k_{\rm B}T}\right)^2+\cdots\right] \\
(\text{p.f.}) &= \frac{e^{-h\nu/2k_{\rm B}T}}{1-e^{-h\nu/k_{\rm B}T}}
\end{aligned}
\tag{2.46}
$$

Now use some thermodynamics. It follows from Eqs. (1.18) and (1.20) that at fixed N

$$
A = E - TS = E + T\left(\frac{\partial A}{\partial T}\right)_V \tag{2.47}
$$

Hence

$$
E = A - T\left(\frac{\partial A}{\partial T}\right)_V \equiv -T^2\frac{\partial}{\partial T}\left(\frac{A}{T}\right)_V \tag{2.48}
$$

where the last relation is an identity. The first law states that

$$
dE = đQ - PdV \tag{2.49}
$$

The constant-volume heat capacity is then[9]

$$
C_V \equiv \left(\frac{đQ}{dT}\right)_V = \left(\frac{dE}{dT}\right)_V \equiv \left(\frac{\partial E}{\partial T}\right)_V \quad ; \text{ heat capacity} \tag{2.50}
$$

where the last relation follows from the definition of a partial derivative.[10]

It follows from Eqs. (2.39) that the Helmholtz free energy for the $3N$ oscillators is

$$
\begin{aligned}
A &= -3Nk_{\rm B}T\ln(\text{p.f.}) \\
&= \frac{3}{2}Nh\nu + 3Nk_{\rm B}T\ln\left(1-e^{-h\nu/k_{\rm B}T}\right)
\end{aligned}
\tag{2.51}
$$

The energy E is then obtained from Eq. (2.48) as

$$
\begin{aligned}
E &= -T^2\left[\frac{3}{2}Nh\nu\left(-\frac{1}{T^2}\right) + 3Nk_{\rm B}\left(-\frac{h\nu}{k_{\rm B}T^2}\right)\frac{e^{-h\nu/k_{\rm B}T}}{1-e^{-h\nu/k_{\rm B}T}}\right] \\
E &= \frac{3}{2}Nh\nu + 3N\frac{h\nu}{e^{h\nu/k_{\rm B}T}-1}
\end{aligned}
\tag{2.52}
$$

---

[9] We use the terms "heat capacity" and "specific heat" interchangeably in this work.
[10] Recall that N is also fixed here.

The first term is the (constant) zero-point energy, and the second term is the *Planck distribution.* The heat capacity now follows from Eq. (2.50) as

$$C_V = 3Nk_\mathrm{B}\left(\frac{h\nu}{k_\mathrm{B}T}\right)^2 \frac{e^{h\nu/k_\mathrm{B}T}}{(e^{h\nu/k_\mathrm{B}T}-1)^2} \quad ; \text{ Einstein} \tag{2.53}$$

Consider the two limiting cases:

- In the high-temperature limit $T \to \infty$, this reproduces the molar specific heat law of Dulong and Petit

$$C_V = 3N_\mathrm{A}k_\mathrm{B} = 3R = 6 \text{ cal/mole-deg} \qquad ; T \to \infty \qquad ; \text{ Dulong-Petit} \tag{2.54}$$

  Here N_A is Avogadro's number and R is the gas constant [see Eqs. (2.137)].
- At low temperature $T \to 0$, this specific heat vanishes

$$C_V = 3Nk_\mathrm{B}\left(\frac{h\nu}{k_\mathrm{B}T}\right)^2 e^{-h\nu/k_\mathrm{B}T} \qquad ; T \to 0 \tag{2.55}$$

This vanishing of the specific heat at low temperature was one of the early successes of quantum theory (see [Walecka (2008)]); however, an exponential decrease is too severe, and experiment indicates a T^3 dependence at low T. We will return to this.

2.4 Method of Steepest Descent

Our initial focus is on *independent, localized systems* (Fig. 2.1) with a set of energy levels $\varepsilon_1, \varepsilon_2, \varepsilon_3, \cdots$ for each system. In the assembly, these levels are occupied by $n_1, n_2, n_3, \cdots$ systems. We denote this set of distribution numbers by $\{n_i\}$. The total number of complexions is then given by the constrained sum in Eqs. (2.18)

$$\Omega = \sum_{\text{all sets of } \{n_i\}} \frac{N!}{n_1!n_2!n_3!\cdots} \equiv \sum_{\{n\}} \frac{N!}{\prod_i n_i!}$$
$$\sum_i n_i = N \qquad ; \sum_i n_i\varepsilon_i = E \qquad ; \text{ fixed} \tag{2.56}$$

The sum is to be carried out for a fixed total number of systems N, and fixed total energy E.

So far, motivated by our introductory example of the two-level system, we have evaluated Ω under the following approximations:

(1) Only the largest term in the sum is retained, so that $\ln \Omega \cong \ln t(n^\star)$;
(2) In addition to $\ln N! \cong N \ln N - N$, Stirling's formula is also used for the distribution numbers, so that $\ln n_i^\star! \cong n_i^\star \ln n_i^\star - n_i^\star$.

While certainly valid, for example, when used in Einstein's theory of the specific heat, we will encounter other situations where it is essential to have an improved evaluation of the sum. In this section we present the *method of steepest descent*, due to [Darwin and Fowler (1922)], which permits an evaluation of Ω in the limit $N \to \infty$, obtained without invoking either of the above assumptions.

Recall the binomial theorem

$$(a_1 + a_2)^N = \sum_{\{n_i\}} \frac{N!}{n_1! n_2!} a_1^{n_1} a_2^{n_2} \qquad ; \; n_1 + n_2 = N \tag{2.57}$$

This is readily generalized to the *multinomial theorem*

$$(a_1 + a_2 + \cdots + a_L)^N = \sum_{\{n_i\}} \frac{N!}{n_1! n_2! \cdots n_L!} a_1^{n_1} a_2^{n_2} \cdots a_L^{n_L} \qquad ; \; n_1 + n_2 + \cdots + n_L = N \tag{2.58}$$

Let $L \to \infty$ in this expression

$$(a_1 + a_2 + a_3 + \cdots)^N = \sum_{\{n_i\}} \frac{N!}{n_1! n_2! n_3! \cdots} a_1^{n_1} a_2^{n_2} a_3^{n_3} \cdots \qquad ; \; n_1 + n_2 + n_3 + \cdots = N \tag{2.59}$$

This allows us to define a *generating function* for Ω

$$f(z) \equiv \sum_{i=1}^{\infty} z^{\varepsilon_i} = z^{\varepsilon_1} + z^{\varepsilon_2} + z^{\varepsilon_3} + \cdots \qquad ; \text{ generating function} \tag{2.60}$$

For reasons that will soon become apparent, we want to work entirely with *integers*. Assume that $\varepsilon_1, \varepsilon_2, \varepsilon_3, \cdots$ form a sequence of non-decreasing integers with no common divisor. This is actually no loss of generality, since

(1) Any common divisor ε_D can be extracted and used in the definition of the unit of energy, so that the energy statement takes the form

$$\sum_i n_i \left(\frac{\varepsilon_i}{\varepsilon_D}\right) = \left(\frac{E}{\varepsilon_D}\right) \tag{2.61}$$

(2) The actual eigenvalues can be approximated to arbitrary accuracy by a sequence of integers, for example

$$\{3.13,\, 4.28,\, 5.15,\, \cdots\} \to \{313,\, 428,\, 515,\, \cdots\} \tag{2.62}$$

where the r.h.s. again follows from a redefinition of the unit of energy. Compute with these values. The result must be a continuous function of these energies, and it can then be changed infinitesimally by going to the correct $\{\varepsilon_i\}$ at the end;[11]

(3) For simplicity, choose the zero of energy so that

$$\varepsilon_1 = 0 \tag{2.63}$$

Then by the multinomial theorem, the generating function in Eq. (2.60), raised to the power N, takes the form

$$[f(z)]^N = \left[\sum_i z^{\varepsilon_i}\right]^N = \sum_{\{n_i\}} \frac{N!}{n_1!n_2!n_3!\cdots} \left(z^{\varepsilon_1}\right)^{n_1} \left(z^{\varepsilon_2}\right)^{n_2} \left(z^{\varepsilon_3}\right)^{n_3} \cdots$$
$$; \sum_i n_i = N \tag{2.64}$$

The last product in the first line can be rewritten as

$$\left(z^{\varepsilon_1}\right)^{n_1} \left(z^{\varepsilon_2}\right)^{n_2} \left(z^{\varepsilon_3}\right)^{n_3} \cdots = z^{\varepsilon_1 n_1 + \varepsilon_2 n_2 + \varepsilon_3 n_3 \cdots} \equiv z^E \tag{2.65}$$

We now observe that *the coefficient of z^E in the expression in Eq. (2.64) is precisely the Ω we are looking for!*

$$\Omega(E,N) = \sum_{\text{all sets of } \{n_i\}} \frac{N!}{n_1!n_2!n_3!\cdots}$$
$$\sum_i n_i = N \qquad ; \sum_i n_i\varepsilon_i = E \tag{2.66}$$

This follows since

$$\sum_{\{n_i\}} = \sum_E \left[\sum_{\{n_i\}}\right]_{\sum_i n_i\varepsilon_i = E} \tag{2.67}$$

[11]For the harmonic oscillator with $\varepsilon/h\nu = (p + 1/2)$, this result holds immediately; it is exact if the ε_i are *rational* multiples of each other.

merely represents a regrouping of the terms in the sum. Hence, in summary,

$$[f(z)]^N = [1 + z^{\varepsilon_2} + z^{\varepsilon_3} + \cdots]^N = \sum_{E'} \Omega(E', N) z^{E'}$$

$$E' = \sum_i n_i \varepsilon_i \qquad ; \text{ an integer} \tag{2.68}$$

We proceed to analyze this result through a series of observations and steps:[12]

1) $[f(z)]^N$ is a *power series* in z, containing integral powers $z^{E'}$. It has a radius of convergence $R \leq 1$, for if one sets $z = 1$ in the series in the first line of Eqs. (2.68), the sum diverges;

2) Within its radius of convergence, a power series is *analytic*;

3) To *solve for* $\Omega(E, N)$, one can divide by z^{E+1}, and then integrate term by term using the integral formula

$$\frac{1}{2\pi i} \oint \frac{dz}{z^n} = \delta_{n,1} \tag{2.69}$$

This gives[13]

$$\Omega(E, N) = \frac{1}{2\pi i} \oint_C dz \frac{[f(z)]^N}{z^{E+1}} \tag{2.70}$$

where the contour C is illustrated in Fig. 2.4.

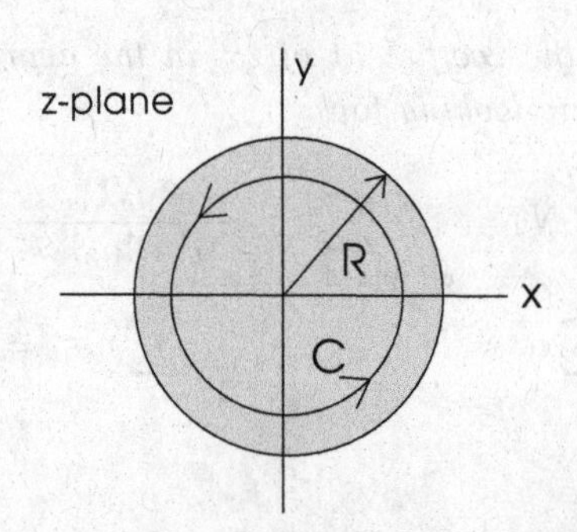

Fig. 2.4 The contour C in Eq. (2.70).

[12] For a review of complex variables, see appendix A of [Fetter and Walecka (2003a)].

[13] There is again a V dependence, which is here suppressed, coming implicitly through the $\{\varepsilon_i\}$.

Equation (2.70) is just the residue theorem for the Laurent series $[f(z)]^N/z^{E+1}$. *This is a truly remarkable result, since it relates the desired sum to the contour integral of an analytic function with an isolated singularity at the origin!*

4) Look along the real x-axis, where the integrand is also real. Since $[f(x)]^N$ is a real, monotonically increasing function of x that diverges at $x = R$, and $1/x^{E+1}$ is a real, monotonically decreasing function of x that diverges at $x = 0$, there is a *minimum* in the integrand in Eq. (2.70) along the real x-axis at a point x_0 that lies between 0 and R (Fig. 2.5).

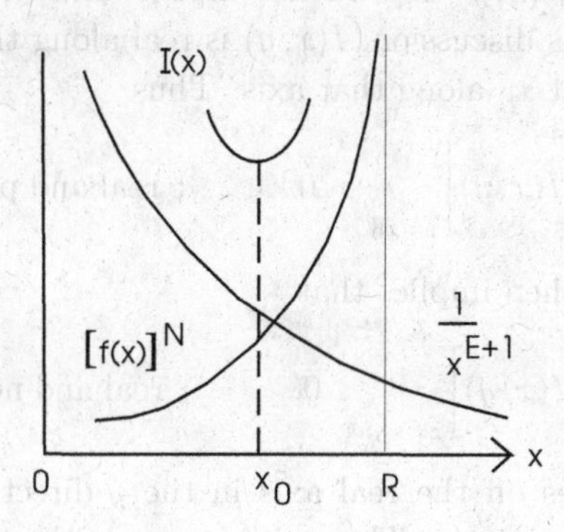

Fig. 2.5 Behavior of $I(z) \equiv [f(z)]^N/z^{E+1}$ along the real x-axis where $z = x$.

5) Away from the origin and inside R, the integrand

$$I(z) \equiv \frac{[f(z)]^N}{z^{E+1}} \quad ; \text{ integrand} \tag{2.71}$$

is *analytic*. This has several implications:

- The function $I(z)$ has a unique derivative with respect to z;
- The Cauchy-Riemann equations hold. If one separates $I(z)$ into its real and imaginary parts, then[14]

$$I(z) = u(x, y) + iv(x, y)$$
$$\frac{\partial u}{\partial x} = \frac{\partial v}{\partial y} \quad ; \quad \frac{\partial u}{\partial y} = -\frac{\partial v}{\partial x} \quad ; \text{ Cauchy-Riemann} \tag{2.72}$$

[14]See Prob. 2.2.

- One more partial derivative then leads to the relations

$$\frac{\partial^2 u}{\partial x^2} + \frac{\partial^2 u}{\partial y^2} = 0 \quad ; \quad \frac{\partial^2 v}{\partial x^2} + \frac{\partial^2 v}{\partial y^2} = 0 \tag{2.73}$$

If $I(z)$ is analytic, and $I(z) = u(x,y)+iv(x,y)$, then both u and v satisfy Laplace's equation. One says that (u,v) are "harmonic functions". Evidently, since it is true for both the real and imaginary parts,

$$\left[\frac{\partial^2}{\partial x^2} + \frac{\partial^2}{\partial y^2}\right] I(x,y) = 0 \quad ; \; I(z) \equiv I(x,y) \tag{2.74}$$

Here we make the (x,y) dependence explicit and write $I(z) \equiv I(x,y)$.

- From the previous discussion, $I(x,y)$ is real along the real x-axis, and has a *minimum* at x_0 along that axis. Thus

$$\left[\frac{\partial^2}{\partial x^2} I(x,y)\right]_{z=x_0} > 0 \quad ; \text{ real and positive} \tag{2.75}$$

Equation (2.74) then implies that

$$\left[\frac{\partial^2}{\partial y^2} I(x,y)\right]_{z=x_0} < 0 \quad ; \text{ real and negative} \tag{2.76}$$

Thus, as one goes off the real axis in the y-direction, the integrand $I(x,y)$ has a *maximum*. The point $z = x_0$ therefore constitutes a *saddle point* of the function $I(x,y)$.

The relations in Eqs. (2.75) and (2.76) also hold for the *modulus* of the integrand $|I(z)| = \sqrt{u^2+v^2}$ at the point x_0 (see Prob. 2.3), which provides a convenient way of plotting things. If the contour C is taken to run through the saddle point, then we have the situation illustrated in Fig. 2.6.

We now want to investigate what happens for very large N. In the vicinity of the saddle point, one can define a new function $g(z)$ by

$$I(z) = \frac{[f(z)]^N}{z^{E+1}} \equiv e^{Ng(z)} \tag{2.77}$$

The logarithm of this relation then gives

$$g(z) = \ln f(z) - \left(\frac{E}{N} + \frac{1}{N}\right) \ln z \tag{2.78}$$

Since E/N is a finite quantity in the limit $N \to \infty$, and $1/N$ is negligible, $g(z)$ is an *intensive quantity independent of N*.

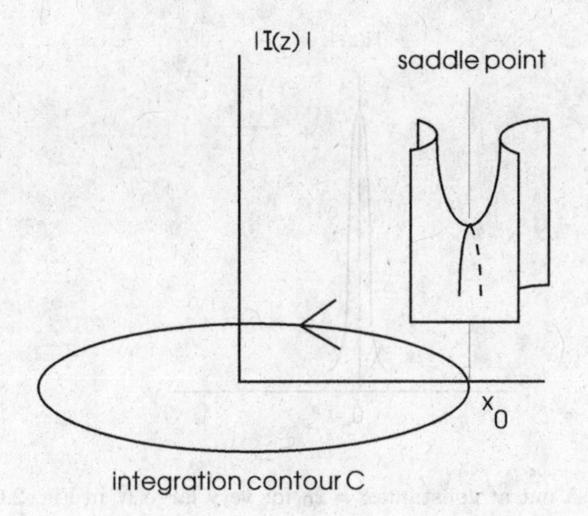

Fig. 2.6 Integration contour and saddle point.

One derivative of the analytic function in Eq. (2.77) along the x-axis, evaluated at the saddle point gives

$$\left(\frac{dI}{dz}\right)_{x_0} = 0$$

$$\Longrightarrow \qquad g'(x_0) = 0 \qquad ; \text{ defines saddle point} \tag{2.79}$$

The second relation defines the location of the saddle point. A second derivative taken along the x-axis then yields

$$\left(\frac{\partial^2 I}{\partial x^2}\right)_{x_0} = Ng''(x_0)e^{Ng(x_0)}$$
$$\to \infty \qquad ; N \to \infty \tag{2.80}$$

One has *infinitely steep peaks and valleys in the limit* $N \to \infty$.[15] For very large N, a cut in the y-direction at fixed $x = x_0$ in Fig. 2.6 thus produces the situation shown in Fig. 2.7.

6) Let us now *do the integral* for very large N, by expanding the exponent about the saddle point. Make the Taylor expansion

$$g(z) = g(x_0) + \frac{1}{2}(z - x_0)^2 g''(x_0) + \cdots \tag{2.81}$$

[15]Since I has a minimum in the x-direction, $g''(x_0) > 0$. Furthermore, since $f(x_0) > 1$, and $x_0 < 1$, one has $g(x_0) = \ln f(x_0) - (E/N) \ln x_0 > 0$.

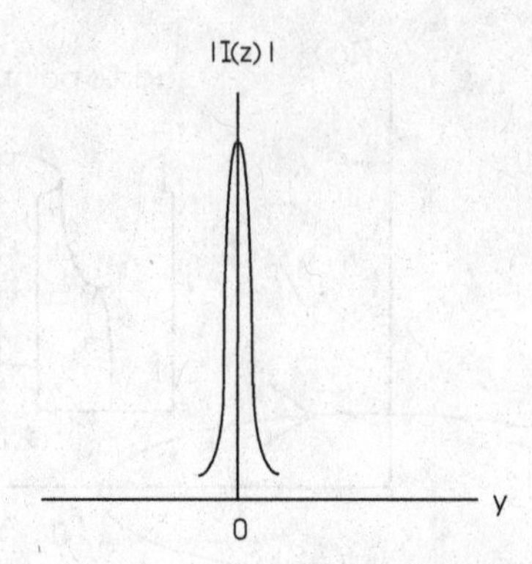

Fig. 2.7 A cut at constant $x = x_0$ for very large N in Fig. 2.6.

where Eq. (2.79) has been employed. Then

$$\begin{aligned}\Omega(E,N) &= \frac{1}{2\pi i}\oint dz\, e^{Ng(z)} \\ &\approx \frac{1}{2\pi}e^{Ng(x_0)}\int_{-\infty}^{\infty} dy\, e^{-\frac{1}{2}Ng''(x_0)y^2} \qquad ; \; z = x_0 + iy \end{aligned} \tag{2.82}$$

Here we have written $z = x_0 + iy$, integrated in the y-direction in Fig. 2.7, and used the exponential decrease of the integrand to extend the limits to $\pm\infty$. Now change variables in the integral

$$t \equiv y\left[\frac{Ng''(x_0)}{2}\right]^{1/2} \tag{2.83}$$

to obtain

$$\Omega(E,N) = \frac{1}{2\pi}\frac{e^{Ng(x_0)}}{[Ng''(x_0)/2]^{1/2}}\int_{-\infty}^{\infty} dt\, e^{-t^2} \tag{2.84}$$

The integral is just $\sqrt{\pi}$, and thus

$$\Omega(E,N) = \frac{e^{Ng(x_0)}}{[2\pi Ng''(x_0)]^{1/2}} \tag{2.85}$$

Let us re-define $g(z)$ so that

$$g(z) \equiv \ln f(z) - \left(\frac{E}{N}\right)\ln z \tag{2.86}$$

The difference between this expression and the one in Eq. (2.78) is of $O(1/N)$ and therefore negligible. First observe that

$$e^{Ng(z)} = \left[e^{g(z)}\right]^N \tag{2.87}$$

It follows from Eq (2.86) that

$$\left[e^{g(z)}\right]^N = \left[\frac{f(z)}{z^{E/N}}\right]^N \equiv \phi(z)^N \tag{2.88}$$

where this expression defines $\phi(z)$. Thus

$$\begin{aligned}\phi(z) &= \frac{f(z)}{z^{E/N}} = e^{g(z)} \\ g'(x_0) &= 0 \qquad ; \text{defines saddle point}\end{aligned} \tag{2.89}$$

Two derivatives then give

$$\begin{aligned}\phi''(x_0) &= g''(x_0)e^{g(x_0)} \\ \Longrightarrow \qquad \frac{\phi''(x_0)}{\phi(x_0)} &= g''(x_0)\end{aligned} \tag{2.90}$$

When written in terms of $\phi(z)$, our final result becomes

$$\begin{aligned}\Omega(E,N) &= \frac{1}{2\pi i}\oint \frac{dz}{z}\,[\phi(z)]^N \\ &= \frac{[\phi(x_0)]^N}{x_0}\left[\frac{\phi(x_0)}{2\pi N\phi''(x_0)}\right]^{1/2}\left[1+O\left(\frac{1}{N}\right)\right]\end{aligned} \tag{2.91}$$

Here

$$\phi(z) = \frac{f(z)}{z^{E/N}} \qquad ; f(z) = \sum_i z^{\varepsilon_i} \tag{2.92}$$

Several comments:

- This is the principal result of the method of steepest descent;
- The expression in Eq. (2.91) provides an evaluation of the total number of complexions $\Omega(E,N)$ by an integration over the high mountain in Fig. 2.7 through the saddle point pass in Fig. 2.6;
- The $1/z$ in the integrand in Eq. (2.91) has been evaluated at the saddle point as $1/x_0$. From the discussion of Eq. (2.86), this is an approximation that holds to $O(1/N)$;[16]

[16] See also Prob. 2.4(a). The saddle point is now defined through Eqs. (2.89).

- If one keeps more terms in the Taylor series in Eq. (2.81), and then goes to the dimensionless variable t in Eq. (2.83), the first non-vanishing correction is a factor[17]

$$\exp\left\{\frac{N}{4!}g''''(x_0)y^4\right\} = \exp\left\{\frac{g''''(x_0)}{6N[g''(x_0)]^2}t^4\right\} = 1 + O\left(\frac{1}{N}\right) \tag{2.93}$$

Thus the result in Eq. (2.91) is indeed accurate to $O(1/N)$, as claimed.

7) Are there any other mountains of comparable height along the contour C? Write $z \equiv x_0 e^{i\theta}$, and then

$$[f(z)]^N = \left[1 + (x_0 e^{i\theta})^{\varepsilon_2} + (x_0 e^{i\theta})^{\varepsilon_3} + \cdots\right]^N \qquad ;\text{ on } C \tag{2.94}$$

The modulus of $[f(z)]^N$ is a maximum when all terms are real and positive. The condition for this is $\theta\varepsilon_i = 2\pi p_i$ where $p_i = 0$ or an integer. If $\theta \neq 0$, then $2\pi/\theta$ is a rational number, and this would mean $\varepsilon_i = (2\pi/\theta)p_i$. This possibility is ruled out by the fact that the integers ε_i are assumed to have no common divisor. Hence, *the absolute maximum of the modulus of* $[f(z)]^N$ *along* C *occurs at* $\theta = 0$.

What about the contribution of the other possible maxima along the contour C? *It doesn't matter!* Call the ratio of the maxima

$$\left|\frac{\phi(x_0, \theta)}{\phi(x_0)}\right| \equiv r \qquad ; r < 1 \tag{2.95}$$

From the previous argument, r is strictly less than 1. To establish the error in Eq. (2.91), it is sufficient to show that

$$r^N < \frac{1}{N^{3/2}} \qquad ; N \to \infty \tag{2.96}$$

However[18]

$$r^N < \frac{1}{N^p} \qquad ;\text{ any integer } p \qquad ; N \to \infty \tag{2.97}$$

Thus, indeed, other maxima on C with $r < 1$ *do not matter.*

8) Now define

$$x_0 \equiv e^{\beta} \qquad ; \beta < 0 \tag{2.98}$$

[17] The first odd correction integrates out; see also Prob. 2.4(c).

[18] Consider $N^p r^N = N^p e^{-N \ln 1/r} \to 0$ as $N \to \infty$.

Since $0 < x_0 < 1$, we must have $\beta < 0$. The relation locating the saddle point in Eq. (2.89) is written in terms of f through Eq. (2.86) as

$$g'(x_0) = \frac{f'(x_0)}{f(x_0)} - \frac{E}{N}\frac{1}{x_0} = 0 \tag{2.99}$$

This gives

$$\frac{E}{N} = \frac{x_0 f'(x_0)}{f(x_0)} = \frac{\sum_i \varepsilon_i e^{\beta\varepsilon_i}}{\sum_i e^{\beta\varepsilon_i}} \tag{2.100}$$

This relation determines $\beta(E, N)$; it is identical to Eq. (2.32).[19]

9) The entropy then follows from Eq. (2.91) as[20]

$$S = k_\mathrm{B} \ln \Omega(E, N) = k_\mathrm{B} N \ln \phi(x_0) + O(\ln N) \tag{2.101}$$

The last term is negligible, and hence from Eqs. (2.92)

$$S = k_\mathrm{B}\left[N \ln\left(\sum_i e^{\beta\varepsilon_i}\right) - \beta E\right] \tag{2.102}$$

This is identical to Eq. (2.31). It then follows, exactly as in the previous discussion, that

$$\begin{aligned}
\beta &= -\frac{1}{k_\mathrm{B}T} && \text{; temperature} \\
(\text{p.f.}) &= \sum_i e^{-\varepsilon_i/k_\mathrm{B}T} && \text{; partition function} \\
A &= -Nk_\mathrm{B}T \ln(\text{p.f.}) && \text{; Helmholtz free energy}
\end{aligned} \tag{2.103}$$

It is quite remarkable that now the only assumption made in evaluating the entropy from Eqs. (2.56) is that N is very large!

2.5 Independent Non-Localized Systems

So far, it has been assumed that the systems are localized, as in a crystal, and that their mean positions are fixed and distinguishable. Thus the two pictures on the left in Fig. 2.8, where the labels (A, B) denote the systems and the numbers $(2, 5)$ denote the energy levels $(\varepsilon_2, \varepsilon_5)$, represent *different distinguishable complexions.*

---

[19] Recall, again, that the V dependence here enters implicitly through the $\{\varepsilon_i\}$.

[20] Note that we have actually calculated the $O(\ln N)$ term here!

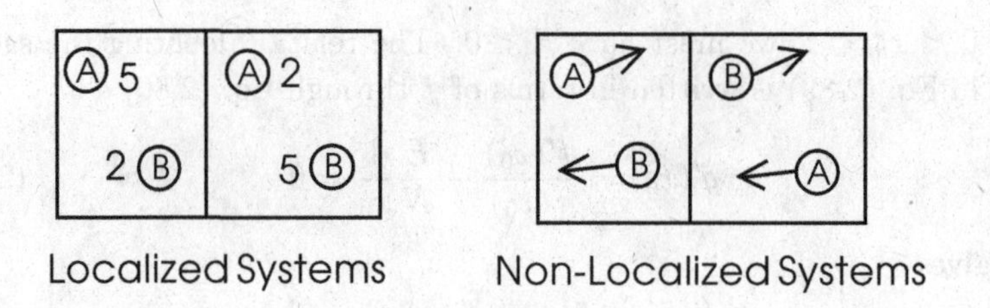

Fig. 2.8 Comparison of complexions with localized and non-localized systems. The labels (A, B) denote the systems, and on the left, the numbers denote the energy levels $(\varepsilon_2, \varepsilon_5)$.

If the systems are now identical and *non-localized*, then in the evaluation of all complexions, say by enumerating all sets of $\{\mathbf{p}_i, \mathbf{q}_i\}$, we have clearly *overcounted*. The two complexions shown on the right in Fig. 2.8, differing only by the interchange of systems (A, B), are *indistinguishable.* The number of identical re-arrangements of the systems is just the number of ways the system labels can be chosen, and that is $N!$.

> *For identical systems, non-localized as in a gas, we must reduce our previous result for the total number of complexions by* $N!$*, the number of re-arrangements of the systems.*

We therefore adopt the following working procedure for identical, independent systems:

(1) Treat the systems as distinguishable and label them (put them on the sites, for example);
(2) Find the number of complexions Ω. With a set of energy levels $\{\varepsilon_i\}$ and distribution numbers $\{n_i\}$, then

$$\Omega = \sum_{\{n_i\}} \frac{N!}{n_1!n_2!n_3!\cdots}$$
$$N = \sum_i n_i \qquad ; E = \sum_i \varepsilon_i n_i \tag{2.104}$$

This is just the problem we have already analyzed.
(3) For identical, *non-localized* systems divide the resulting Ω by $N!$.

We return to a discussion of the validity of this procedure at the end of this section, after giving an example to show that one really *must* do this to get meaningful, extensive, results for the entropy and Helmholtz free energy of an assembly of non-localized systems.[21] It is useful first to provide a *summary* of the implied results for independent, non-localized systems.

$$\begin{aligned} S &= k_{\rm B} \ln \Omega(E,V,N) \\ &= k_{\rm B}\left[N \ln\left(\sum_i e^{\beta\varepsilon_i}\right) - \beta E\right] - k_{\rm B}\ln N! \\ \frac{E}{N} &= \frac{\sum_i \varepsilon_i e^{\beta\varepsilon_i}}{\sum_i e^{\beta\varepsilon_i}} \qquad ; \quad \frac{n_s^\star}{N} = \frac{e^{\beta\varepsilon_s}}{\sum_i e^{\beta\varepsilon_i}} \end{aligned} \tag{2.105}$$

Since the additional, energy-independent term in the entropy does not affect the determination of the temperature through Eq. (2.34), it again follows that

$$\beta = -\frac{1}{k_{\rm B}T} \tag{2.106}$$

Thus

$$\begin{aligned} S &= k_{\rm B}N \ln{\rm (p.f.)} + \frac{E}{T} - k_{\rm B}\ln N! \quad &&; \text{ independent non-localized systems} \\ A &= -k_{\rm B}T \ln \frac{{\rm (p.f.)}^N}{N!} &&; \text{ Helmholtz free energy} \\ {\rm (p.f.)} &= \sum_i e^{-\varepsilon_i/k_{\rm B}T} &&; \text{ partition function} \end{aligned} \tag{2.107}$$

Note that, based on these arguments, the entire effect of going from localized to non-localized systems in the microcanonical ensemble is to replace $({\rm p.f.})^N \rightarrow ({\rm p.f.})^N/N!$ in the argument of the logarithm in the Helmholtz free energy.

2.5.1 *Perfect Gas of Structureless Particles*

As an example, let us evaluate the partition function (p.f.) for a structureless particle of mass m confined to a cubical box with sides (a, b, c) (Fig. 2.9). The stationary state Schrödinger equation is

$$T\psi = \varepsilon\psi$$

$$\text{or;} \qquad -\frac{\hbar^2\boldsymbol{\nabla}^2}{2m}\psi = \frac{\hbar^2 k^2}{2m}\psi \tag{2.108}$$

[21] In contrast to our discussion of localized systems, the volume V now plays a central role.

Here the eigenvalue has been parameterized as

$$\varepsilon \equiv \frac{\hbar^2 k^2}{2m} \tag{2.109}$$

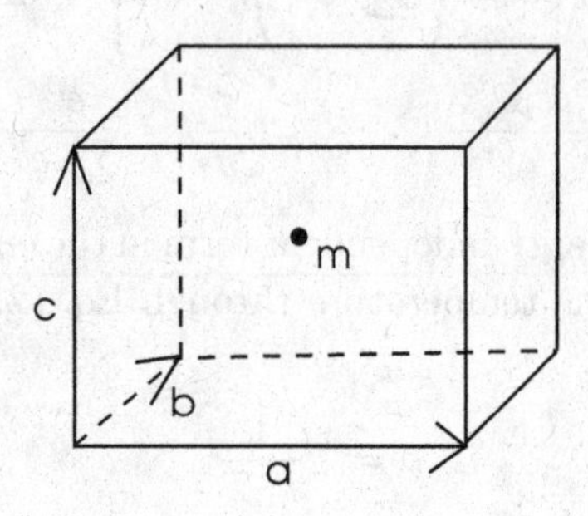

Fig. 2.9 Particle in a box.

Equation (2.108) is just the scalar Helmholtz equation

$$\left(\boldsymbol{\nabla}^2 + k^2\right)\psi = 0 \tag{2.110}$$

The boundary conditions are that the wave function must vanish on all the walls

$$\psi = 0 \qquad ; \text{ on all walls} \tag{2.111}$$

The solution is evidently

$$\begin{aligned} \psi &= \mathcal{N} \sin\left(k_x x\right) \sin\left(k_y y\right) \sin\left(k_z z\right) \\ k_x a = n_x \pi \qquad &; \; k_y b = n_y \pi \qquad ; \; k_z c = n_z \pi \\ (n_x, n_y, n_z) &= 1, 2, 3, \cdots \end{aligned} \tag{2.112}$$

Substitution into Eq. (2.110), and identification of k^2, then gives the eigenvalue spectrum

$$\begin{aligned} \varepsilon &= \frac{\hbar^2 k^2}{2m} = \frac{\hbar^2}{2m}\left(k_x^2 + k_y^2 + k_z^2\right) \\ &= \frac{\hbar^2 \pi^2}{2m}\left(\frac{n_x^2}{a^2} + \frac{n_y^2}{b^2} + \frac{n_z^2}{c^2}\right) \qquad ; \; (n_x, n_y, n_z) = 1, 2, 3, \cdots \end{aligned} \tag{2.113}$$

The partition function is now given by

$$
\begin{aligned}
(\text{p.f.}) &= \sum_{n_x}\sum_{n_y}\sum_{n_z} \exp\left\{-\frac{\hbar^2\pi^2}{2mk_{\rm B}T}\left(\frac{n_x^2}{a^2}+\frac{n_y^2}{b^2}+\frac{n_z^2}{c^2}\right)\right\} \\
&= \left[\sum_{n_x}\exp\left(-\frac{\hbar^2\pi^2 n_x^2}{2ma^2k_{\rm B}T}\right)\right]\left[\sum_{n_y}\exp\left(-\frac{\hbar^2\pi^2 n_y^2}{2mb^2k_{\rm B}T}\right)\right] \\
&\quad\times\left[\sum_{n_z}\exp\left(-\frac{\hbar^2\pi^2 n_z^2}{2mc^2k_{\rm B}T}\right)\right]
\end{aligned}
\tag{2.114}
$$

where it has been observed that the exponentials, and hence the sums, *factor*. It is therefore sufficient to evaluate just one of them[22]

$$
\sum_{n=1}^{\infty}\exp\left(-\frac{\hbar^2\pi^2 n^2}{2ma^2k_{\rm B}T}\right) = \sum_{n=1}^{\infty} e^{-\alpha^2 n^2} \tag{2.115}
$$

where we have defined[23]

$$
\alpha^2 \equiv \frac{\hbar^2\pi^2}{2ma^2k_{\rm B}T} = \frac{h^2}{8ma^2k_{\rm B}T} \tag{2.116}
$$

What sort of values are we interested in for α? Take

$$
\begin{aligned}
k_{\rm B} &= 1.38\times 10^{-16}\,\text{erg}/^{\rm o}\text{K} \qquad &;\ h = 6.63\times 10^{-27}\,\text{erg-sec} \\
m_{\rm H} &= 1.67\times 10^{-24}\,\text{gm} \qquad &;\ T = 300\,^{\rm o}\text{K} \\
a &= 1\,\text{cm}
\end{aligned}
\tag{2.117}
$$

where we use the mass of the hydrogen atom. This gives

$$
\alpha^2 \approx 10^{-16} \tag{2.118}
$$

which is a *very small number.*

In this case, the sum can be replaced by an *integral*

$$
\sum_{n=1}^{\infty} e^{-\alpha^2 n^2} \to \int_0^{\infty} dt\, e^{-\alpha^2 t^2} \qquad ;\ \alpha \to 0 \tag{2.119}
$$

[22] This is the partition function for a particle in a one-dimensional box.

[23] Recall $\hbar \equiv h/2\pi$, where h is Planck's constant.

This is demonstrated as follows. Since one is summing over integers, the sum can be written

$$\sum_{n=1}^{\infty} e^{-\alpha^2 n^2} = \sum_{n=1}^{\infty} \Delta n \, e^{-\alpha^2 n^2} \qquad ; \Delta n = 1 \tag{2.120}$$

where we have defined $\Delta n \equiv 1$. For very small α, the function $e^{-\alpha^2 n^2}$ changes only infinitesimally as n increases by an integer and one is computing the area under the curve, as illustrated in Fig. 2.10.

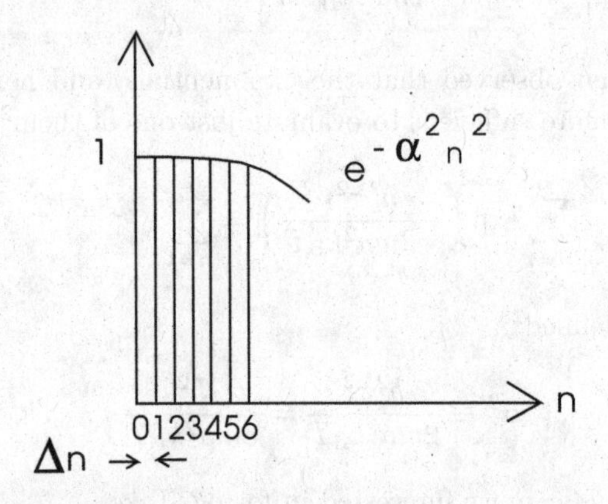

Fig. 2.10 Conversion of sum in Eq. (2.120) to an integral. Here $\Delta n = 1$.

To see this in detail, rewrite Eq. (2.120) as

$$\sum_{n=1}^{\infty} \Delta n \, e^{-\alpha^2 n^2} = \frac{1}{\alpha} \sum_{n=1}^{\infty} (\alpha \Delta n) e^{-\alpha^2 n^2} \tag{2.121}$$

Call

$$\alpha \Delta n = \Delta x_n \qquad ; \alpha n = x_n \tag{2.122}$$

Then

$$\sum_{n=1}^{\infty} \Delta x_n \, f(x_n) \rightarrow \int_0^{\infty} f(x) dx \qquad ; \alpha \rightarrow 0 \tag{2.123}$$

by the definition of the definite integral. Thus, in the limit $\alpha \rightarrow 0$,

Eqs. (2.120) and (2.121) becomes

$$\sum_{n=1}^{\infty} e^{-\alpha^2 n^2} \to \frac{1}{\alpha}\int_0^{\infty} e^{-x^2}\,dx \qquad ;\ \alpha \to 0$$
$$= \int_0^{\infty} e^{-\alpha^2 t^2}\,dt \tag{2.124}$$

which is the result that was to be established.

The integral is immediately done with the aid of a trick. Consider its square, which can be written as a double integral

$$I \equiv \int_0^{\infty} e^{-\alpha^2 t^2}\,dt$$
$$I^2 = \int_0^{\infty} dx \int_0^{\infty} dy\, e^{-\alpha^2(x^2+y^2)} \tag{2.125}$$

Go to polar coordinates, and define $r^2 \equiv u$

$$I^2 = \int_0^{\infty} r dr\, e^{-\alpha^2 r^2} \int_0^{\pi/2} d\phi = \frac{\pi}{4}\int_0^{\infty} du\, e^{-\alpha^2 u} = \frac{\pi}{4\alpha^2} \tag{2.126}$$

Hence[24]

$$I = \int_0^{\infty} e^{-\alpha^2 t^2}\,dt = \frac{\sqrt{\pi}}{2\alpha} \tag{2.127}$$

In the limit of small α, the sum in Eq. (2.115) thus becomes

$$\sum_{n=1}^{\infty} e^{-\alpha^2 n^2} \to \frac{\sqrt{\pi}}{2\alpha} \qquad ;\ \alpha \to 0$$
$$= a\left(\frac{2\pi m k_B T}{h^2}\right)^{1/2} \tag{2.128}$$

The partition function in Eq. (2.114) involves the product of three of these sums and takes the form

$$\text{(p.f.)} = abc\left(\frac{2\pi m k_B T}{h^2}\right)^{3/2} \tag{2.129}$$

Since $abc = V$ is the volume of the box, this is

$$\text{(p.f.)} = V\left(\frac{2\pi m k_B T}{h^2}\right)^{3/2} \qquad ;\ \text{particle in box} \tag{2.130}$$

---

[24] We will subsequently frequently use $\int_{-\infty}^{\infty} e^{-\alpha^2 t^2}\,dt = \sqrt{\pi}/\alpha$.

The Helmholtz free energy for an assembly of these independent, non-localized systems in a box then follows from Eq. (2.107) as[25]

$$A = -Nk_{\rm B}T\ln(\text{p.f.}) + k_{\rm B}T(N\ln N - N)$$
$$A = -Nk_{\rm B}T\left[\ln V - \ln N + \frac{3}{2}\ln T + \frac{3}{2}\ln\left(\frac{2\pi m k_{\rm B}}{h^2}\right) + 1\right] \qquad (2.131)$$

We make several comments:

- This free energy is *extensive*, as it should be. If one lets $V \to 2V$ and $N \to 2N$, then $\ln V - \ln N$ remains unchanged, and the free energy scales with the overall N. Note that it is essential to have the $1/N!$ in the argument of the logarithm in Eq. (2.107) to achieve this;
- We now have the Helmholtz free energy $A(T, V, N)$ for the perfect gas, and all the thermodynamics follows from this. The first of Eqs. (1.26) states that

$$dA = -SdT - PdV + \mu dN \qquad (2.132)$$

  Thus, for example,

$$S = -\left(\frac{\partial A}{\partial T}\right)_{V,N} = -\frac{A}{T} + \frac{3}{2}Nk_{\rm B}$$
$$A + TS = E$$
$$E = \frac{3}{2}Nk_{\rm B}T \qquad ;\ \text{energy} \qquad (2.133)$$

  This expression for the energy of a perfect gas

  - Is independent of V;
  - Depends only on T;
  - Follows from the equipartition theorem;[26]

- The constant volume heat capacity is immediately obtained from Eq. (2.133) as

$$C_V = \left(\frac{\partial E}{\partial T}\right)_N = \frac{3}{2}Nk_{\rm B} \qquad ;\ \text{heat capacity} \qquad (2.134)$$

- The pressure follows from Eqs. (2.131) and (2.132) as

$$P = -\left(\frac{\partial A}{\partial V}\right)_{N,T} = \frac{Nk_{\rm B}T}{V} \qquad (2.135)$$

[25] We use Stirling's formula $\ln N! = N\ln N - N$, which is certainly justified here.

[26] The equipartition theorem (see later) here assigns $k_{\rm B}T/2$ units of energy for each translational degree of freedom.

This yields the *equation of state* of the perfect gas

$$PV = Nk_{\rm B}T \qquad ; \text{ equation of state} \qquad (2.136)$$

- The properties of the perfect gas can be used to determine Boltzmann's constant. It is related to Avogadro's number and the gas constant by

$$\begin{aligned} k_{\rm B} &= 1.381 \times 10^{-23}\ \text{J/}^{\rm o}\text{K} && ; \text{ Boltzmann's constant} \\ N_{\rm A} &= 6.022 \times 10^{23}\ \text{/mole} && ; \text{ Avogadro's number} \\ R = N_{\rm A}k_{\rm B} &= 1.987\ \text{cal/mole-}^{\rm o}\text{K} && ; \text{ gas constant} \qquad (2.137) \end{aligned}$$

It is a striking fact that all these properties of the perfect gas follow from our two basic statistical hypotheses!

- The entropy is given by the first of Eqs. (2.133)

$$\begin{aligned} S &= Nk_{\rm B}\left[\ln V - \ln N + \frac{3}{2}\ln T + \frac{3}{2}\ln\left(\frac{2\pi m k_{\rm B}}{h^2}\right) + \frac{5}{2}\right] \\ S &= Nk_{\rm B}\left\{\frac{5}{2}\ln T - \ln P + \ln\left[\left(\frac{2\pi m}{h^2}\right)^{3/2} k_{\rm B}^{5/2}\right] + \frac{5}{2}\right\} \\ & \qquad ; \text{ Sackur-Tetrode equation} \qquad (2.138) \end{aligned}$$

Here the equation of state has been used in the second line. This is the *Sackur-Tetrode equation* for the translational entropy of a perfect gas. It provides the *absolute entropy.* Since a crystal at $T = 0$ has $S = 0$, we now know the change in entropy between that state and the state at high T and low P where it (and everything) becomes a perfect gas, no matter how complicated the intermediates states (liquids, for example) and intermediate heat flows (vaporization, *etc.*).

- The chemical potential is given by

$$\begin{aligned} \mu &= \left(\frac{\partial A}{\partial N}\right)_{V,T} \\ &= -k_{\rm B}T\left[\ln V - \ln N + \frac{3}{2}\ln T + \frac{3}{2}\ln\left(\frac{2\pi m k_{\rm B}}{h^2}\right) + 1\right] + k_{\rm B}T \\ \mu &= k_{\rm B}T\left\{\ln P - \frac{5}{2}\ln T - \ln\left[\left(\frac{2\pi m}{h^2}\right)^{3/2} k_{\rm B}^{5/2}\right]\right\} \qquad (2.139) \end{aligned}$$

This is the absolute chemical potential for a perfect gas. It provides a very useful result for our subsequent discussions of chemical reactions, and of quantum gases.

- Note that both S and μ contain Planck's constant

$$h = 6.626 \times 10^{-34}\,\text{Joule-sec} \qquad ;\ \text{Planck's constant} \tag{2.140}$$

We will shortly return to this.

2.5.2 *Validity*

Let us examine the validity of what we have done for many identical, independent, non-localized systems. The starting point is the number of complexions given by

$$\Omega = \frac{1}{N!}\sum_{\{n_i\}} \frac{N!}{n_1!n_2!\cdots} = \sum_{\{n_i\}} \frac{1}{n_1!n_2!\cdots} \tag{2.141}$$

where we have divided by the $N!$. Since this must be a sum of integers, it only makes sense if $n_i = 0, 1$. *The levels should not be multiply occupied.* We want many more levels than particles [Fig. 2.11(a)].

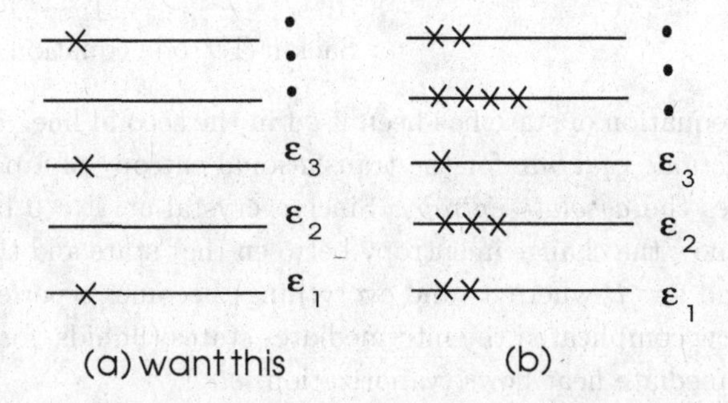

Fig. 2.11 Want many more levels than particles (a); Multiply occupied levels (b).

In the situation in Fig. 2.11(a), the counting has been done correctly. Since we have *already* corrected for the rearrangement of the systems in the same energy levels, one cannot correct for it *again* through the $1/N!$ in the situation in Fig. 2.11(b).

If $n_i = 0, 1$, one clearly cannot use Stirling's formula for the distribution numbers. This invalidates the initial approach using Lagrange's method of undetermined multipliers to maximize the logarithm of the largest term in the first sum in Eq. (2.141); however, *the method of steepest descent still holds.*

When the states are not multiply occupied, and the previous analysis applies for identical, independent, non-localized systems, one is in the domain of *classical, or Boltzmann, statistics.* Classical statistics thus holds when

$$n_i^\star \ll 1 \qquad ; \text{ classical statistics} \tag{2.142}$$

Now

$$n_i^\star = \frac{N}{(\text{p.f.})} e^{-\varepsilon_i/k_B T} \tag{2.143}$$

Since the ε_i can be quite small, the condition in Eq. (2.142) becomes

$$\frac{N}{(\text{p.f.})} = \left(\frac{N}{V}\right)\left(\frac{h^2}{2\pi m k_B T}\right)^{3/2} \ll 1 \qquad ; \text{ classical statistics} \tag{2.144}$$

where Eq. (2.130) has been employed. This relation is satisfied for

- High T;
- Low density (N/V);
- Large m;
- $h \to 0$.

If the condition in Eq. (2.144) is *not* satisfied, the gas is said to be *degenerate*, and one is in the domain of *quantum statistics.* The correct counting procedure in this case of quantum statistics is to take

$$\Omega = \sum_{\{n_i\}} 1 \qquad ; \text{ quantum statistics}$$

$$\sum_i n_i = N \qquad ; \sum_i \varepsilon_i n_i = E \tag{2.145}$$

For each set of distribution numbers $\{n_i\}$, there is just one many-body wave function:

- For identical *bosons*, this total wave function must be symmetric under the interchange of any two systems, and the n_i can take any value;
- For identical *fermions*, this total wave function must be antisymmetric under the interchange of any two systems, and the $n_i = 0, 1$.

We will later analyze quantum statistics in detail. We will find that when the condition in Eq. (2.144) is satisfied, classical statistics is recovered.[27]

[27] This is, ultimately, the real justification for the classical statistics of identical, independent, non-localized systems! See the discussion following Eq. (7.18).

For now, we focus on classical statistics.

2.6 Transition to Classical Dynamics

So far, all the counting has been done from the point of view of quantum mechanics, because it is simple to enumerate and sum complexions. We now work back to classical statistical mechanics, starting with individual systems, and then working back to assemblies. Historically, of course, the development was in exactly the *opposite* direction.

2.6.1 *Classical Mechanics*

Consider a single system. Pick a set of independent, *generalized coordinates* q_i. The lagrangian is then given by

$$L(q_i, \dot{q}_i) = T - V \tag{2.146}$$

where $i = 1, \cdots, s$ labels the degrees of freedom, and $\dot{q}_i \equiv dq_i(t)/dt$. The canonical momenta and hamiltonian are given by

$$\begin{aligned} p_i &= \frac{\partial L}{\partial \dot{q}_i} \qquad ; \; i = 1, 2, \cdots, s \\ H(p_i, q_i) &= \sum_i p_i \dot{q}_i - L = T + V \end{aligned} \tag{2.147}$$

It is assumed for the present purposes that the hamiltonian is the total energy. Hamilton's equations are

$$\dot{q}_i = \frac{\partial H}{\partial p_i} \qquad ; \; \dot{p}_i = -\frac{\partial H}{\partial q_i} \qquad ; \; i = 1, 2, \cdots, s \tag{2.148}$$

Classically, one specifies (p_i, q_i), a representative point in *phase space*, and then uses Hamilton's equations to follow the motion. For a system with s degrees of freedom, this is a $2s$-dimensional space. Each set of values of (p_i, q_i) represents a different, possible configuration of the system. For *equal a priori weighting* of the possible states of the system we would take

$$d\omega \propto dp_1 \cdots dp_s \, dq_1 \cdots dq_s \qquad ; \text{ equal a priori weighting} \tag{2.149}$$

This phase space volume has two important properties in this regard:[28]

- It is unchanged along a phase trajectory; this is Liouville's theorem;

---

[28] See [Walecka (2000)].

- It is invariant under a canonical transformation to new generalized coordinates and momenta.

2.6.2 *Quantum Mechanics*

The Heisenberg uncertainty principle, however, says that

$$\Delta p \Delta q \geq h \qquad ; \text{Heisenberg} \tag{2.150}$$

It is therefore a meaningless question to ask just where the system is within this small cell in phase space. That cannot be determined. This suggests that, at best, all we can do is find out how many of these cells we have, and then give each cell equal weighting as *each cell now represents a distinguishable configuration of the system.*

To see this a little more quantitatively, go back to the problem of a particle of mass m in a one-dimensional box of size L (Fig. 2.12).

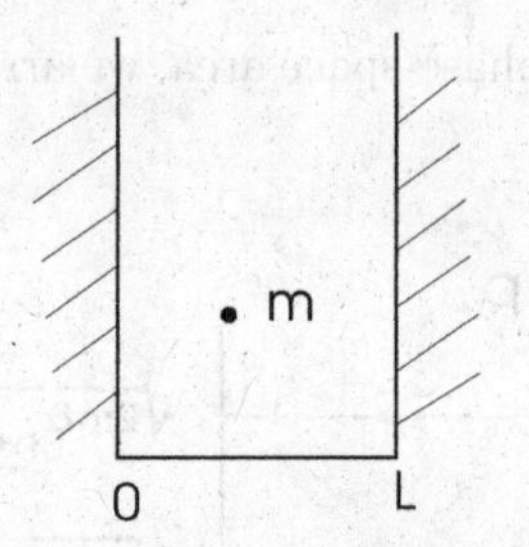

Fig. 2.12 Particle of mass m in a one-dimensional box of size L.

The energy levels are given by[29]

$$\varepsilon = \frac{p^2}{2m} = \frac{\hbar^2\pi^2}{2m}\frac{n^2}{L^2} \qquad ; n = 1, 2, 3, \cdots, \infty \tag{2.151}$$

Classically, constant energy gives a constant $p = \pm(2m\varepsilon)^{1/2}$, which can take any value. In quantum mechanics, only discrete values of $p = \pm(2m\varepsilon_n)^{1/2}$ are allowed. These values of p are separated by

$$\sqrt{2m\varepsilon_{n+1}} - \sqrt{2m\varepsilon_n} = \frac{\hbar\pi}{L}[(n+1) - n] = \frac{\hbar\pi}{L} \tag{2.152}$$

We illustrate the phase space for this system in Fig. 2.13.

[29]See Eq. (2.113); here $p = \hbar k$.

With each *new state* of the system in quantum mechanics, one can associate an *area in phase space*. From Fig. 2.13 and Eq. (2.152), that area is

$$\mathcal{A}_{\text{p.s.}} = 2\left[L\left(\frac{\hbar\pi}{L}\right)\right] = h \qquad ; \text{ area in phase space} \tag{2.153}$$

An identical result is obtained for the simple harmonic oscillator (see Prob. 2.13). Thus, *finding the number of cells of area h is equivalent to counting the number of quantum mechanical states.*

We also recall the Sommerfeld-Wilson quantization condition, an early attempt at quantization, which is asymptotically correct for large quantum numbers

$$\oint p dq = nh \qquad ; n = 0, 1, 2, \cdots \qquad \text{Sommerfeld-Wilson} \tag{2.154}$$

Since the l.h.s. is just the phase-space area, we arrive more generally at the previous result.

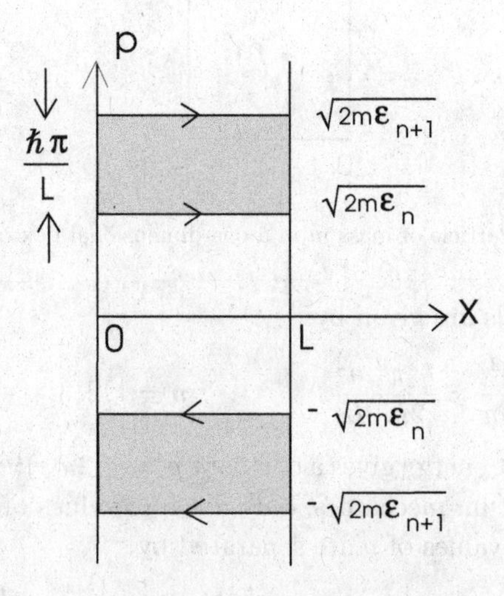

Fig. 2.13 Phase space for particle in a one-dimensional box. Constant energy implies constant $p = \pm\sqrt{2m\varepsilon}$. The phase space orbits are thus straight lines, and classically, p can take any value. In quantum mechanics, the energies are quantized, and hence so are the momenta. We show $p = \pm\sqrt{2m\varepsilon_{n+1}}$ and $p = \pm\sqrt{2m\varepsilon_n}$.

Thus, for equal a priori weighting of the states in classical statistical mechanics, one just counts the number of cells of volume h^s in $2s$-dimensional phase space (see Fig. 2.14)

$$d\omega = \frac{dp_1 \cdots dp_s dq_1 \cdots dq_s}{h^s} \quad ; \text{ number of states} \tag{2.155}$$

This weighting allows us to produce the *classical limit* in statistical mechanics.[30] For example, the classical limit of the partition function is given by

$$\begin{aligned}(\text{p.f.}) &= \sum_i e^{-\varepsilon_i/k_B T} \\ &\to \frac{1}{h^s}\int \cdots \int e^{-\varepsilon(p_1,\cdots,p_s,\,q_1,\cdots,q_s)/k_B T}\, dp_1 \cdots dp_s\, dq_1 \cdots dq_s \\ &\qquad ; \text{ classical limit}\end{aligned} \tag{2.156}$$

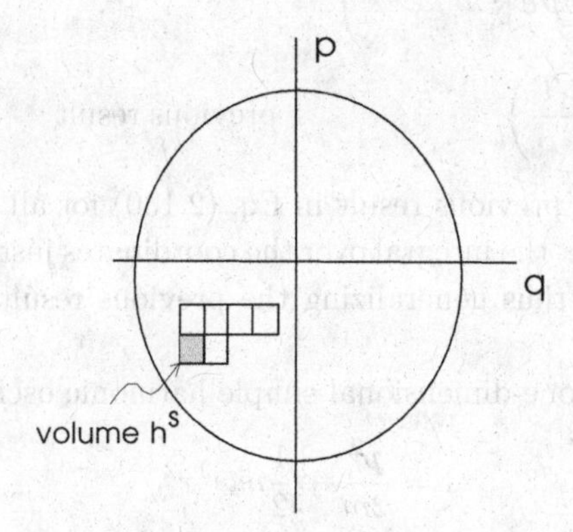

Fig. 2.14 The classical limit is obtained by dividing $2s$-dimensional phase space into cells of volume h^s, and then counting the number of cells.

Some comments:

- We note that it is only correct to use an integral if there are *many cells*, that is, *many quantum states*;
- At high T many states contribute to the partition function, and the classical limit is also, quite generally, the *high-temperature* limit;

[30]We distinguish between the *classical limit*, where classical dynamics is applicable, and *classical statistics*, where $n_i^\star \ll 1$.

- The factor of $1/h^s$ was required classically, although one did not know just why;[31]
- Most of the time the factor of $1/h^s$ will not affect the answer, for example, in computing *differences* such as $(\Delta S, C_V,$ *etc.*$)$;
- The factor of $1/h^s$ is only necessary in evaluating *absolute* quantities such as (S, μ, A).

We give two examples of the classical limit. For a system in a perfect gas, the energy is

$$\varepsilon = \frac{1}{2m}\left(p_1^2 + p_2^2 + p_3^2\right) \tag{2.157}$$

The classical partition function is then

$$\begin{aligned}
(\text{p.f.})_{\text{cl}} &= \frac{1}{h^3}\int \cdots \int dx_1 dx_2 dx_3\, dp_1 dp_2 dp_3 \exp\left\{-\frac{1}{2mk_{\text{B}}T}\left(p_1^2 + p_2^2 + p_3^2\right)\right\} \\
&= \frac{V}{h^3}\left(\int_{-\infty}^{\infty} dp\, e^{-p^2/2mk_{\text{B}}T}\right)^3 \\
&= V\left(\frac{2\pi m k_{\text{B}}T}{h^2}\right)^{3/2} \qquad ; \text{ previous result}
\end{aligned} \tag{2.158}$$

This reproduces our previous result in Eq. (2.130) for all T!

In the second line, the integral over the coordinates just gives the volume V of the container, thus generalizing the previous result to containers of any *shape*.[32]

The energy of a one-dimensional simple harmonic oscillator is

$$\varepsilon = \frac{p^2}{2m} + \frac{1}{2}m\omega^2 x^2 \tag{2.159}$$

The classical partition function is then

$$\begin{aligned}
(\text{p.f.})_{\text{cl}} &= \frac{1}{h}\int_{-\infty}^{\infty} dp \int_{-\infty}^{\infty} dx\, e^{-p^2/2mk_{\text{B}}T}\, e^{-m\omega^2 x^2/2k_{\text{B}}T} \\
&= \frac{\pi}{h}\left(2mk_{\text{B}}T\right)^{1/2}\left(\frac{2k_{\text{B}}T}{m\omega^2}\right)^{1/2} = \frac{k_{\text{B}}T}{\hbar\omega} \\
&= \frac{k_{\text{B}}T}{h\nu} \qquad ; \text{ previous result at high } T
\end{aligned} \tag{2.160}$$

This is just the high-T limit of our previous result in Eq. (2.46).

---

[31] It is required to make $d\omega$ dimensionless, if for no other reason.

[32] A study of container shapes can be found in [Gutiérrez and Yáñez (1997)].

These two examples also serve to illustrate the classical equipartition theorem, which is stated and proven in Prob. 2.14.

2.6.3 *Compute* $\Omega(E, V, N)$

Let us use these results to calculate the total number of complexions $\Omega(E, V, N)$ in the classical limit. Recall from Eq. (2.107) that the entropy of an assembly of independent non-localized systems is given by

$$\begin{aligned} S &= k_B \ln \frac{(\text{p.f.})^N}{N!} + \frac{E}{T} \\ &= k_B \ln \left[\frac{(\text{p.f.})^N}{N!} e^{E/k_B T} \right] \\ &= k_B \ln \Omega \end{aligned} \tag{2.161}$$

Thus we can identify

$$\Omega = \frac{(\text{p.f.})^N}{N!} e^{E/k_B T} \qquad ; \text{non-localized systems} \tag{2.162}$$

The $N!$ in the denominator is absent for localized systems, and therefore

$$\Omega = (\text{p.f.})^N e^{E/k_B T} \qquad ; \text{localized systems} \tag{2.163}$$

As an example, let us focus on the perfect gas.[33] With the insertion of the classical partition function, Eq. (2.162) becomes

$$\begin{aligned} \Omega &= \frac{e^{E/k_B T}}{h^{3N} N!} \left(\int \cdots \int e^{-\varepsilon(p_x, p_y, p_z, x, y, z)/k_B T} \, dp_x dp_y dp_z \, dx \, dy \, dz \right)^N \\ &\equiv \frac{e^{E/k_B T}}{h^{3N} N!} \bar{\Omega} \end{aligned} \tag{2.164}$$

The quantity $\bar{\Omega}$ can be written as a multiple integral

$$\begin{aligned} \bar{\Omega} &= \left(\int \cdots \int e^{-\varepsilon[\mathbf{p}^{(1)}, \mathbf{x}^{(1)}]/k_B T} dp_x^{(1)} dp_y^{(1)} dp_z^{(1)} dx^{(1)} dy^{(1)} dz^{(1)} \right) \\ &\quad \times \left(\int \cdots \int e^{-\varepsilon[\mathbf{p}^{(2)}, \mathbf{x}^{(2)}]/k_B T} dp_x^{(2)} dp_y^{(2)} dp_z^{(2)} dx^{(2)} dy^{(2)} dz^{(2)} \right) \times \cdots \\ &= \int \cdots \int e^{-[\varepsilon(1) + \varepsilon(2) + \cdots + \varepsilon(N)]/k_B T} \left(d^3p \, d^3x\right)_1 \left(d^3p \, d^3x\right)_2 \cdots \left(d^3p \, d^3x\right)_N \end{aligned} \tag{2.165}$$

[33] The argument is easily generalized.

This, in turn, can be recast as an integral over the 6-N dimensional phase space for the entire assembly

$$\bar{\Omega} = \int \cdots \int e^{-[\varepsilon(1)+\varepsilon(2)+\cdots+\varepsilon(N)]/k_{\rm B}T}\, dp_1 \cdots dp_{3N}\, dq_1 \cdots dq_{3N} \qquad (2.166)$$

where the coordinates are now labeled by $(p_1, p_2, \cdots, p_{3N}, q_1, q_2, \cdots, q_{3N})$.

We proceed to make a *heuristic* connection to the full classical analysis.[34] The exponent in the integral in Eq. (2.166) involves

$$\varepsilon(1) + \varepsilon(2) + \cdots + \varepsilon(N) = E \qquad (2.167)$$

This is the total energy, and the total energy is to be held *fixed* in our development! Thus the factors of $e^{E/k_{\rm B}T}e^{-E/k_{\rm B}T}$ cancel, and Eqs. (2.164) and (2.166) reduce to

$$\Omega = \frac{1}{h^{3N}N!}\int \cdots \int dp_1 \cdots dp_{3N}\, dq_1 \cdots dq_{3N}$$
$$;\ E(p_1, \cdots, p_{3N}, q_1, \cdots, q_{3N}) = E \quad ;\ \text{fixed} \qquad (2.168)$$

This expression is clearly meaningless as it stands. In our simple example in Fig. 2.13 it amounts to computing the *area of a line*, and more generally, it amounts to computing the *volume of a surface.* The best we can do is to restrict the energy to some interval[35]

$$E \leq E(p_1, \cdots, p_{3N}, q_1, \cdots, q_{3N}) \leq E + \Delta E \qquad (2.169)$$

Thus

$$\Omega = \frac{1}{h^{3N}N!}\int \cdots \int dp_1 \cdots dp_{3N}\, dq_1 \cdots dq_{3N}$$
$$;\ E \leq E(p_1, \cdots, p_{3N}, q_1, \cdots, q_{3N}) \leq E + \Delta E \qquad (2.170)$$

The situation is illustrated in Fig. 2.15. We have thus come full circle. The number of complexions in the classical limit is given by the number of cells of size h^{3N} contained in the 6-N dimensional phase-space volume of the assembly lying between the energy surfaces E and $E + \Delta E$.

For the perfect gas in a box, the energy is

$$E(p, q) = H = \sum_{i=1}^{N} \frac{1}{2m}\mathbf{p}_i^2 \qquad (2.171)$$

[34] This is done properly in Chapter 4.

[35] We will return shortly to a discussion of ΔE.

To evaluate Ω, first define $\omega(E)$ as the result arising from the *total phase space volume contained inside the surface of constant* E[36]

$$\omega(E) \equiv \frac{1}{h^{3N}N!}\int\cdots\int dp_1\cdots dp_{3N}dq_1\cdots dq_{3N}$$

$$;\ 0 \leq \sum_{i=1}^{N}\mathbf{p}_i^2 \leq 2mE \qquad (2.172)$$

Then the number of complexions is the *difference*

$$\Omega(E,V,N) = \omega(E+\Delta E) - \omega(E)\,. \qquad (2.173)$$

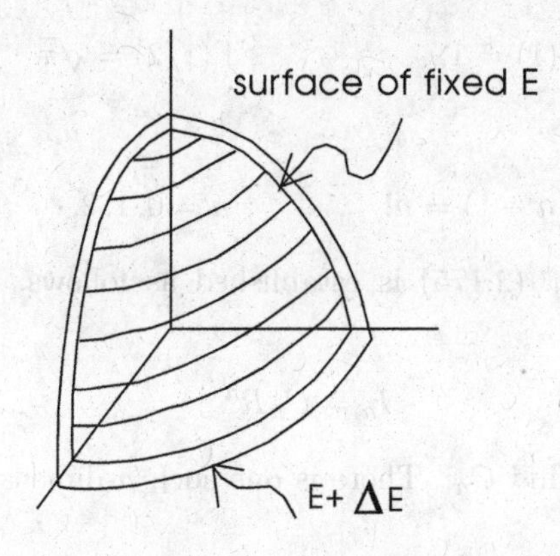

Fig. 2.15 Integration region in phase space $E \leq E(p_1,\cdots,p_{3N},\,q_1,\cdots,q_{3N}) \leq E+\Delta E$.

An integration over all the spatial coordinates produces a factor of V^N, and Eq. (2.172) becomes

$$\omega(E) = \frac{V^N}{h^{3N}N!}\int\cdots\int dp_1\cdots dp_{3N} \qquad (2.174)$$

$$;\ 0 \leq p_1^2 + p_2^2 + \cdots + p_{3N}^2 \leq 2mE$$

The goal, then, is to evaluate this $\omega(E)$.

The volume of a sphere in n-dimensional euclidian space is given by the

[36]For clarity, we here suppress the dependence of $\omega(E,V,N)$ on the latter two variables.

Dirichlet integral

$$
\begin{aligned}
I_n(R) &\equiv \int \cdots \int dx_1 \cdots dx_n \qquad ; \; x_1^2 + x_2^2 + \cdots + x_n^2 \leq R^2 \\
&= \frac{(\sqrt{\pi}\,)^n}{\Gamma(1+n/2)} R^n
\end{aligned} \tag{2.175}
$$

Here $\Gamma(z)$ is the gamma function with the properties[37]

$$
\begin{aligned}
\Gamma(z) &= \int_0^\infty e^{-t}\, t^{z-1} dt \qquad ; \; \mathrm{Re} z > 0 \\
\Gamma(z+1) &= z\Gamma(z)
\end{aligned} \tag{2.176}
$$

Furthermore

$$
\Gamma(1) = 1 \qquad ; \; \Gamma(1/2) = \sqrt{\pi} \tag{2.177}
$$

and

$$
\Gamma(n+1) = n! \qquad ; \; n = 0, 1, 2, \cdots \tag{2.178}
$$

The result in Eq. (2.175) is established as follows. By dimensional arguments

$$
I_n = C_n R^n \tag{2.179}
$$

The challenge is to find C_n. There is one such n-dimensional integral we know how to do

$$
\int_{-\infty}^{\infty} dx_1 \cdots \int_{-\infty}^{\infty} dx_n \, e^{-(x_1^2 + \cdots x_n^2)} = \left(\int_{-\infty}^{\infty} dx \, e^{-x^2} \right)^n = (\sqrt{\pi}\,)^n \tag{2.180}
$$

Now introduce polar-spherical angles in n-dimensions. No matter what the details, after all the angular integrations have been carried out, the result must be of the form[38]

$$
\int \cdots \int dx_1 \cdots dx_n = n C_n r^{n-1} dr \qquad ; \text{ over all angles} \tag{2.181}
$$

The remaining radial integral then gives

$$
\int_0^R n C_n r^{n-1} dr = C_n R^n \tag{2.182}
$$

---

[37] See, for example, [Fetter and Walecka (2003a)].

[38] Compare Prob. 2.12.

Now transform the integral on the l.h.s. of Eq. (2.180) to polar-spherical coordinates, do the integrals over all angles, and introduce $t = r^2$

$$\begin{aligned} nC_n \int_0^\infty e^{-r^2} r^{n-1} dr &= \frac{n}{2} C_n \int_0^\infty e^{-t}\, t^{n/2-1} dt \\ &= \frac{n}{2} C_n\, \Gamma\left(\frac{n}{2}\right) = C_n\, \Gamma\left(\frac{n}{2}+1\right) \end{aligned} \tag{2.183}$$

A comparison with Eq. (2.180) then gives

$$C_n = \frac{(\sqrt{\pi}\,)^n}{\Gamma(1+n/2)} \tag{2.184}$$

which is the stated result.

The use of the Dirichlet integral for the momentum integrations in Eq. (2.174) then produces the following expression

$$\omega(E) = \frac{V^N}{h^{3N} N!} \frac{\pi^{3N/2}(2mE)^{3N/2}}{\Gamma(1+3N/2)} \tag{2.185}$$

We proceed to analyze this result.

The energy dependence in Eq. (2.185) is now explicit, and it can be exhibited as

$$\omega(E) = K(N,V)E^{\,3N/2} \tag{2.186}$$

Equation (2.173) for $\Omega(E,V,N)$ then becomes

$$\begin{aligned} \Omega &= K(N,V)\left[(E+\Delta E)^{\,3N/2} - E^{\,3N/2}\right] \\ &= \omega(E)\left[\left(1+\frac{\Delta E}{E}\right)^{3N/2} - 1\right] \end{aligned} \tag{2.187}$$

The quantity of interest, the logarithm of Ω, is then

$$\ln\Omega(E,V,N) = \ln\omega(E) + \ln\left[\left(1+\frac{\Delta E}{E}\right)^{3N/2} - 1\right] \tag{2.188}$$

We must now confront the issue of just what is ΔE. We first observe that the energy of the assembly can never be determined *exactly*. There will always be some experimental error. Recall that for a perfect gas

$$E = \frac{3N}{2} k_{\rm B}T \qquad ; \text{ perfect gas} \tag{2.189}$$

Try

$$\Delta E = k_{\rm B}T \tag{2.190}$$

The experimental error is certainly greater than $k_{\rm B}T$ in this case! Then

$$\left(1+\frac{2}{3N}\right)^{3N/2} \to e \qquad ; N \to \infty \tag{2.191}$$

where e is the natural-logarithm base.[39] The second term in Eq. (2.188) then becomes $\ln(e-1)$. Now $\ln \omega(E)$ is of order N, and hence the second term in Eq. (2.188) is *completely negligible.*

Take $\Delta E = xk_{\rm B}T$ where x is *any* finite number. In this case, the second term in Eq. (2.188) becomes $\ln(e^x - 1)$, and it is again *completely negligible* as $N \to \infty$.

Thus the following provide *completely equivalent expressions for the entropy as $N \to \infty$, provided $\Delta E \neq 0$:*

$$\begin{aligned} S &= k_{\rm B} \ln \Omega(E,V,N) \\ &= k_{\rm B} \ln \omega(E) \qquad\qquad ; N \to \infty \qquad ; \Delta E \neq 0 \\ &= k_{\rm B} \ln \frac{\partial \omega(E)}{\partial E} \end{aligned} \tag{2.192}$$

The last relation follows since

$$\begin{aligned} \frac{\partial \omega(E)}{\partial E} &= \frac{3N}{2E}\omega(E) \\ \ln \frac{\partial \omega(E)}{\partial E} &= \ln \omega(E) + \ln \frac{3N}{2} - \ln E \end{aligned} \tag{2.193}$$

The last two terms are of $O(\ln N)$ and thus are again negligible.

Therefore the following expressions for the required phase-space integral in Eq. (2.170) are all *equivalent for very large N, provided $\Delta E \neq 0$,*

$$\begin{aligned} &\int \cdots \int \qquad ; E \leq E(p,q) \leq E + \Delta E \qquad ; \Delta E \neq 0 \\ &\int \cdots \int \qquad ; 0 \leq E(p,q) \leq E \\ \frac{\partial}{\partial E} &\int \cdots \int \qquad ; 0 \leq E(p,q) \leq E \end{aligned} \tag{2.194}$$

The situation is illustrated in Fig. 2.16. As was the case in picking out

[39] Define $(1+\epsilon)^{1/\epsilon} \equiv e^u$. Then $u = (1/\epsilon)\ln(1+\epsilon) \to 1$ as $\epsilon \to 0$.

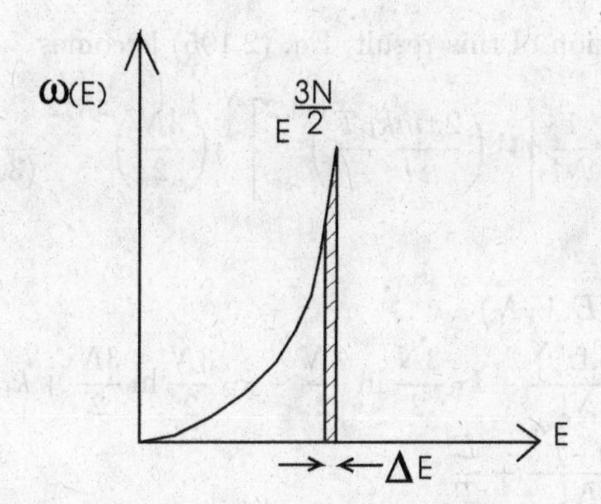

Fig. 2.16 The function $\omega(E)$ grows very rapidly with energy. As a result, the logarithms $\ln[\omega(E+\Delta E)-\omega(E)] = \ln\omega(E) = \ln\partial\omega(E)/\partial E$ all give equivalent results for very large N and $\Delta E \neq 0$. The reader is asked to contemplate $E^{3N/2}$ for $N \approx 10^{23}$.

the largest term in the sum, these non-intuitive results arise from the fact that N is very large and we are dealing with logarithms.

Let us show that we reproduce our previous results for the perfect gas with the present analysis. The entropy of the assembly is now

$$S(E,V,N) = k_{\rm B}\ln\omega(E,V,N)$$
$$\omega(E,V,N) = \frac{V^N}{N!}\left(\frac{2\pi mE}{h^2}\right)^{3N/2}\frac{1}{(3N/2)!} \tag{2.195}$$

where we restore the full dependence to $\omega(E,V,N)$. For simplicity, we assume here that $3N/2$ is an integer.[40]

The first and second laws of thermodynamics give

$$dE = TdS - PdV + \mu dN \tag{2.196}$$

Thus

$$\left(\frac{\partial S}{\partial E}\right)_{N,V} = \frac{1}{T}$$
$$= \frac{\partial}{\partial E}\left(\frac{3Nk_{\rm B}}{2}\ln E\right) = \frac{3Nk_{\rm B}}{2E} \tag{2.197}$$

We therefore recover the equation of state for the perfect gas

$$E = \frac{3}{2}Nk_{\rm B}T \qquad ;\ \text{perfect gas} \tag{2.198}$$

[40]So that we can use $\Gamma(1+n) = n!$; see, however, Prob. 4.1(c).

With the substitution of this result, Eq. (2.195) becomes

$$\omega(E,V,N) = \frac{1}{N!}\left[V\left(\frac{2\pi m k_{\rm B}T}{h^2}\right)^{3/2}\right]^N \left(\frac{3N}{2}\right)^{3N/2} \frac{1}{(3N/2)!} \tag{2.199}$$

Then

$$\begin{aligned} S &= k_{\rm B} \ln \omega(E,V,N) \\ &= k_{\rm B} \ln \frac{(\text{p.f.})^N}{N!} + k_{\rm B}\frac{3N}{2} \ln \frac{3N}{2} - k_{\rm B}\frac{3N}{2} \ln \frac{3N}{2} + k_{\rm B}\frac{3N}{2} \\ &= k_{\rm B} \ln \frac{(\text{p.f.})^N}{N!} + \frac{E}{T} \end{aligned} \tag{2.200}$$

Note that the second and third terms cancel in the second line, and the fourth term is just E/T. Since $A = E - TS$, we now recover our previous result

$$\begin{aligned} A(T,V,N) &= -k_{\rm B}T \ln \frac{(\text{p.f.})^N}{N!} \qquad ; \text{ perfect gas} \\ (\text{p.f.}) &= V\left(\frac{2\pi m k_{\rm B}T}{h^2}\right)^{3/2} \end{aligned} \tag{2.201}$$

Chapter 3

Applications of the Microcanonical Ensemble

3.1 Internal Partition Function

Suppose that for a given system, one can separate the energy into several *independent contributions*[1]

$$\varepsilon_{i,j,k,\cdots} = \varepsilon_i^{(1)} + \varepsilon_j^{(2)} + \varepsilon_k^{(3)} + \cdots \quad \text{; single system independent contributions} \tag{3.1}$$

Here the superscript denotes the mode, and the subscript denotes the energy level within that mode. The partition function then becomes

$$\begin{aligned}
(\text{p.f.}) &= \sum_{i,j,k,\cdots} e^{-\varepsilon_{i,j,k,\cdots}/k_{\text{B}}T} \\
&= \sum_{i,j,k,\cdots} e^{-\left(\varepsilon_i^{(1)}+\varepsilon_j^{(2)}+\varepsilon_k^{(3)}+\cdots\right)/k_{\text{B}}T} \\
&= \left(\sum_i e^{-\varepsilon_i^{(1)}/k_{\text{B}}T}\right)\left(\sum_j e^{-\varepsilon_j^{(2)}/k_{\text{B}}T}\right)\left(\sum_k e^{-\varepsilon_k^{(3)}/k_{\text{B}}T}\right)\cdots
\end{aligned} \tag{3.2}$$

The partition function *factors* into the product of partition functions for each of the independent modes of excitation

$$(\text{p.f.}) = (\text{p.f.})^{(1)}(\text{p.f.})^{(2)}(\text{p.f.})^{(3)}\cdots \qquad \text{; factors} \tag{3.3}$$

This has two immediate consequences:

[1] We shall see many examples of this, when the systems themselves have internal structure. Recall also the result in classical mechanics that for small oscillations about a stable equilibrium configuration, the introduction of *normal modes* reduces the energy to a sum of uncoupled simple harmonic oscillators (see [Fetter and Walecka (2003a)]).

(1) Since the Helmholtz free energy, and then the entropy, follow from the logarithm of the partition function, these independent modes make *additive* contributions to A and S

$$A = A^{(1)} + A^{(2)} + A^{(3)} + \cdots \qquad ; \text{ additive}$$
$$S = S^{(1)} + S^{(2)} + S^{(3)} + \cdots \tag{3.4}$$

(2) The mean number of particles in the level i in the mode (1), for example, is obtained by summing over all the other possibilities

$$\frac{n_i^{(1)\star}}{N} = \sum_{j,k,\cdots} \frac{n_{i,j,k,\cdots}^{\star}}{N} = \sum_{j,k,\cdots} \frac{e^{-\varepsilon_{i,j,k,\cdots}/k_{\rm B}T}}{(\text{p.f.})}$$
$$= \sum_{j,k,\cdots} \frac{e^{-\varepsilon_i^{(1)}/k_{\rm B}T} e^{-\varepsilon_j^{(2)}/k_{\rm B}T} e^{-\varepsilon_k^{(3)}/k_{\rm B}T} \cdots}{(\text{p.f.})^{(1)}(\text{p.f.})^{(2)}(\text{p.f.})^{(3)} \cdots}$$
$$= \frac{e^{-\varepsilon_i^{(1)}/k_{\rm B}T}}{(\text{p.f.})^{(1)}} \left(\frac{\sum_j e^{-\varepsilon_j^{(2)}/k_{\rm B}T}}{(\text{p.f.})^{(2)}} \right) \left(\frac{\sum_k e^{-\varepsilon_k^{(3)}/k_{\rm B}T}}{(\text{p.f.})^{(3)}} \right) \cdots$$
$$\frac{n_i^{(1)\star}}{N} = \frac{e^{-\varepsilon_i^{(1)}/k_{\rm B}T}}{(\text{p.f.})^{(1)}} \tag{3.5}$$

Since the sum over all the other modes factors and cancels in the ratio, the mean occupation number in a given mode can be calculated all by itself, independent of the other modes of excitation of a system.

3.2 Molecular Spectroscopy

As an initial application of the microcanonical ensemble to physical assemblies, we consider molecular spectroscopy, and we start with diatomic molecules.[2]

3.2.1 *Diatomic Molecules*

Introduce the center-of-mass (C-M) and relative coordinates for the diatomic molecule illustrated in Fig. 3.1. It is a general result from classical mechanics that the kinetic energy is a sum of the kinetic energy of the center-of-mass and the kinetic energy of motion about the center-of-mass.

[2]See [Herzberg (2008)].

The *hamiltonian* for the system in Fig. 3.1 thus takes the form

$$H = \frac{1}{2}M\dot{\mathbf{R}}^2 + \left(\frac{1}{2}m_A\dot{\mathbf{r}}_A^2 + \frac{1}{2}m_B\dot{\mathbf{r}}_B^2 + \frac{1}{2}m\sum_i \dot{\mathbf{r}}_i^2\right) + \frac{1}{2}\sum_{i\neq j}\frac{e^2}{|\mathbf{r}_i - \mathbf{r}_j|}$$
$$+ \frac{Z_a Z_B e^2}{|\mathbf{r}_A - \mathbf{r}_B|} - Z_A e^2 \sum_i \frac{1}{|\mathbf{r}_i - \mathbf{r}_A|} - Z_B e^2 \sum_i \frac{1}{|\mathbf{r}_i - \mathbf{r}_B|} \tag{3.6}$$

Here (m_A, m_B, m) are the masses of the nuclei and electrons, M is the total mass, and $\mathbf{R}$ locates the position of the center-of-mass. The vectors $\mathbf{r}_A$ and $\mathbf{r}_B$ locate the two nuclei with respect to the C-M, and the vectors $\mathbf{r}_i$ locate the electrons with respect to the C-M. The charges on the nuclei are (Z_A, Z_B), and the Coulomb interactions between the constituent nuclei and electrons have all been included in Eq. (3.6).

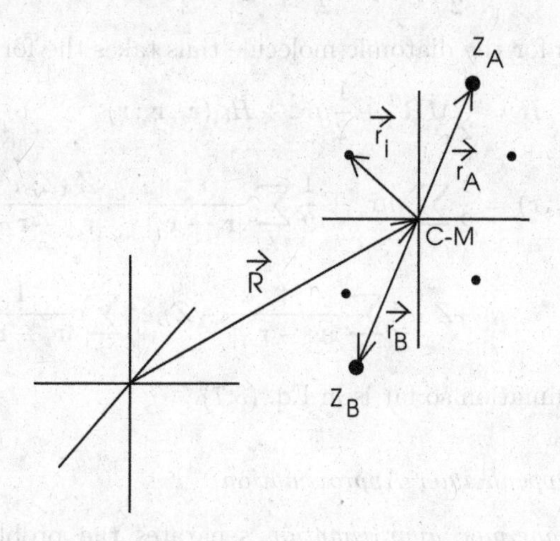

Fig. 3.1 Center-of-mass (C-M) and relative coordinates for a diatomic molecule. $\mathbf{R}$ locates the position of the center-of-mass, $\mathbf{r}_A$ and $\mathbf{r}_B$ locate the two nuclei with respect to the C-M; the charges on the nuclei are (Z_A, Z_B) respectively. The vectors $\mathbf{r}_i$ locate the electrons with respect to the C-M.

Since all the mass is in the nuclei, one has

$$m_A\mathbf{r}_A + m_B\mathbf{r}_B \approx 0 \qquad ; \text{ C-M system} \tag{3.7}$$

Define the relative nuclear coordinate by

$$\mathbf{r}_A \equiv \frac{m_B}{m_A + m_B}\mathbf{r} \tag{3.8}$$

It follows from Eq. (3.7) that

$$\mathbf{r}_B = -\frac{m_A}{m_A + m_B}\mathbf{r}$$

$$\mathbf{r}_A - \mathbf{r}_B = \mathbf{r} \qquad ; \text{ relative coordinate } \mathbf{r} \tag{3.9}$$

The reduced mass is defined by

$$\frac{1}{\mu} \equiv \frac{1}{m_A} + \frac{1}{m_B}$$

$$\mu = \frac{m_A m_B}{m_A + m_B} \qquad ; \text{ reduced mass} \tag{3.10}$$

Then

$$\frac{1}{2}m_A\dot{\mathbf{r}}_A^2 + \frac{1}{2}m_B\dot{\mathbf{r}}_B^2 = \frac{1}{2}\mu\dot{\mathbf{r}}^2 \tag{3.11}$$

The hamiltonian for the diatomic molecule thus takes the form

$$H = \frac{1}{2}M\dot{\mathbf{R}}^2 + \frac{1}{2}\mu\dot{\mathbf{r}}^2 + H_{\text{el}}(\dot{\mathbf{r}}_i, \mathbf{r}_i; \mathbf{r})$$

$$H_{\text{el}}(\dot{\mathbf{r}}_i, \mathbf{r}_i; \mathbf{r}) = \frac{1}{2}\sum_i m\dot{\mathbf{r}}_i^2 + \frac{1}{2}\sum_{i \neq j}\frac{e^2}{|\mathbf{r}_i - \mathbf{r}_j|} + \frac{Z_A Z_B e^2}{|\mathbf{r}_A - \mathbf{r}_B|}$$

$$-Z_A e^2 \sum_i \frac{1}{|\mathbf{r}_i - \mathbf{r}_A|} - Z_B e^2 \sum_i \frac{1}{|\mathbf{r}_i - \mathbf{r}_B|} \tag{3.12}$$

The only approximation so far is in Eq. (3.7).

3.2.1.1 *Born-Oppenheimer Approximation*

The *Born-Oppenheimer approximation* separates the problem into two parts:

(1) The electrons are very light, and they move very fast. The nuclei are heavy, and their motion is much slower. Consider the nuclei first as *fixed*, and solve the electron problem[3]

$$H_{\text{el}}(\dot{\mathbf{r}}_i, \mathbf{r}_i; \mathbf{r})\psi_{el}(\mathbf{r}_i; \mathbf{r}) = U(r)\psi_{\text{el}}(\mathbf{r}_i; \mathbf{r}) \qquad ; \text{ electron problem} \tag{3.13}$$

Here the energy eigenvalue, which will depend on r, has been labeled as $\varepsilon^{\text{el}}(r) = U(r)$;

[3] As usual for the electrons, $\dot{\mathbf{r}}_i = \mathbf{p}_i/m$.

(2) Then solve for the much slower nuclear motion using the hamiltonian[4]

$$H = \frac{1}{2}M\dot{\mathbf{R}}^2 + \frac{1}{2}\mu\dot{\mathbf{r}}^2 + U(r)$$
$$= H_{\rm CM} + H_{\rm int} \qquad ; \text{ nuclear motion} \tag{3.14}$$

The energy clearly separates into C-M and internal contributions

$$\varepsilon = \varepsilon^{\rm CM} + \varepsilon^{\rm int} \tag{3.15}$$

and we know the partition function for the free C-M motion from our previous analysis.

Let us then concentrate on the internal configuration in the C-M frame. There, for the two nuclei, one has an *equivalent one-body problem* for the motion of a particle of reduced mass μ and relative coordinate $\mathbf{r}$ moving in the potential $U(r)$ generated by the solution for the electrons. The general shape of $U(r)$ is sketched in Fig. 3.2.

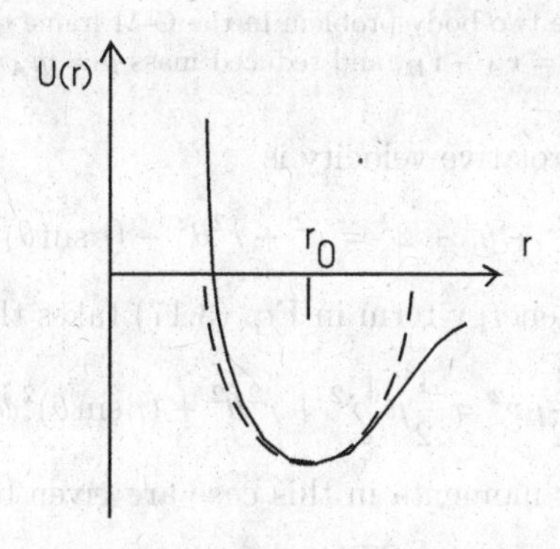

Fig. 3.2 Inter-nuclear potential in a diatomic molecule in the Born-Oppenheimer approximation. The potential is approximately quadratic about the equilibrium point r_0.

For small oscillations, the potential is approximately quadratic about the minimum

$$U(r) = U(r_0) + \frac{1}{2}(r - r_0)^2 U''(r_0) + \cdots \tag{3.16}$$

[4]See Prob. 3.12.

and we retain just the first two terms. Then

$$H_{\text{int}} = \frac{1}{2}\mu\dot{\mathbf{r}}^2 + U(r_0) + \frac{1}{2}(r - r_0)^2 U''(r_0) \tag{3.17}$$

The standard spherical coordinates (r, θ, ϕ) for this equivalent one-body central-force problem are shown in Fig. 3.3.

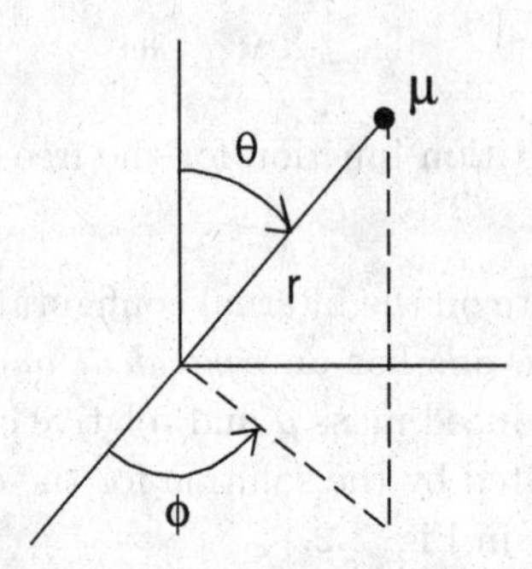

Fig. 3.3 Reduction of the two-body problem in the C-M frame to a one-body problem with relative coordinate $\mathbf{r} = \mathbf{r}_A - \mathbf{r}_B$, and reduced mass $\mu = m_A m_B/(m_A + m_B)$.

The square of the relative velocity is

$$\dot{\mathbf{r}}^2 = \dot{x}^2 + \dot{y}^2 + \dot{z}^2 = \dot{r}^2 + r^2\dot{\theta}^2 + (r\sin\theta)^2\dot{\phi}^2 \tag{3.18}$$

and hence the kinetic energy term in Eq. (3.17) takes the form

$$T = \frac{1}{2}\mu\dot{\mathbf{r}}^2 = \frac{1}{2}\mu\left[\dot{r}^2 + r^2\dot{\theta}^2 + (r\sin\theta)^2\dot{\phi}^2\right] \tag{3.19}$$

The canonical angular momenta in this case are given by

$$\begin{aligned} p_\theta &= \frac{\partial T}{\partial\dot{\theta}} = \mu r^2\dot{\theta} \\ p_\phi &= \frac{\partial T}{\partial\dot{\phi}} = \mu r^2\sin^2\theta\,\dot{\phi} \end{aligned} \tag{3.20}$$

The kinetic energy is therefore re-written in terms of these canonical variables as

$$T = \frac{1}{2}\mu\dot{r}^2 + \frac{1}{2\mu r^2}\left[p_\theta^2 + \frac{1}{\sin^2\theta}p_\phi^2\right] \tag{3.21}$$

The equivalent one-body angular momentum is defined by

$$\mathbf{L} \equiv \mathbf{r} \times (\mu\dot{\mathbf{r}}) \tag{3.22}$$

Its square is

$$\begin{aligned}\mathbf{L}^2 &= \mu^2(\mathbf{r}\times\dot{\mathbf{r}})\cdot(\mathbf{r}\times\dot{\mathbf{r}}) \\ &= \mu^2\left[\mathbf{r}^2\,\dot{\mathbf{r}}^2 - (\mathbf{r}\cdot\dot{\mathbf{r}})^2\right] = \mu^2 r^2\left(\dot{\mathbf{r}}^2 - \dot{r}^2\right) \\ &= \mu^2 r^2\left[r^2\dot{\theta}^2 + (r\sin\theta)^2\dot{\phi}^2\right]\end{aligned} \tag{3.23}$$

Thus

$$\mathbf{L}^2 = p_\theta^2 + \frac{1}{\sin^2\theta}p_\phi^2 \tag{3.24}$$

Equations (3.21) and (3.17) can therefore be re-written as

$$\begin{aligned}T &= \frac{1}{2}\mu\dot{r}^2 + \frac{1}{2\mu r^2}\mathbf{L}^2 \\ H_{\text{int}} &= \frac{1}{2}\mu\dot{r}^2 + \frac{1}{2\mu r^2}\mathbf{L}^2 + U(r_0) + \frac{1}{2}(r-r_0)^2U''(r_0)\end{aligned} \tag{3.25}$$

To a first approximation, the low-lying vibrational motion of the molecule will not significantly change the mean inter-nuclear separation in the rotational motion, and these motions *decouple* if we set $r \approx r_0$ in the second term on the r.h.s. of Eqs. (3.25).[5] The *moment of inertia* of the molecule is then defined as

$$I \equiv \mu r_0^2 \qquad \text{; moment of inertia} \tag{3.26}$$

Define the radial displacement from equilibrium as $r - r_0 \equiv x$. Then

$$\begin{aligned}r - r_0 &\equiv x \\ \frac{1}{2}\mu\dot{r}^2 &= \frac{1}{2}\mu\dot{x}^2 \\ \frac{1}{2}(r-r_0)^2U''(r_0) &= \frac{1}{2}\mu\omega^2x^2\end{aligned} \tag{3.27}$$

where the oscillator frequency has been defined through

$$U''(r_0) \equiv \mu\omega^2 \tag{3.28}$$

Then

$$H_{\text{int}} = \frac{1}{2I}\mathbf{L}^2 + \frac{1}{2}\mu\dot{x}^2 + \frac{1}{2}\mu\omega^2x^2 + U(r_0) \tag{3.29}$$

[5] See Prob. 3.11.

A combination of these results reduces the hamiltonian in Eq. (3.14) to the separated form

$$\begin{aligned}
H &= H_{\rm CM} + H_{\rm rot} + H_{\rm vib} + H_{\rm el} \\
H_{\rm CM} &= \frac{1}{2}M\dot{\mathbf{R}}^2 \qquad ; H_{\rm rot} = \frac{1}{2I}\mathbf{L}^2 \\
H_{\rm vib} &= \frac{1}{2}\mu\dot{x}^2 + \frac{1}{2}\mu\omega^2 x^2 \qquad ; H_{\rm el} = U(r_0)
\end{aligned} \tag{3.30}$$

The energy eigenvalues correspondingly separate into

$$\varepsilon = \varepsilon^{\rm CM} + \varepsilon^{\rm rot} + \varepsilon^{\rm vib} + \varepsilon^{\rm el} \tag{3.31}$$

The internal energy is now reduced to the form in Eq. (3.1), the internal partition function factors as in Eq. (3.3), and the contributions can be analyzed mode by mode.

3.2.1.2 *Partition Function*

(1) *Translation*: For the free C-M motion, one needs $({\rm p.f.})_{\rm trans}^N/N!$ where from Eq. (2.201)[6]

$$({\rm p.f.})_{\rm trans} = V\left[\frac{2\pi(m_A+m_B)k_{\rm B}T}{h^2}\right]^{3/2} \qquad ; \text{translation} \tag{3.32}$$

(2) *Vibration*: The vibrational hamiltonian is just that of a one-dimensional simple harmonic oscillator, and thus the vibrational energies are

$$\varepsilon^{\rm vib} = h\nu\left(n+\frac{1}{2}\right) \qquad ; n = 0,1,2,\cdots \tag{3.33}$$

The vibrational partition function follows as in Eq. (2.46)

$$\begin{aligned}
({\rm p.f.})_{\rm vib} &= \frac{e^{-\theta_{\rm V}/2T}}{1-e^{-\theta_{\rm V}/T}} \qquad ; \text{vibration} \\
\theta_{\rm V} &\equiv \frac{h\nu}{k_{\rm B}}
\end{aligned} \tag{3.34}$$

(3) *Rotation*: The eigenvalues of $\mathbf{L}^2$, the square of the one-body angular momentum, are

$$\begin{aligned}
\mathbf{L}^2 &= \hbar^2 l(l+1) \qquad ; l = 0,1,2,\cdots \\
\varepsilon^{\rm rot} &= \frac{\hbar^2}{2I}l(l+1)
\end{aligned} \tag{3.35}$$

[6] We now label the various partition functions with a subscripted name.

The eigenfunctions are just the spherical harmonics[7]

$$\psi_{\text{rot}}(\theta,\phi) = Y_{lm}(\theta,\phi) \qquad ; l = 0,1,2,\cdots$$
$$-l \le m \le l \tag{3.36}$$

The quantum number m runs from $-l \le m \le l$ in integer steps, and the rotational eigenvalue is independent of m. Thus there is a $(2l+1)$-fold *degeneracy* of the energy eigenvalues in Eq. (3.35).
The rotational partition function therefore becomes [recall Eq. (2.40)]

$$(\text{p.f.})_{\text{rot}} = \sum_{l=0}^{\infty}(2l+1)e^{-\hbar^2 l(l+1)/2Ik_{\text{B}}T} \qquad ; \text{rotation} \tag{3.37}$$

Although this cannot be evaluated in closed form, it is easy to get some limits:

- At *low temperature*, the terms in the sum decrease very rapidly and

$$(\text{p.f.})_{\text{rot}} = 1 + 3e^{-2\theta_{\text{R}}/T} + 5e^{-6\theta_{\text{R}}/T} + \cdots \qquad ; T \to 0$$
$$\theta_{\text{R}} \equiv \frac{\hbar^2}{2Ik_{\text{B}}} \tag{3.38}$$

- At *high temperature*, many terms contribute to the sum, and it can be converted to an integral as before[8]

$$(\text{p.f.})_{\text{rot}} = \int_0^{\infty} e^{-\theta_{\text{R}}u/T}\,du = \frac{T}{\theta_{\text{R}}} \qquad ; T \to \infty$$
$$= \frac{8\pi^2 I k_{\text{B}} T}{h^2} \tag{3.39}$$

This latter result can also be obtained from the classical rotation par-

[7]See, for example, [Walecka (2008)].
[8]See Prob. 3.2; note that if $l(l+1) = u$, then in the sum over integers $(2l+1)\Delta l = du$

tition function[9]

$$
\begin{aligned}
(\text{p.f.})_{\text{class}} &= \frac{1}{h^2}\int_0^{2\pi} d\phi \int_0^{\pi} d\theta \int_{-\infty}^{\infty} dp_\phi \int_{-\infty}^{\infty} dp_\theta \\
&\quad \times \exp\left\{-\frac{1}{2Ik_{\text{B}}T}\left(p_\theta^2 + \frac{1}{\sin^2\theta}p_\phi^2\right)\right\} \\
&= \frac{2\pi}{h^2}\left(2\pi I k_{\text{B}}T\right)^{1/2}\left(2\pi I k_{\text{B}}T\right)^{1/2}\int_0^{\pi} \sin\theta\, d\theta \\
&= \frac{8\pi^2 I k_{\text{B}}T}{h^2}
\end{aligned}
\tag{3.40}
$$

(4) *Electronic*: The energy of excitation of the electronic states is generally much larger than those of the molecular states considered above (we shall later review typical sizes of these quantities). For the present purposes, we assume that the electron configuration remains in the ground state (compare Prob. 3.3)

$$
\begin{aligned}
&\varepsilon^{\text{el}}(r_0) = U(r_0) \equiv \varepsilon_0 \\
&(\text{p.f.})_{\text{el}} = e^{-\varepsilon_0/k_{\text{B}}T} \qquad ; \text{ electronic ground state}
\end{aligned}
\tag{3.41}
$$

3.2.1.3 *Heat Capacity*

The Helmholtz free energy follows from the partition function through Eqs. (2.107) and (3.3)

$$
\begin{aligned}
A(T,V,N) &= -k_{\text{B}}T\ln\left[\frac{(\text{p.f.})_{\text{trans}}^N}{N!}(\text{p.f.})_{\text{vib}}^N(\text{p.f.})_{\text{rot}}^N(\text{p.f.})_{\text{el}}^N\right] \\
&= -k_{\text{B}}T\left[\ln\frac{(\text{p.f.})_{\text{trans}}^N}{N!} + \ln(\text{p.f.})_{\text{vib}}^N + \ln(\text{p.f.})_{\text{rot}}^N + \ln(\text{p.f.})_{\text{el}}^N\right]
\end{aligned}
\tag{3.42}
$$

The energy is immediately obtained from the Helmholtz free energy through Eq. (2.48)

$$
E = -T^2\frac{\partial}{\partial T}\left(\frac{A}{T}\right)_{N,V} \qquad ; \text{ energy} \tag{3.43}
$$

The constant-volume heat capacity is then given by Eq. (2.50)

$$
C_V = \left(\frac{\partial E}{\partial T}\right)_{N,V} \qquad ; \text{ heat capacity} \tag{3.44}
$$

[9]Note, in particular, the use of the canonical momenta in this calculation.

The molar constant-volume heat capacity $\mathcal{C}_V$, in units of R, for a gas of diatomic molecules in the electronic ground state is sketched in Fig. 3.4.

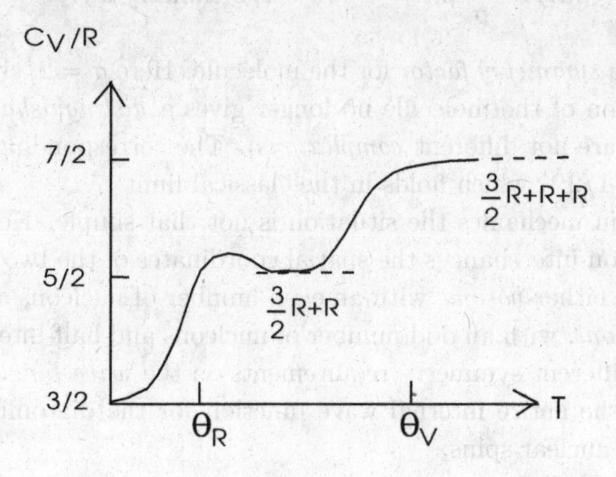

Fig. 3.4 Sketch of the molar constant-volume heat capacity $\mathcal{C}_V$, in units of R, for a gas of diatomic molecules in the electronic ground state (see Prob. 3.5).

Several comments:

- Given $(\theta_\mathrm{R}, \theta_\mathrm{V})$, this is a *universal curve*;
- In general $\theta_\mathrm{R} \ll \theta_\mathrm{V}$;
- At very low temperature, one sees only the translational contribution of $\mathcal{C}_V = (3/2)R$; however, since θ_R is so small, this is actually only observed for the lightest gases $\mathrm{H}_2, \mathrm{D}_2$, and HD;
- The high-temperature partition functions for both rotation and vibration are proportional to T, and therefore one finds a step increase in $\mathcal{C}_V$ of R for $T \gg \theta$ in each case;[10]
- Note the actual bump in $\mathcal{C}_V$ after θ_R;
- A good numerical calculation of the curve in Fig. 3.4 for representative values of $(\theta_\mathrm{R}, \theta_\mathrm{V})$, is assigned as Prob. 3.5.

3.2.1.4 *Symmetry of the Wave Function*

The above discussion is for diatomic molecules with distinct nuclei. If the nuclei are the *same* ("homonuclear molecules"), then there is an additional symmetry in the problem.

[10]See Prob. 3.4.

Classically, one takes

$$(\text{p.f.}) \to \frac{1}{\sigma}(\text{p.f.}) \qquad ; \ \sigma \text{ is symmetry factor} \tag{3.45}$$

where σ is the *symmetry factor* for the molecule. Here $\sigma = 2$, since an end-for-end rotation of the molecule no longer gives a *distinguishable configuration* (these are not different *complexions*). The corresponding correction factor in Ω is $1/2^N$, which holds in the classical limit.

In quantum mechanics the situation is not that simple. Here the end-for-end rotation interchanges the spatial coordinates of the two nuclei, and the nuclei are either *bosons*, with an even number of nucleons and integral spin, or *fermions*, with an odd number of nucleons and half-integral spin.[11] This places different symmetry requirements on the *wave function*.

Consider the entire internal wave function for the diatomic molecule, including the nuclear spins,

$$\Psi_{\text{tot}} = \psi_{\text{el}}(\mathbf{r}_i; \mathbf{r})\, \psi_{\text{vib}}(x)\, \psi_{\text{rot}}(\theta, \phi)\, \psi_{\text{nuclear spins}} \tag{3.46}$$

Now carry out the transformation ($\mathbf{r} \to -\mathbf{r}$) by letting $(\theta, \phi) \to (\pi - \theta, \phi + \pi)$, which rotates the molecule end-for-end. Then

(1) The behavior of the rotational wave function under the end-for-end rotation follows from that of the spherical harmonics in Eq. (3.36)

$$\psi_{\text{rot}}(\pi - \theta, \phi + \pi) = (-1)^l \psi_{\text{rot}}(\theta, \phi) \tag{3.47}$$

(2) Since $x = r - r_0$, The vibrational wave function is unchanged;

$$\psi_{\text{vib}}(x) \to \psi_{\text{vib}}(x) \tag{3.48}$$

(3) What about the electronic wave function? Let us ask what sort of single-electron wave functions we might expect to see for the molecule. The Hartree-Fock single-particle wave functions[12] will be even or odd about the C-M, reflecting the symmetry of the overall Coulomb potential. In the linear combination of atomic orbitals (LCAO) method, tbe molecular one-electron wave functions formed from the ground-state 1s

[11] A nucleus is said to be "odd" or "even" depending on whether A, the total number of nucleons in it, is odd or even.

[12] See, for example, [Fetter and Walecka (2003)].

wave functions for each atom are given by[13]

$$\psi_{1s}^{\rm sym}(\mathbf{r}_1;\mathbf{r}) = \mathcal{N}\left[\psi_{1s}\left(\left|\mathbf{r}_1 - \frac{\mathbf{r}}{2}\right|\right) + \psi_{1s}\left(\left|\mathbf{r}_1 + \frac{\mathbf{r}}{2}\right|\right)\right] \quad ; \text{ground state } \mathrm{H}_2^+$$

$$\psi_{1s}^{\rm anti}(\mathbf{r}_1;\mathbf{r}) = \tilde{\mathcal{N}}\left[\psi_{1s}\left(\left|\mathbf{r}_1 - \frac{\mathbf{r}}{2}\right|\right) - \psi_{1s}\left(\left|\mathbf{r}_1 + \frac{\mathbf{r}}{2}\right|\right)\right] \qquad \text{LCAO} \qquad (3.49)$$

These wave functions satisfy

$$\psi_{1s}^{\rm sym}(\mathbf{r}_1;-\mathbf{r}) = \psi_{1s}^{\rm sym}(\mathbf{r}_1;\mathbf{r})$$
$$\psi_{1s}^{\rm anti}(\mathbf{r}_1;-\mathbf{r}) = -\psi_{1s}^{\rm anti}(\mathbf{r}_1;\mathbf{r}) \qquad (3.50)$$

They are sketched in Fig. 3.5.

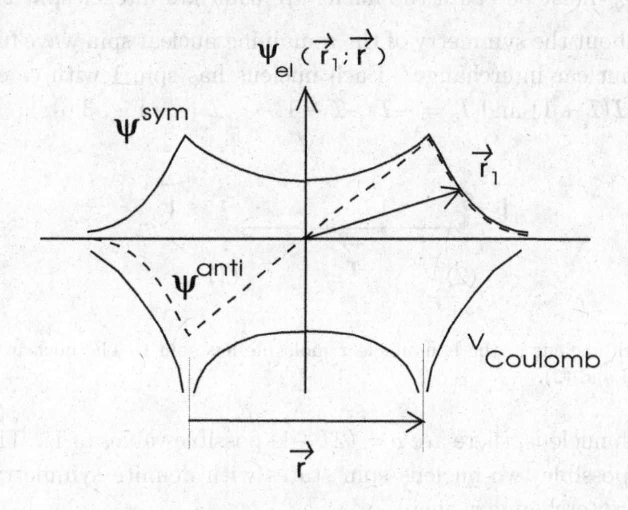

Fig. 3.5 Sketch of the even and odd single-electron wave functions $\psi_{1s}^{\rm sym}(\mathbf{r}_1;\mathbf{r})$ and $\psi_{1s}^{\rm anti}(\mathbf{r}_1;\mathbf{r})$ in Eqs. (3.49), obtained from the atomic 1s states with the LCAO method in the homonuclear diatomic molecule. The wave functions have cylindrical symmetry about the horizontal axis.

The *binding orbitals* are generally the even functions, since they make maximum use of the attractive Coulomb potential in the mid-plane coming from both nuclei; the odd orbitals have a node in the mid-plane. In fact, the LCAO in the first line of Eq. (3.49) provides an excellent variational estimate for the ground-state of the molecule H_2^+. Thus, in general, the ground-state electronic wave function of homonuclear diatomic molecules is *even*, or *unchanged*, under the transformation

[13]These levels can each be filled with two electrons of opposite spin.

$\mathbf{r} \to -\mathbf{r}$

$$\psi_{\text{g.s.}}(\mathbf{r}_i; -\mathbf{r}) = \psi_{\text{g.s.}}(\mathbf{r}_i; \mathbf{r}) \qquad ; \text{ground state} \tag{3.51}$$

We will assume this to be the case for the homonuclear diatomic molecules of interest.[14]

(4) The whole effect of the transformation $\mathbf{r} \to -\mathbf{r}$ is now to interchange the spatial coordinates of the two nuclei. We then have the following conditions on Ψ_{tot} when nuclear spin exchange is included, which follow from either Bose or Fermi statistics for the two identical nuclei:[15]

- Ψ_{tot} must be *even* if the nuclei are *even* (integer nuclear spin $\mathcal{I}$);
- Ψ_{tot} must be *odd* if the nuclei are *odd* (half-integer spin $\mathcal{I}$).

(5) What about the symmetry of the remaining nuclear spin wave functions under nuclear interchange? Each nucleus has spin $\mathbf{I}$ with eigenvalues $\mathbf{I}^2 = \hbar^2\mathcal{I}(\mathcal{I}+1)$ and $I_z = -\mathcal{I}, -\mathcal{I}+1, \cdots, \mathcal{I}$ (see Fig. 3.6).

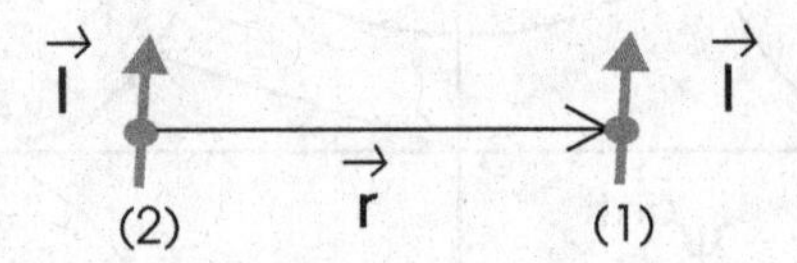

Fig. 3.6 Each nucleus in the homonuclear molecule has spin **I**. The nuclear spins are labeled by (1) and (2).

For each nucleus, there are $\rho = (2\mathcal{I}+1)$ possible values of I_z. The number of possible two-nucleus spin states with definite symmetry under nuclear interchange is then

- Number of *antisymmetric* combinations is $\rho(\rho-1)/2$;
- Number of *symmetric* combinations is $\rho(\rho+1)/2$;
- *Total* number of two-nucleus spin states is ρ^2.

For example, for nuclear spin 1/2 with the two spin states $\alpha = \psi_\uparrow$ and $\beta = \psi_\downarrow$, one has the wave functions shown in Table 3.1. Molecules with symmetric nuclear spin wave functions and total spin $\mathcal{I} = 1$ in this case are said to be in the "ortho" state, while those with antisymmetric

[14]There are exceptions.

[15]The Pauli principle states that the wave function must be *unchanged* under the interchange of the space and spin coordinates for two identical bosons, while it must *change sign* under the interchange of the space and spin coordinates for two identical fermions.

nuclear spin wave functions and total spin $\mathcal{I} = 0$ are in the "para" state.

Table 3.1 Two-nucleus spin wave functions for nuclear spin $\mathcal{I} = 1/2$ with $\rho = 2\mathcal{I} + 1 = 2$. The total nuclear spin $\mathcal{I}$, and the symmetry of these wave functions under nuclear spin interchange $(1) \rightleftharpoons (2)$ is indicated.

wave function	$\mathcal{I}$	symmetry	name
$\alpha(1)\alpha(2)$	1	symmetric	"ortho"
$[\alpha(1)\beta(2) + \alpha(2)\beta(1)]/\sqrt{2}$	1	symmetric	"ortho"
$\beta(1)\beta(2)$	1	symmetric	"ortho"
$[\alpha(1)\beta(2) - \alpha(2)\beta(1)]/\sqrt{2}$	0	antisymmetric	"para"

A combination of the above observations then leads to the following results for homonuclear diatomic molecules with an even electronic wave function:

- If the nuclei are *odd*, then they are *fermions*. The overall wave function must then *change sign* under end-for-end rotation and nuclear spin exchange. It follows from Eq. (3.47) that
  - If the angular momentum l of the molecule is even, the nuclear spin wave function must be antisymmetric. There are $\rho(\rho - 1)/2$ of these antisymmetric wave functions, where $\rho = 2\mathcal{I} + 1$;
  - If the angular momentum l of the molecule is odd, the nuclear spin wave function must be symmetric. There are $\rho(\rho + 1)/2$ of these symmetric wave functions.
- If the nuclei are *even*, then they are *bosons*. The overall wave function must then be *unchanged* under end-for-end rotation and nuclear spin exchange. Thus
  - If the angular momentum l of the molecule is even, the nuclear spin wave function must be symmetric. There are $\rho(\rho + 1)/2$ of these symmetric wave functions;
  - If the angular momentum l of the molecule is odd, the nuclear spin wave function must be antisymmetric. There are $\rho(\rho - 1)/2$ of these antisymmetric wave functions.

With the inclusion of the appropriate degeneracy factor in each case, the rotational partition functions for homonuclear diatomic molecules with an even electronic wave function are then as follows:

$$(\text{p.f.})_{\text{rot}} = \frac{1}{2}\rho(\rho-1)\sum_{\text{even } l}(2l+1)e^{-l(l+1)\theta_{\text{R}}/T} + \frac{1}{2}\rho(\rho+1)\sum_{\text{odd } l}(2l+1)e^{-l(l+1)\theta_{\text{R}}/T} \quad ; \text{ odd A} \tag{3.52}$$

and

$$(\text{p.f.})_{\text{rot}} = \frac{1}{2}\rho(\rho+1)\sum_{\text{even } l}(2l+1)e^{-l(l+1)\theta_{\text{R}}/T} + \frac{1}{2}\rho(\rho-1)\sum_{\text{odd } l}(2l+1)e^{-l(l+1)\theta_{\text{R}}/T} \quad ; \text{ even A} \tag{3.53}$$

A few comments:

- The number of states, and hence the counting in the microcanonical ensemble, is directly determined by the *nuclear statistics.* The nuclear statistics enter even though there are no nuclear *interactions* in this problem;
- In the high-temperature limit many states contribute to the partition function, and the sums over even and odd l become *identical*

$$\sum_{\text{even } l} = \sum_{\text{odd } l} = \frac{1}{2}\sum_{l} \quad ; T \to \infty \tag{3.54}$$

  In this case, the two partition functions reduce to the same expression

$$(\text{p.f.})_{\text{rot}} = \frac{1}{\sigma}\left[\rho^2\sum_{l}(2l+1)e^{-l(l+1)\theta_{\text{R}}/T}\right] \quad ; T \to \infty$$
$$\sigma = 2 \tag{3.55}$$

  This is just the result in Eq. (3.45) with the correct symmetry factor of $\sigma = 2$. Here the spin degeneracy ρ of each nucleus is also included;[16]
- *If the nuclei are spin-zero bosons ($\rho = 1$), then only even l are allowed.* The overall wave function must be *invariant* under end-for-end rotation of the molecule in this case. It is a striking fact that the rotational spectrum of molecular O_2, for example, is missing all the odd-l states!

[16]See Probs. 3.7–3.8; notice, in particular, the factor of 1/2 in the last of Eqs. (3.54).

3.2.1.5 *Ortho- and Para-Hydrogen* H_2

The result for the molar heat capacity C_V/R for molecular H_2 calculated from Eq. (3.52), with $A = 1, \mathcal{I} = 1/2$, and $\rho = 2$, is shown in Fig. 3.7. It is clear from a comparison with the observed values indicated by the dots that the above expression is *completely wrong* at low temperature!

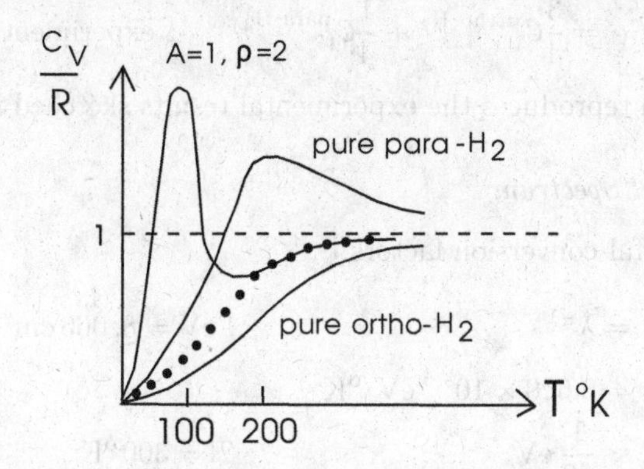

Fig. 3.7 Sketch of the molar heat capacity C_V/R for molecular H_2. The result calculated from Eq. (3.52) is labeled with the appropriate values of $A = 1$ and $\rho = 2$ ($\mathcal{I} = 1/2$). The corresponding results for pure para-H_2 and pure ortho-H_2 are also shown (see Prob. 3.6). The behavior of the observed value of C_V/R for molecular H_2 is indicted by the dots (see, for example, [Wannier (1987)]).

The solution to this problem was given by [Dennison (1927)]. The above analysis assumes *statistical equilibrium* at each temperature. In fact, at low temperature Eq. (3.52) gives

$$\begin{aligned} \frac{N_{\text{ortho-H}_2}}{N_{\text{para-H}_2}} &\rightarrow \frac{9e^{-2\theta_R/T}}{1+5e^{-6\theta_R/T}} \qquad ; T \rightarrow 0 \\ &\rightarrow 0 \end{aligned} \tag{3.56}$$

This says that as $T \rightarrow 0$, all the molecules should be in the para-state. Now statistical equilibrium relies on collisions to equilibrate the ortho- and para-species. These collisions must be able to change the *nuclear spin states*, and ordinary molecular collisions are extremely inefficient in this regard. As a result, the statistical occupation of the nuclear spin states that is obtained over long times at room temperature is *frozen in* as the temperature is lowered, and what one has is a *metastable* assembly that retains the high-temperature ratio of

$$\frac{N_{\text{ortho-H}_2}}{N_{\text{para-H}_2}} = \frac{\rho(\rho+1)}{\rho(\rho-1)} = \frac{3}{1} \qquad ; \text{ “frozen in”} \tag{3.57}$$

In this case, the heat capacity is obtained from an assembly that is 3/4 ortho-H_2 and 1/4 para-H_2

$$\mathcal{C}_V = \frac{3}{4}\mathcal{C}_V^{\text{ortho-H}_2} + \frac{1}{4}\mathcal{C}_V^{\text{para-H}_2} \qquad ; \text{ experiment} \tag{3.58}$$

This expression reproduces the experimental results sketched in Fig. 3.7.[17]

3.2.1.6 *Typical Spectrum*

First, some useful conversion factors

$$\begin{aligned} \frac{\Delta\varepsilon}{hc} &= \lambda^{-1} & &; 1\,\text{eV} = 8{,}066\,\text{cm}^{-1} \\ k_{\text{B}} &= 8.620 \times 10^{-5}\,\text{eV}/^{\text{o}}\text{K} & & \\ k_{\text{B}}T &\approx \frac{1}{40}\,\text{eV} & &; T = 300\,^{\text{o}}\text{K} \end{aligned} \tag{3.59}$$

For orientation, the excitation energies of the various modes for a typical diatomic molecule are then indicated in Fig. 3.8.

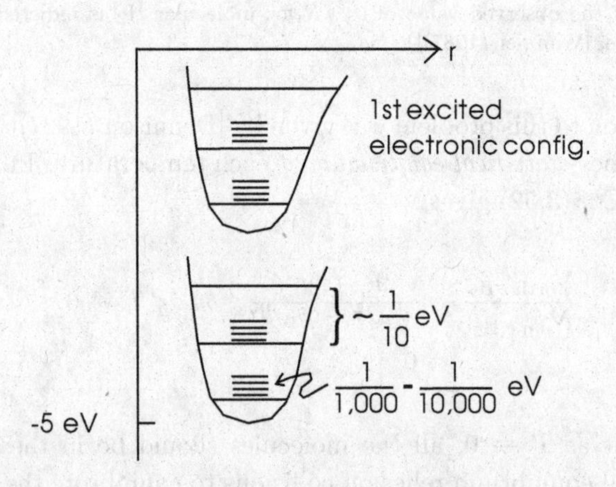

Fig. 3.8 Typical energies for rotational, vibrational, and electronic excitations of a diatomic molecule.

[17]See Prob. 3.6. It is worth re-emphasizing the very valuable lesson learned here: *statistical mechanics assumes statistical equilibrium at each temperature.*

A few comments:

- Rotation-vibration spectra, arising from transitions between vibrational and rotational states for a given electronic state (Fig. 3.9) are in the *infrared*;

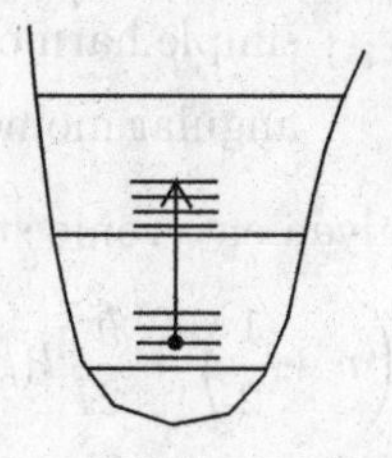

Fig. 3.9 Rotation-vibration spectra arising from transitions between vibrational and rotational states for a given electronic state.

- Electronic spectra, arising from transitions between electronic states are in the *visible* and *ultraviolet*;
- At room temperature, while the electronic and vibrational modes are in their lowest states, the rotational states are *thermally populated*;
- Typically

$$\theta_{\rm R} \ll T_{\rm room} \qquad ; \ \theta_{\rm V} \gg T_{\rm room} \tag{3.60}$$

3.2.1.7 *Selection Rules*

Some interesting results follow when the above considerations are combined with the selection rules for the dominant electric-dipole radiation.

Rotation-Vibration Spectra: The rate for transitions between the vibrational and rotational states in Fig. 3.9 is governed by the transition matrix elements of the electric dipole operator

$$M_{fi} \propto \langle \Psi_f | \mathbf{r} | \Psi_i \rangle \qquad ; \text{ electric dipole operator} \tag{3.61}$$

where $\mathbf{r}$ is the relative nuclear coordinate.[18] In spherical coordinates, the components of this vector can be written

$$r_{1m} = (r_0 + x)\left(\frac{4\pi}{3}\right)^{1/2} Y_{1m}(\theta, \phi) \qquad ; \ m = 0, \pm 1 \tag{3.62}$$

[18] See [Schiff (1968)] and Prob. 3.10.

Consider the rotation-vibration spectra arising from the transitions illustrated in Fig. 3.9. The selection rules for the transition matrix elements of the operator in Eq. (3.62) taken between the vibrational simple-harmonic-oscillator states in Eq. (3.33), and the rotational spherical harmonic eigenstates in Eq. (3.36), are[19]

$$\begin{aligned} \Delta n &= \pm 1 \qquad &&; \text{ simple harmonic oscillator} \\ \Delta l &= \pm 1 \qquad &&; \text{ angular momentum and parity} \end{aligned} \tag{3.63}$$

The energy eigenvalues for a given electronic state of the molecule are

$$\varepsilon_{nl} = h\nu_{\rm osc}\left(n + \frac{1}{2}\right) + \frac{\hbar^2}{2I}l(l+1) + \varepsilon_0 \tag{3.64}$$

The photon absorption frequency for a rotation-vibration transition with $n \to n+1$ and $l \to l+1$, for example, is therefore

$$\begin{aligned} h\nu &= \varepsilon_{n+1,l+1} - \varepsilon_{nl} \\ &= h\nu_{\rm osc} + \frac{\hbar^2}{2I}[(l+1)(l+2) - l(l+1)] \\ &= h\nu_{\rm osc} + \frac{h^2}{4\pi^2 I}(l+1) \qquad ; \; l = 0, 1, 2, 3, \cdots \end{aligned} \tag{3.65}$$

For a rotation-vibration transition with $n \to n+1$ and $l \to l-1$, the photon absorption frequency is

$$\begin{aligned} h\nu &= \varepsilon_{n+1,l-1} - \varepsilon_{nl} \\ &= h\nu_{\rm osc} + \frac{\hbar^2}{2I}[(l-1)(l) - l(l+1)] \\ &= h\nu_{\rm osc} + \frac{h^2}{4\pi^2 I}(-l) \qquad ; \; l = 1, 2, 3, \cdots \end{aligned} \tag{3.66}$$

In both cases the photon frequency is linear in l, and one sees *rotational bands* in the rotation-vibration absorption spectra with a uniform spacing (see Fig. 3.10)

$$\begin{aligned} \Delta\nu &= \frac{h}{4\pi^2 I} \qquad &&; \text{ spacing in rotational band} \\ \nu &= \nu_{\rm osc} \qquad &&; \text{ missing} \end{aligned} \tag{3.67}$$

Since the rotational state must change, the central line with $\nu = \nu_{\rm osc}$ is *missing*. This provides a useful identification feature in the spectra.

---

[19]See Prob. 3.9.

Since the initial rotational states are statistically populated at room temperature, one sees a variation in the strength of the absorption lines which reflects that population. Our statistical arguments say that the population of the initial rotational states is given by

$$\frac{n_l^\star}{N} = \frac{(2l+1)e^{-l(l+1)\hbar^2/2Ik_BT}}{(\text{p.f.})_{\text{rot}}} \quad ; \text{ initial states} \qquad (3.68)$$

With the insertion of Eq. (3.39), this becomes

$$n_l^\star = N\left(\frac{\theta_R}{T}\right)(2l+1)e^{-l(l+1)\theta_R/T} \quad ; \; T \gg \theta_R$$

$$\theta_R = \frac{\hbar^2}{2Ik_B} = \frac{h^2}{8\pi^2 Ik_B} \qquad (3.69)$$

The implied modulation of the absorption strengths is sketched in Fig. 3.10. Note that the band spacing measures the moment of inertia I, while the strength distribution then measures the temperature T.

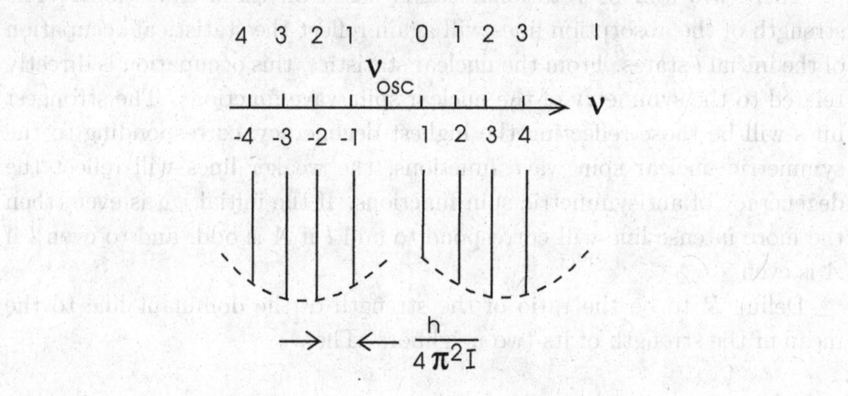

Fig. 3.10 Rotation-vibration rotational band in the photon absorption spectra as given by Eqs. (3.65) and (3.66). The spacing within the band is given by $\Delta\nu = h/4\pi^2 I$ where I is the moment of inertia. Note that the central photon frequency $\nu = \nu_{\text{osc}}$ is *missing.* The initial value of l is also shown. The strength modulation due to the statistical population of the initial low-lying rotational states at room temperature in Eq. (3.69) is indicated.

Homonuclear Diatomic Molecules: Since there is no interaction with the nuclear spins, electric dipole radiation cannot change the symmetry of the nuclear spins states. However, since $\Delta l = \pm 1$, this symmetry *must* change with a given electronic state due to the nuclear statistics [recall Eqs. (3.52)–

(3.53)]. As a consequence, electric dipole transitions in a homonuclear diatomic molecule must involve a *change in the electronic state* (Fig. 3.11).[20]

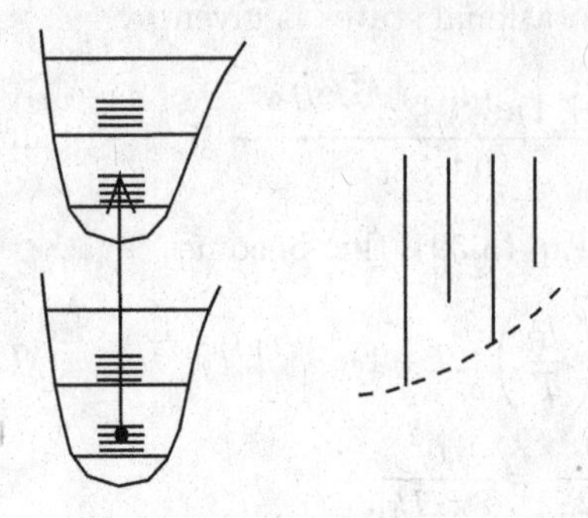

Fig. 3.11 Rotation-electronic band with strength distribution of absorption lines reflecting the statistical occupation of the nuclear spin states in a homonuclear molecule.

There will still be rotational bands based on these transitions. The strength of the absorption lines will again reflect the statistical occupation of the initial l states. From the nuclear statistics, this occupation is directly related to the symmetry of the nuclear spin wave functions. The strongest lines will be those reflecting the highest degeneracy, corresponding to the symmetric nuclear spin wave functions; the weaker lines will reflect the degeneracy of antisymmetric spin functions. If the initial $\psi_{\rm el}$ is even, then the more intense line will correspond to odd l if A is odd, and to even l if A is even.

Define $\mathcal{R}$ to be the ratio of the strength of the dominant line to the mean of the strength of its two neighbors. Then

$$\mathcal{R} \equiv \frac{\text{strength of dominant line}}{\text{mean of two side lines}} = \frac{\rho(\rho+1)}{\rho(\rho-1)} = \frac{\mathcal{I}+1}{\mathcal{I}} \tag{3.70}$$

This strength ratio measures the nuclear spin $\mathcal{I}$! Note that all slowly varying factors (transition probability, Boltzmann factor, *etc.*) cancel if $\mathcal{I}$ is determined in this manner.

[20]A homonuclear diatomic molecule in a given electronic state has no electric dipole moment (see Prob. 3.10).

3.2.2 *Polyatomic Molecules*

We start with the simplest extension to polyatomic molecules, and that is the *symmetric top*, examples of which are shown in Fig. 3.12.

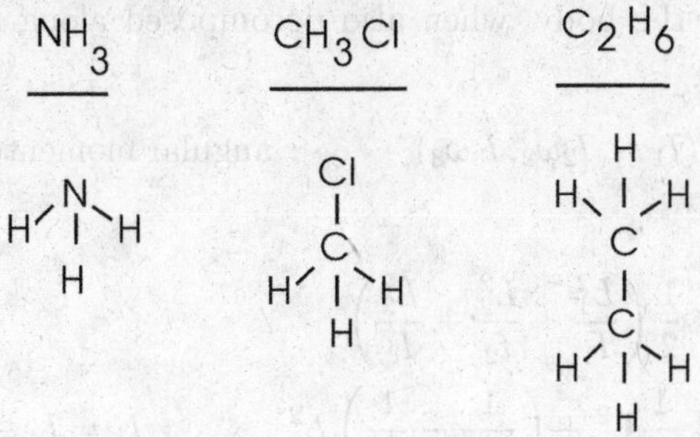

NH_3 $\quad$ CH_3Cl $\quad$ C_2H_6

Fig. 3.12 Examples of a symmetric top.

3.2.2.1 *Symmetric Top*

The moment of inertia tensor for an arbitrary rigid body can be put on principal axes, where it is diagonal with moments of inertia (I_1, I_2, I_3). A symmetric top is a body where two of the principal moments are equal, and unequal to the third (see Fig. 3.13)

$$I_1 = I_2 \neq I_3 \qquad ; \text{ symmetric top} \tag{3.71}$$

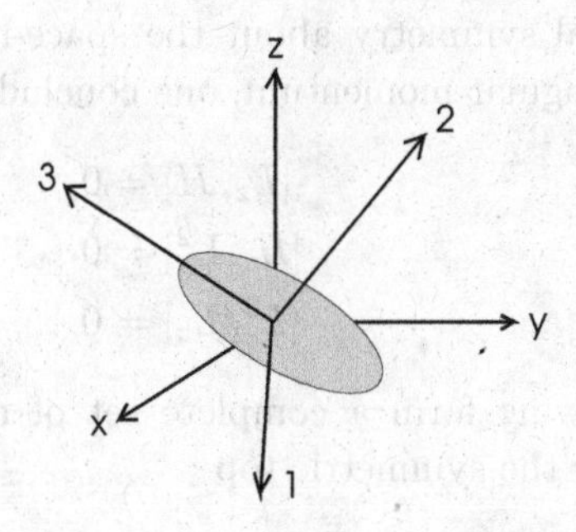

Fig. 3.13 Principal axes for the symmetric top where $I_1 = I_2 \neq I_3$.

When the instantaneous angular velocity $\boldsymbol{\omega} = (\omega_1, \omega_2, \omega_3)$ is decomposed along principal axes in the body-fixed frame, the kinetic energy for a symmetric top takes the form[21]

$$T = H = \frac{1}{2}\left(I_1\omega_1^2 + I_2\omega_2^2 + I_3\omega_3^2\right) \qquad ; \boldsymbol{\omega} = (\omega_1, \omega_2, \omega_3) \qquad (3.72)$$

For a free top, the kinetic energy is also the hamiltonian. The angular momentum $\mathbf{L}$ of the body, when also decomposed along prinicipal axes, takes the form

$$\mathbf{L} = (I_1\omega_1, I_2\omega_2, I_3\omega_3) \qquad ; \text{angular momentum} \qquad (3.73)$$

Hence

$$\begin{aligned} H &= \frac{1}{2}\left(\frac{L_1^2}{I_1} + \frac{L_2^2}{I_2} + \frac{L_3^2}{I_3}\right) \\ &= \frac{1}{2I}\mathbf{L}^2 + \left(\frac{1}{2I_3} - \frac{1}{2I}\right)L_3^2 \qquad ; I_1 = I_2 \equiv I \end{aligned} \qquad (3.74)$$

where we have identified $\mathbf{L}^2 = L_1^2 + L_2^2 + L_3^2$ and defined $I_1 = I_2 \equiv I$.

The spectrum of eigenvalues of this H can be obtained without doing any actual calculations, and this is accomplished as follows:

(1) For a free top the total angular momentum must be a constant of the motion, and therefore[22]

$$\begin{aligned} & [\mathbf{L}^2, H] = 0 \\ \Longrightarrow \quad & [L_3^2, \mathbf{L}^2] = 0 \\ \Longrightarrow \quad & [L_3^2, H] = 0 \end{aligned} \qquad (3.75)$$

(2) From the rotational symmetry about the space-fixed z-axis, and the properties of the angular momentum, one concludes that

$$\begin{aligned} & [L_z, H] = 0 \\ & [L_z, \mathbf{L}^2] = 0 \\ \Longrightarrow \quad & [L_3^2, L_z] = 0 \end{aligned} \qquad (3.76)$$

Therefore the following form a complete set of mutually commuting hermitian operators for the symmetric top

$$\{H, \mathbf{L}^2, L_z, L_3^2\} \quad \text{form complete set of mutually commuting operators} \qquad (3.77)$$

[21] See [Fetter and Walecka (2003a)]; it is also shown there that the unit vectors $(\mathbf{e}_1, \mathbf{e}_2, \mathbf{e}_3)$ defining the principal axes are orthonormal, with $\mathbf{e}_i \cdot \mathbf{e}_s = \delta_{rs}$.

[22] Recall $[\mathbf{L}^2, \mathbf{L}^2] \equiv 0$.

In fact, one can actually use just L_3, rather than L_3^2. This corresponds to the fact that the projection of the angular momentum along the figure axis is *also* a constant of the motion for a symmetric top.

The eigenfunctions for the symmetric top are actually the rotation matrices (see [Edmonds (1974)])

$$\psi^L_{L_z L_3}(\alpha,\beta,\gamma) = \mathcal{N}\mathcal{D}^l_{m\kappa}(\alpha,\beta,\gamma) \qquad ; \; l = 0,1,2,\cdots \qquad -l \leq (m,\kappa) \leq l \tag{3.78}$$

where (α, β, γ) are the Euler angles. As usual, the eigenvalues of $\mathbf{L}^2$ are $\hbar^2 l(l+1)$ with $l = 0, 1, 2, \cdots$. The eigenvalues of L_z are $\hbar m$ where $m = -l, -l+1, \cdots, l$, and the eigenvalues of the projection of the angular momentum along the figure axis are $L_3 = \hbar\kappa$ where *also* $\kappa = -l, -l+1, \cdots, l$. Thus

$$\begin{aligned}
\mathbf{L}^2 &= \hbar^2 l(l+1) \qquad &&; \; l = 0,1,2,\cdots \\
L_z &= \hbar m &&; \; m = -l, -l+1, \cdots, l \\
L_3 &= \hbar\kappa &&; \; \kappa = -l, -l+1, \cdots, l
\end{aligned} \tag{3.79}$$

The energy eigenvalues of the hamiltonian in Eq. (3.74) are therefore

$$\varepsilon^{\text{top}} = \frac{\hbar^2}{2I} l(l+1) + \hbar^2 \left(\frac{1}{2I_3} - \frac{1}{2I}\right)\kappa^2 \qquad ; \text{ symmetric top} \tag{3.80}$$

where the quantum numbers take the values in Eqs. (3.79). These eigenvalues are again independent of the overall spatial orientation of the top, and thus there is again a $(2l+1)$*-fold degeneracy in* m.

3.2.2.2 *Partition Function*

The partition function for the symmetric top follows as

$$(\text{p.f.})_{\text{top}} = \sum_{l=0}^{\infty} \sum_{\kappa=-l}^{l} (2l+1) \exp\left\{-\frac{\hbar^2 l(l+1)}{2Ik_{\text{B}}T}\right\} \exp\left\{-\frac{\hbar^2\kappa^2}{2k_{\text{B}}T}\left(\frac{1}{I_3} - \frac{1}{I}\right)\right\} \qquad ; \text{ symmetric top} \tag{3.81}$$

This partition function cannot be evaluated in closed form, but it is again relatively easy to extract the high-temperature limit. First interchange the orders of summation, making use of the diagram in Fig. 3.14.

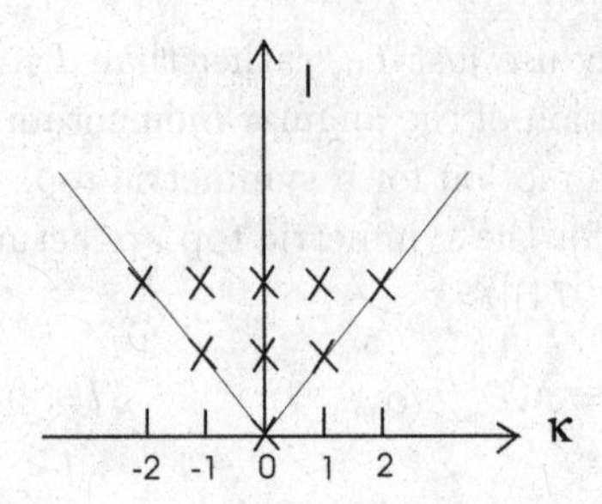

Fig. 3.14 Terms in the partition function for the symmetric top. l runs from 0 to ∞, and κ runs from $-l$ to l.

$$(\text{p.f.})_{\text{top}} = \sum_{\kappa=-\infty}^{\infty} \exp\left\{-\frac{\hbar^2\kappa^2}{2k_BT}\left(\frac{1}{I_3}-\frac{1}{I}\right)\right\}\sum_{l=|\kappa|}^{\infty}(2l+1)\exp\left\{-\frac{\hbar^2 l(l+1)}{2Ik_BT}\right\} \tag{3.82}$$

Now change the sum to an integral in the second factor, exactly as in Eq. (3.39)

$$\begin{aligned}\sum_{l=|\kappa|}^{\infty}(2l+1)\exp\left\{-\frac{\hbar^2 l(l+1)}{2Ik_BT}\right\} &= \int_{|\kappa|(|\kappa|+1)}^{\infty} e^{-\theta_R u/T}\,du \qquad ; T\to\infty \\ &= \frac{T}{\theta_R}e^{-\theta_R|\kappa|(|\kappa|+1)/T} \qquad \theta_R = \frac{\hbar^2}{2Ik_B}\end{aligned} \tag{3.83}$$

Thus

$$\begin{aligned}(\text{p.f.}) &= \frac{T}{\theta_R}\sum_{\kappa=-\infty}^{\infty}\exp\left\{-\frac{\hbar^2\kappa^2}{2k_BT}\left(\frac{1}{I_3}-\frac{1}{I}\right)\right\}\exp\left\{-\frac{\hbar^2}{2k_BT}\left(\frac{|\kappa|(|\kappa|+1)}{I}\right)\right\} \\ &= \frac{T}{\theta_R}\sum_{\kappa=-\infty}^{\infty}\exp\left\{-\frac{\hbar^2}{2I_3k_BT}\left(\kappa^2+\frac{I_3}{I}|\kappa|\right)\right\}\end{aligned} \tag{3.84}$$

This last result can, in turn, be written

$$\begin{aligned}(\text{p.f.})_{\text{top}} &= \frac{T}{\theta_R}\exp\left\{\frac{\hbar^2}{2I_3k_BT}\left(\frac{I_3}{2I}\right)^2\right\}\mathcal{S} \qquad ; T\to\infty \\ \mathcal{S} &\equiv \sum_{\kappa=-\infty}^{\infty}\exp\left\{-\frac{\hbar^2}{2I_3k_BT}\left(|\kappa|+\frac{I_3}{2I}\right)^2\right\}\end{aligned} \tag{3.85}$$

Conversion of the sum to an integral in the high-T limit gives

$$\mathcal{S} = 2\int_0^\infty d\kappa \exp\left\{-\frac{\hbar^2}{2I_3 k_\mathrm{B}T}\left(\kappa + \frac{I_3}{2I}\right)^2\right\}$$
$$= 2\int_0^\infty dt \exp\left\{-\frac{\hbar^2 t^2}{2I_3 k_\mathrm{B}T}\right\} - 2\int_0^{I_3/2I} dt \exp\left\{-\frac{\hbar^2 t^2}{2I_3 k_\mathrm{B}T}\right\} \qquad (3.86)$$

where we have changed variables to $t = \kappa + I_3/2I$ and then explicity subtracted the contribution from 0 to $I_3/2I$. The last term is of $O(1)$ as $T \to \infty$, while the first term is our familiar gaussian integral, which is of $O(T^{1/2})$. Thus

$$\mathcal{S} = \sqrt{\pi}\left(\frac{2I_3 k_\mathrm{B}T}{\hbar^2}\right)^{1/2} \quad ; T \to \infty \qquad (3.87)$$

Hence Eqs. (3.85) give[23]

$$(\text{p.f.})_\text{top} = \sqrt{\pi}\left(\frac{8\pi^2 k_\mathrm{B}T}{h^2}\right)^{3/2}\left(I^2 I_3\right)^{1/2} \quad ; T \to \infty \qquad (3.88)$$

We comment on these results:

- For a *diatomic molecule*, which is *also* a symmetric top, the third moment of inertia becomes vanishingly small, In this case the energy of excitation in Eq. (3.80) for anything but the $\kappa = 0$ mode becomes prohibitively large, and thus

$$I_3 \to 0 \quad \Longrightarrow \kappa = 0 \qquad ; \text{ diatomic molecule} \qquad (3.89)$$

  One then recovers the energy spectrum in Eq. (3.35);
- A *spherical top* has *all* of its moments of inertia equal

$$I_3 = I \qquad ; \text{ spherical top} \qquad (3.90)$$

  Examples of a spherical top are shown in Fig. 3.15. The condition $I_3 = I$ again reduces the energy spectrum to Eq. (3.35);
- For an *asymmetric top*, one makes the following replacement in Eq. (3.88)[24]

$$(I^2 I_3)^{1/2} \to (I_1 I_2 I_3)^{1/2} \qquad ; \text{ asymmetric top} \qquad (3.91)$$

---

[23]Note that $(\text{p.f.})_\text{top} \to (T/\theta_R)\mathcal{S}$ as $T \to \infty$.
[24]See, for example, [Davidson (2003)].

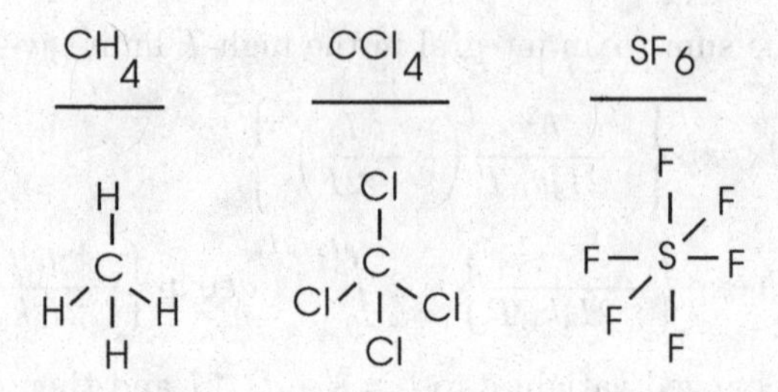

Fig. 3.15 Examples of a spherical top.

- To get the full internal partition function, it is necessary to include the vibrational partition function $(\text{p.f.})_{\text{vib}}$. To compute this, one must carry out a normal-mode analysis of the vibrations of the molecule, making use of its symmetry properties;[25]
- It may be necessary to also include $(\text{p.f.})_{\text{el}} = \omega_1 e^{-\varepsilon_1/k_{\text{B}}T} + \omega_2 e^{-\varepsilon_2/k_{\text{B}}T} + \cdots$, if there are a few low-lying electronic levels;
- It is a striking fact that all of the above features of molecular spectra, including intrinsic, vibrational, and rotational excitations, are also observed in the spectra of *highly-deformed nuclei!* (see [Bohr and Mottelson (1975)]).

3.2.2.3 *Hindered Rotation*

Molecules also exhibit *hindered rotation* of subunits. As a simple model for this, consider the one-dimensional rotor illustrated in Fig. 3.16.

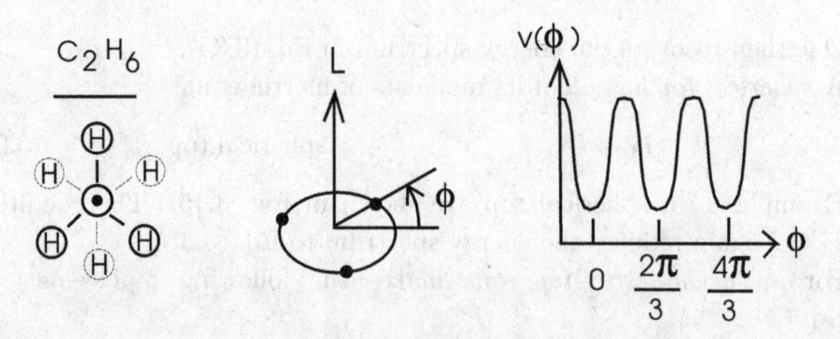

Fig. 3.16 One-dimensional rotor as a model for the hindered internal rotation of the two carbon groups in ethane. The figure on the left looks down the axis of the molecule.

[25]See, for example, [Davidson (2003); Wilson, Decius, and Cross (1980)].

The hamiltonian is

$$H = \frac{1}{2}I\dot{\phi}^2 + v(\phi) \qquad ; \text{ one-dimensional rotor}$$
$$= \frac{p_\phi^2}{2I} + v(\phi) \qquad ; p_\phi = I\dot{\phi} \tag{3.92}$$

The corresponding classical partition function is given by

$$(\text{p.f.})_{\text{class}} = \frac{1}{\sigma h}\int_{-\infty}^{\infty} dp_\phi \int_0^{2\pi} d\phi\, e^{-p_\phi^2/2Ik_BT}\, e^{-v(\phi)/k_BT} \tag{3.93}$$

Here σ is a symmetry factor for the internal rotation; in this case $\sigma = 3$ (Fig. 3.16).[26] The momentum integral is a standard one, and

$$(\text{p.f.})_{\text{class}} = \frac{1}{\sigma h}(2\pi I k_BT)^{1/2}\int_0^{2\pi} d\phi\, e^{-v(\phi)/k_BT} \tag{3.94}$$

If $k_BT \gg v(\phi)$, this gives

$$(\text{p.f.})_{\text{class}} = \frac{2\pi}{\sigma h}(2\pi I k_BT)^{1/2} \qquad ; k_BT \gg v(\phi) \tag{3.95}$$

If $k_BT \ll v(\phi)$, then the system performs small oscillations in the minima of $v(\phi)$. Consider the first small ϕ region, and expand the potential as

$$v(\phi) = v(0) + \frac{1}{2}\phi^2 v''(0) + \cdots \tag{3.96}$$

Keep just these first two terms, and choose the zero of energy so that $v(0) = 0$. Now define

$$I \equiv \mu r_0^2 \qquad ; r_0\phi \equiv \eta \qquad ; v''(0) \equiv \mu\omega^2 r_0^2 \tag{3.97}$$

The hamiltonian in Eq. (3.92) takes the form

$$H = \frac{1}{2}\mu\dot{\eta}^2 + \frac{1}{2}\mu\omega^2\eta^2 \qquad ; \text{s.h.o.} \tag{3.98}$$

This is just the hamiltonian for a one-dimensional harmonic oscillator, and the partition function for this was calculated in Eq. (2.46)

$$(\text{p.f.}) = \frac{e^{-\theta_V/2T}}{1 - e^{-\theta_V/T}} \qquad ; k_BT \ll v(\phi)$$
$$\theta_V = h\nu/k_B \tag{3.99}$$

---

[26] The symmetry factor produces $(1/3)\int_0^{2\pi} d\phi$, which effectively confines the problem to one cell of the periodic potential $v(\phi)$.

The full heat capacity for the one-dimensional hindered rotor interpolates between these two limits; it is sketched in Fig. 3.17.

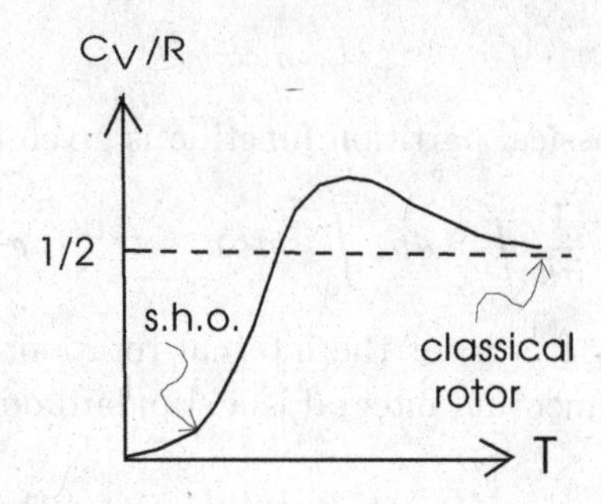

Fig. 3.17 Sketch of molar heat capacity C_V/R for the one-dimensional hindered rotor.

3.2.3 *Comparison of Spectroscopic and Calorimetric Entropies*

So far, we have computed $S_{\rm spec} = k_{\rm B} \ln \Omega$ in the gas phase from all the spectrum information for a constituent system. Statistical mechanics states that at zero temperature $S(0) = k_{\rm B} \ln 1 = 0$, which is the third law of thermodynamics. $S_{\rm spec}(T)$ is thus the absolute entropy difference between $[0, T]$.

$$S_{\rm spec}(T) = S(T) - S(0) = S(T) \tag{3.100}$$

Let us now ask, "What do we measure calorimetrically?" Since most experiments are carried out at constant pressure, the heat transfer is obtained from the constant-pressure heat capacity, and the calorimetric change in entropy is given by[27]

$$S(T) - S(T_{\rm min}) = \int_{T_{\rm min}}^{T} \frac{dQ_R}{T} = \int_{T_{\rm min}}^{T} C_P \frac{dT}{T} \tag{3.101}$$

The measurement is necessarily carried out between two finite temperatures $[T_{\rm min}, T]$. The Debye extrapolation, with $C_P \propto T^3$ (see later), can be used to extrapolate the vibrational modes of the crystal down to $T = 0$, and the

[27]Any additional heat of melting or vaporization at the phase transitions is to be included the integral.

entropy at T_{min} can then be written

$$S(T_{\text{min}}) = S_{\text{Debye}}(T_{\text{min}}) + k_{\text{B}} \ln \Omega_0 \tag{3.102}$$

There may be some additional entropy present, $k_{\text{B}} \ln \Omega_0$, that has not yet been explored calorimetrically. A possible behavior of C_P below T_{min} is sketched in Fig. 3.18.[28]

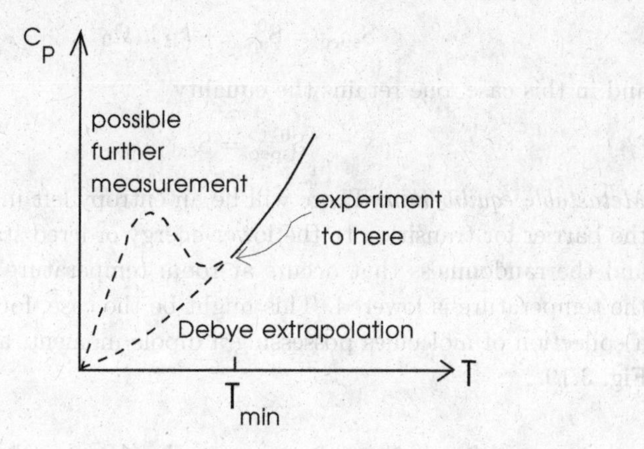

Fig. 3.18 Possible behavior of the very low temperature constant-pressure heat capacity.

Thus

$$S(T) = \int_{T_{\text{min}}}^{T} C_P \frac{dT}{T} + S_{\text{Debye}}(T_{\text{min}}) + k_{\text{B}} \ln \Omega_0 \tag{3.103}$$

While the spectroscopic entropy gives the absolute entropy difference between $[0, T]$, this may be only partially explored in the calorimetric measurement, and one can write

$$\begin{aligned} S_{\text{spec}}(T) &= S_{\text{cal}}(T) + k_{\text{B}} \ln \Omega_0 \\ S_{\text{cal}}(T) &\equiv \int_{T_{\text{min}}}^{T} C_P \frac{dT}{T} + S_{\text{Debye}}(T_{\text{min}}) \end{aligned} \tag{3.104}$$

Note that $k_{\text{B}} \ln \Omega_0$ is independent of T.

3.2.3.1 *Sources of $k_{\text{B}} \ln \Omega_0$*

One can identify possible sources of an additional entropy term $k_{\text{B}} \ln \Omega_0$:

---

[28]Compare Prob. 1.5.

(1) *Degeneracy*: Any degeneracy left in the problem will lead to such an additional entropy. This might come, for example, from[29]

- Nuclear spins
- Isotope mixing, *etc.*

If these degeneracies have also been left out of $S^0_{\rm spec}$, then its does not matter, for then

$$S_{\rm spec} = S^0_{\rm spec} + k_{\rm B} \ln \Omega_0 \tag{3.105}$$

and in this case, one retains the equality

$$S^0_{\rm spec} = S_{\rm cal} \tag{3.106}$$

(2) *Metastable equilibrium*: There will be an entropy left in the problem if the barrier for transition to the lower-energy ordered state is too high, and the randomness that occurs at room temperature is *frozen in* as the temperature is lowered. This might be the case, for example, with a collection of molecules possessing a dipole moment, as illustrated in Fig. 3.19.

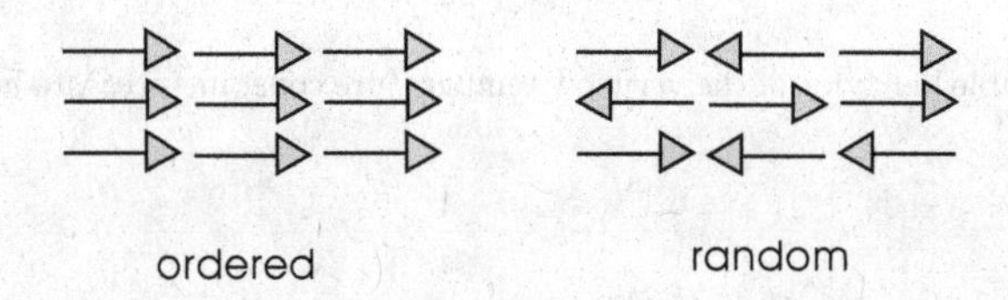

Fig. 3.19 Orientation of dipoles at low temperature.

Since each dipole has two orientations, the additional entropy for random mixing is given by

$$\begin{aligned} \Omega_0 &= 2^N \\ k_{\rm B} \ln \Omega_0 &= k_{\rm B} N \ln 2 \end{aligned} \tag{3.107}$$

For one mole this gives

$$k_{\rm B} \ln \Omega_0 = R \ln 2 = 1.38 \, \text{cal/mole-}^\circ\text{K} \tag{3.108}$$

A comparison between spectroscopic and calorimetric entropies is shown in Table 3.2.

[29]Note Prob. 3.8.

Table 3.2 Comparison of spectroscopic and calorimetric entropies in units of cal/mole-°K. See [Rushbrooke (1949)].

Gas (298 °K, 1 atm)	S_{spect}	S_{cal}	$k_B \ln \Omega_0$
He	30.11	$30.13 \pm .05$	
A	36.96	$36.9 \pm .10$	
Zn	38.46	38.5	
HCl	44.64	44.5	
O_2	49.03	49.1	
CO	47.31	46.2	1.11
H_2O	45.09	44.28	0.81
CH_3Cl	55.98	55.94	
C_2H_4	52.47	52.48	

We make several comments on this table:

- The spectroscopic entropy is obtained from [see Eq. (3.4)]

$$S_{\text{spec}} = S_{\text{trans}} + S_{\text{rot}} + S_{\text{vib}} + S_{\text{el}} \tag{3.109}$$

- The calorimetric entropy is measured into the perfect-gas regime, and then corrected to standard conditions of (298 °K, 1 atm) using the relations for a perfect gas;
- The truly impressive aspect of this table is the overall *agreement* between the entropy obtained from the two basic statistical assumptions, plus the quantum mechanics of a single system, and the entropy obtained from a series of heat-transfer measurements over the entire range of temperature!
- The result for carbon monoxide, CO, can be understood in terms of the above discussion of metastable equilibrium. CO has an electric dipole moment, so the situation in Fig. 3.19 applies; however, this dipole moment aids in the ordering, so that $k_B \ln \Omega_0$ is somewhat less than the completely random result in Eq. (3.108);
- That ubiquitous substance *ice*, solid H_2O, is actually a very complicated material. The additional $k_B \ln \Omega_0$ in this case was first explained by [Pauling (1935)]. Ice is held together by hydrogen bonds pointing in the directions of the corners of a tetrahedron. Two of the H-atoms lie closer to the central oxygen, corresponding to molecular H_2O, and two lie closer to neighboring O-atoms (see Fig. 3.20). The variety of possibilities for *which* two H-atoms lie closer to the central oxygen leads to many complexions for the assembly. The number of complexions can be evaluated with a simple probability argument, which should be valid

for large N.

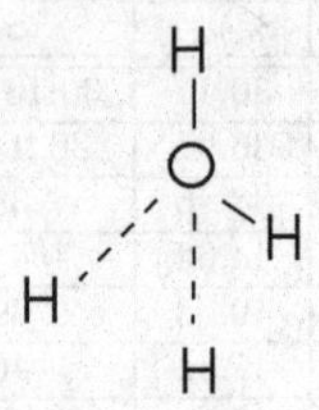

Fig. 3.20 Location of the H-atoms in the hydrogen bonds in ice.

The total number of ways of putting in the H-atoms (there are 2N of them) is

$$\Omega_{\text{tot}} = 2^{2N} \quad ; \text{ total number of ways of putting in H-atoms} \tag{3.110}$$

Only a subset of these provide acceptable configurations. The total number of configurations of H-atoms around one oxygen is $2^4 = 16$. The number of these configurations with two-near and two-far H-atoms is 6. Thus the acceptable fraction of configurations is

$$f_{\text{acc}} = \left(\frac{6}{16}\right)^N \quad ; \text{ acceptable fraction} \tag{3.111}$$

Therefore

$$\Omega_0 = 2^{2N}\left(\frac{6}{16}\right)^N = \left(\frac{3}{2}\right)^N$$

$$k_{\text{B}} \ln \Omega_0 = k_{\text{B}} N \ln\left(\frac{3}{2}\right) \tag{3.112}$$

For one mole this gives

$$k_{\text{B}} \ln \Omega_0 = R \ln\left(\frac{3}{2}\right) = 0.81 \text{ cal/mole-}^{\text{o}}\text{K} \tag{3.113}$$

reproducing the result for ice in Table 3.2.

3.3 Paramagnetic and Dielectric Assemblies

As the next application of the microcanonical ensemble, we discuss paramagnetic and dielectric assemblies, and we start with an analysis of a classical gas of permanent dipoles.

3.3.1 *Classical Gas of Permanent Dipoles*

The interaction energy of a permanent dipole in an external field is given quite generally by

$$E_{\text{int}} = -\boldsymbol{\mu} \cdot \mathbf{F} \tag{3.114}$$

Where $\boldsymbol{\mu}$ is the dipole moment and $\mathbf{F}$ is the external field. More specifically, for a magnetic dipole $\boldsymbol{\mu}$ in an external magnetic field $\mathbf{B}$ one has

$$E_{\text{int}} = -\boldsymbol{\mu} \cdot \mathbf{B} \qquad ; \text{ magnetic} \tag{3.115}$$

while for an electric dipole $\mathbf{d}$ in an electric field $\mathbf{E}$ one has

$$E_{\text{int}} = -\mathbf{d} \cdot \mathbf{E} \qquad ; \text{ electric} \tag{3.116}$$

The configuration is illustrated in Fig. 3.21.

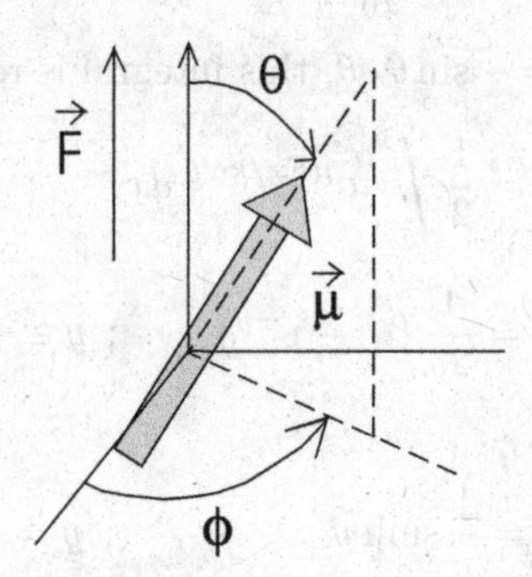

Fig. 3.21 Permanent dipole $\boldsymbol{\mu}$ in an external field $\mathbf{F}$, with an interaction energy $E_{\text{int}} = -\boldsymbol{\mu} \cdot \mathbf{F}$.

With the aid of Eqs. (3.30) and (3.24), one can write the hamiltonian for a rigid rotor in the external field as[30]

$$H = \frac{1}{2I}\left(p_\theta^2 + \frac{1}{\sin^2\theta}p_\phi^2\right) - \mu F \cos\theta \qquad ; \; |\boldsymbol{\mu}| \equiv \mu \tag{3.117}$$

[30] Note $I_3 \to 0$.

Here, and throughout this section, $|\boldsymbol{\mu}| \equiv \mu$ denotes the generic dipole moment. The classical partition function is given by

$$(\text{p.f.})_{\text{class}} = \frac{1}{h^2}\int_0^{\pi} d\theta \int_0^{2\pi} d\phi \int_{-\infty}^{\infty} dp_\theta \int_{-\infty}^{\infty} dp_\phi$$

$$\times \exp\left\{-\frac{1}{2Ik_{\text{B}}T}\left(p_\theta^2 + \frac{1}{\sin^2\theta}p_\phi^2\right) + \frac{\mu F}{k_{\text{B}}T}\cos\theta\right\} \qquad (3.118)$$

The analysis in Eqs. (3.40) reduces this to

$$(\text{p.f.})_{\text{class}} = (\text{p.f.})_{\text{rot}}\,\frac{1}{2}\int_0^{\pi} e^{\mu F\cos\theta/k_{\text{B}}T}\sin\theta\, d\theta$$

$$\equiv (\text{p.f.})_{\text{rot}}(\text{p.f.})_F \qquad (3.119)$$

where $(\text{p.f.})_{\text{rot}}$ is the classical partition function for the rigid rotor in the last of Eqs. (3.40), and[31]

$$(\text{p.f.})_F = \frac{1}{2}\int_0^{\pi} e^{\mu F\cos\theta/k_{\text{B}}T}\sin\theta\, d\theta \qquad (3.120)$$

With $x = \cos\theta$ and $dx = -\sin\theta\, d\theta$, this integral is readily evaluated as

$$(\text{p.f.})_F = \frac{1}{2}\int_{-1}^{1} e^{\mu F x/k_{\text{B}}T}\, dx$$

$$= \frac{1}{2y}\left(e^y - e^{-y}\right) \qquad ; \; y \equiv \frac{\mu F}{k_{\text{B}}T} \qquad (3.121)$$

Hence

$$(\text{p.f.})_F = \frac{1}{y}\sinh y \qquad ; \; y = \frac{\mu F}{k_{\text{B}}T} \qquad (3.122)$$

Our basic result on distribution numbers in Eqs. (3.5) gives[32]

$$\frac{n^\star(\theta)\, d\theta}{N} = \frac{1}{(\text{p.f.})_F}\frac{1}{2}e^{\mu F\cos\theta/k_{\text{B}}T}\sin\theta\, d\theta \qquad (3.123)$$

There will be a dipole moment induced by the external field, which points in the direction of the field (the z-direction). The magnitude of this induced dipole moment for a single system in thermal equilibrium in the assembly

[31] The weighting in $(\text{p.f.})_F$ is just the fractional solid angle $d\Omega/4\pi = \sin\theta\, d\theta d\phi/4\pi$.

[32] Note that $(\text{p.f.})_{\text{rot}}$ has cancelled in this expression.

is evaluated as

$$\begin{aligned}
\langle \mu_F \rangle &= \int_0^{\pi} \mu \cos\theta \, \frac{n^{\star}(\theta)\, d\theta}{N} \\
&= \mu \frac{y}{\sinh y} \frac{1}{2} \int_0^{\pi} e^{y\cos\theta} \cos\theta \sin\theta \, d\theta \\
&= \mu \frac{y}{\sinh y} \frac{d}{dy} \left(\frac{1}{y} \sinh y \right) \qquad ; \; y = \frac{\mu F}{k_{\rm B}T}
\end{aligned} \tag{3.124}$$

Thus

$$\begin{aligned}
\langle \mu_F \rangle &= \mu \frac{d}{dy} \ln\,({\rm p.f.})_F \\
&= k_{\rm B}T \frac{\partial}{\partial F} \ln\,({\rm p.f.})_F
\end{aligned} \tag{3.125}$$

This is the induced magnetic moment per system. The induced dipole moment $\mathcal{M}$ of the *assembly* is given by

$$\mathcal{M} = N\langle \mu_F \rangle = N k_{\rm B}T \frac{\partial}{\partial F} \ln\,({\rm p.f.})_F \tag{3.126}$$

The derivative required in Eq. (3.124) is evaluated as

$$\frac{d}{dy} \left(\frac{1}{y} \sinh y \right) = \frac{\cosh y}{y} - \frac{\sinh y}{y^2} \tag{3.127}$$

The final result in Eq. (3.124) then takes the form

$$\begin{aligned}
\langle \mu_F \rangle &= \mu \, L\left(\frac{\mu F}{k_{\rm B}T} \right) \\
L(y) &\equiv \coth y - \frac{1}{y} \qquad ; \text{ Langevin function}
\end{aligned} \tag{3.128}$$

Here $L(y)$ is known as the *Langevin function.* It is plotted as L_∞ in Fig. 3.22.

The Langevin function has the following limits:

(1) As $y \to \infty$

$$\begin{aligned}
L(y) &= \frac{e^y + e^{-y}}{e^y - e^{-y}} - \frac{1}{y} \\
&\to 1 \qquad ; \; y \to \infty
\end{aligned} \tag{3.129}$$

This is the strong-field, low-temperature limit $\mu F \gg k_{\rm B}T$, where the dipoles are completely aligned with the field;

(2) As $y \to 0$

$$\begin{aligned}
L(y) &\to \frac{2 + y^2 + \cdots}{2y + y^3/3 + \cdots} - \frac{1}{y} \\
&\approx \frac{1}{y}\left(1 + \frac{y^2}{2}\right)\left(1 + \frac{y^2}{6}\right)^{-1} - \frac{1}{y} \\
&\to \frac{y}{3} \qquad\qquad ; \; y \to 0
\end{aligned} \tag{3.130}$$

This is the weak-field, high-temperature limit $\mu F \ll k_{\rm B}T$. In this limit, the classical value of the induced dipole moment per system, in an assembly in thermal equilibrium at temperature T, is given by

$$\langle \mu_F \rangle = \frac{1}{3}\frac{\mu^2 F}{k_{\rm B}T} \qquad ; \; \mu F \ll k_{\rm B}T \tag{3.131}$$

This is an important result.

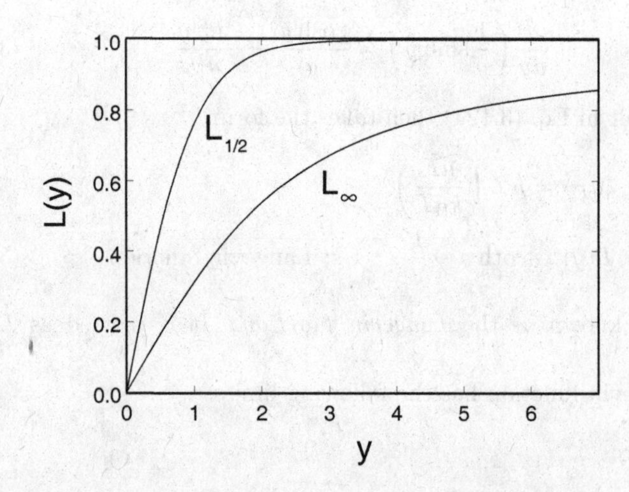

Fig. 3.22 Limiting cases of the response function $L_J(y)$ in Eq. (3.146) for a dipole system in the microcanonical ensemble placed in an external field, where $y = \mu F/k_{\rm B}T$ and J is the maximum value of J_z. Here $L_\infty = L(y)$ is the Langevin function in Eq. (3.128), which is the classical result.

3.3.2 *Magnetic Moments in Quantum Mechanics*

The magnetic dipole moment of a system is proportional to its angular momentum[33]

$$\boldsymbol{\mu} = g\mu_0 \mathbf{J} \tag{3.132}$$

The Bohr magnetons for the most familiar systems are given by

$$\begin{aligned} \mu_0 &= \frac{e\hbar}{2m_e c} & &; \text{ electrons} \\ \mu_0 &= \frac{e_p\hbar}{2m_p c} & &; \text{ nucleons} \end{aligned} \tag{3.133}$$

Here $e = -e_p$. In quantum mechanics, the angular momentum is given by

$$\begin{aligned} \mathbf{J}^2 &= J(J+1) & &; J = 0, \frac{1}{2}, 1, \frac{3}{2}, \cdots \\ J_z &= m & &; m = -J, -J+1, \cdots, J \end{aligned} \tag{3.134}$$

The experimental value of the magnetic moment is obtained by lining the system up as well as possible along the z-axis, so that $m = J$

$$\mu = g\mu_0 J \qquad ; \text{ experimental value} \tag{3.135}$$

The partition function for a point system with spin J and magnetic moment $\boldsymbol{\mu}$ in the external field $\mathbf{F} = F\hat{\mathbf{e}}_z$ is given by

$$(\text{p.f.})_F = \sum_{J_z} \exp\left\{\frac{g\mu_0 F J_z}{k_B T}\right\} = \sum_{m=-J}^{J} \exp\left\{\frac{g\mu_0 F m}{k_B T}\right\} \tag{3.136}$$

This finite sum is easily evaluated. Define

$$z \equiv \frac{g\mu_0 F}{k_B T} \qquad ; \; e^z \equiv x \tag{3.137}$$

Then

$$(\text{p.f.})_F = x^J + x^{J-1} + x^{J-2} + \cdots + x^{-J+1} + x^{-J} \tag{3.138}$$

Clearly

$$x(\text{p.f.})_F - (\text{p.f.})_F = x^{J+1} - x^{-J} \tag{3.139}$$

---

[33]We use c.g.s. units, and the angular momentum is measured in units of $\hbar$. The various sets of units are compared in appendix K of [Walecka (2008)].

The solution is

$$
\begin{aligned}
(\text{p.f.})_F &= \frac{x^{J+1} - x^{-J}}{x-1} \\
&= e^{-Jz}\left[\frac{e^{(2J+1)z} - 1}{e^z - 1}\right] \\
&= \frac{e^{(2J+1)z/2} - e^{-(2J+1)z/2}}{e^{z/2} - e^{-z/2}}
\end{aligned} \tag{3.140}
$$

Thus

$$
(\text{p.f.})_F = \frac{\sinh(2J+1)z/2}{\sinh z/2} \quad ; \; z = \frac{g\mu_0 F}{k_B T} \tag{3.141}
$$

The induced magnetic moment in the z-direction is again evaluated as in Eq. (3.124)

$$
\begin{aligned}
\langle \mu_F \rangle &= \frac{1}{(\text{p.f.})_F} \sum_{m=-J}^{J} (g\mu_0 m) \exp\left\{\frac{g\mu_0 F m}{k_B T}\right\} \\
&= k_B T \frac{\partial}{\partial F} \ln(\text{p.f.})_F \\
&= g\mu_0 \frac{\partial}{\partial z} \ln(\text{p.f.})_F
\end{aligned} \tag{3.142}
$$

Differentiation of Eq. (3.141) leads to

$$
\begin{aligned}
\langle \mu_F \rangle = g\mu_0 &\frac{\sinh z/2}{\sinh(2J+1)z/2} \\
&\times \left[\frac{(2J+1)}{2}\frac{\cosh(2J+1)z/2}{\sinh z/2} - \frac{\sinh(2J+1)z/2}{(\sinh z/2)^2}\frac{1}{2}\cosh\frac{z}{2}\right]
\end{aligned} \tag{3.143}
$$

Thus

$$
\frac{1}{g\mu_0}\langle \mu_F \rangle = \left(\frac{2J+1}{2}\right)\coth\left(\frac{2J+1}{2}\right)z - \frac{1}{2}\coth\frac{z}{2} \tag{3.144}
$$

Recall that

$$
g\mu_0 J = \mu \qquad ; \; zJ = \frac{\mu F}{k_B T} = y \tag{3.145}
$$

where μ is the experimental magnetic moment, and y is now the same expression as in Eq. (3.122). The final result for the induced magnetic moment for a quantum system in an assembly in thermal equilibrium is therefore [compare Eqs. (3.128)]

$$\langle \mu_F \rangle = \mu L_J\left(\frac{\mu F}{k_{\rm B}T}\right)$$

$$L_J(y) \equiv \left(\frac{2J+1}{2J}\right)\coth\left(\frac{2J+1}{2J}\right)y - \frac{1}{2J}\coth\frac{y}{2J} \qquad (3.146)$$

Here we have introduced the generalized Langevin response function $L_J(y)$. It has the following limits:

- For $J \to \infty$ one has[34]

$$L_\infty(y) = \coth y - \frac{1}{y} \qquad ; \text{ Langevin function } L(y) \qquad (3.147)$$

  This is the Langevin function $L(y)$, and one recovers the classical result in Eqs. (3.128);
- For $J = 1/2$ one has

$$\begin{aligned} L_{1/2}(y) &= 2\coth 2y - \coth y \\ &= 2\frac{\cosh^2 y + \sinh^2 y}{2\sinh y \cosh y} - \frac{\cosh y}{\sinh y} \\ &= \frac{\sinh y}{\cosh y} \end{aligned} \qquad (3.148)$$

  Thus one obtains the simple result[35]

$$L_{1/2}(y) = \tanh y \qquad (3.149)$$

- As $y \to 0$, one finds

$$L_J(y) \to \frac{J+1}{3J}y \qquad ; \; y \to 0 \qquad (3.150)$$

  This is the weak-field, high-temperature limit where $\mu F \ll k_{\rm B}T$, and this result reproduces Eq. (3.131) in the classical limit where $J \to \infty$.

The function $L_J(y)$ is graphed in Fig. 3.22 for $J = 1/2$ and $J = \infty$; the other $L_J(y)$ lie between these limiting cases.

---

[34] Recall that $\coth \epsilon \to (1/\epsilon)(1+\epsilon^2/3)$ as $\epsilon \to 0$. The limit here is taken at fixed μ.
[35] See Prob. 3.15.

3.3.3 *Polarization in a Dielectric Medium*

Recall some definitions from electrostatics. Let $\mathbf{P}$ be the induced dipole moment per unit volume in a dielectric, and $\mathbf{E}$ the electric field in the material, then

$$\mathbf{P} = \kappa_e \mathbf{E} \qquad ; \text{ electric susceptibility } \kappa_e \tag{3.151}$$

where κ_e is the *electric susceptibility.*

Consider a dielectric slab in a capacitor as illustrated in Fig. 3.23. A

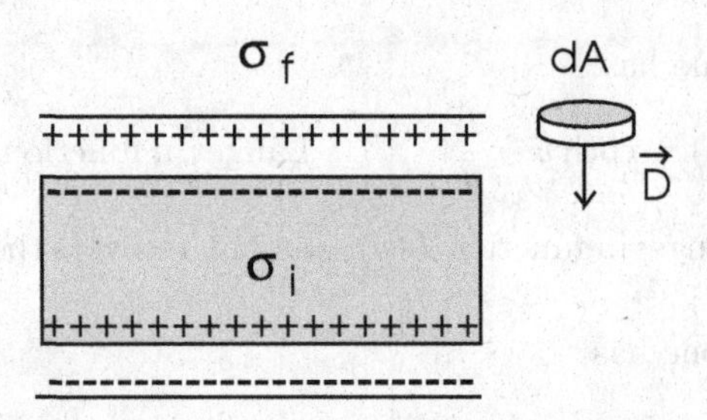

Fig. 3.23 Polarization of a dielectric slab in a capacitor with surface charge density σ_f on the capacitor producing an electric field $\mathbf{D}$ between the plates. A surface charge density σ_i is induced on the dielectric. Also shown is a gaussian pillbox of area dA on the upper plate.

charge will be induced on the surface of the dielectric, and the situation can be analyzed as follows:

(1) The electric field $\mathbf{D}$ is defined to be that arising from the free charge on the condenser plate, so that Maxwell's equation gives

$$\boldsymbol{\nabla} \cdot \mathbf{D} = 4\pi\rho_f \tag{3.152}$$

The field is between the plates and perpendicular to them. It is produced by a surface charge density (charge/area) σ_f. The integration of Eq. (3.152) over a gaussian pillbox on the surface (Fig. 3.23) then gives the magnitude of $\mathbf{D}$

$$\begin{aligned} DdA &= 4\pi\sigma_f dA \\ \Longrightarrow \qquad D &= 4\pi\sigma_f \end{aligned} \tag{3.153}$$

(2) There will be an induced surface charge density σ_i on the dielectric, and the *actual* electric field $\mathbf{E}$ in the dielectric arises from the *total* charge

density, so that a similar argument gives

$$
\begin{aligned}
EdA &= 4\pi(\sigma_f - \sigma_i)dA \\
\Longrightarrow \qquad E &= 4\pi(\sigma_f - \sigma_i)
\end{aligned}
\tag{3.154}
$$

(3) The total electric dipole moment of the dielectric slab arises from the total separated charge multiplied by the distance of separation of that charge, as illustrated in Fig. 3.24. This, in turn, is the dipole moment per unit volume in the dielectric times its volume. If ρ_i is the volume charge density of the displaced charge, and $\mathbf{l}$ is the charge displacement, then

$$
\begin{aligned}
(\rho_i A\mathbf{l})t &= \mathbf{P}(At) \qquad ; \text{ total dipole moment} \\
\Longrightarrow \qquad \mathbf{P} &= \rho_i\mathbf{l} = \sigma_i\hat{\mathbf{n}}
\end{aligned}
\tag{3.155}
$$

where $\hat{\mathbf{n}}$ is a unit vector normal to the surface in the direction of $\mathbf{D}$. The last equality follows from the definition of surface charge density $\sigma_i = l\rho_i$. Note that the dipole moment per unit volume $\mathbf{P}$ in the slab is indeed an intensive quantity depending only on the induced surface charge density σ_i; it is independent of the overall dimensions of the slab.

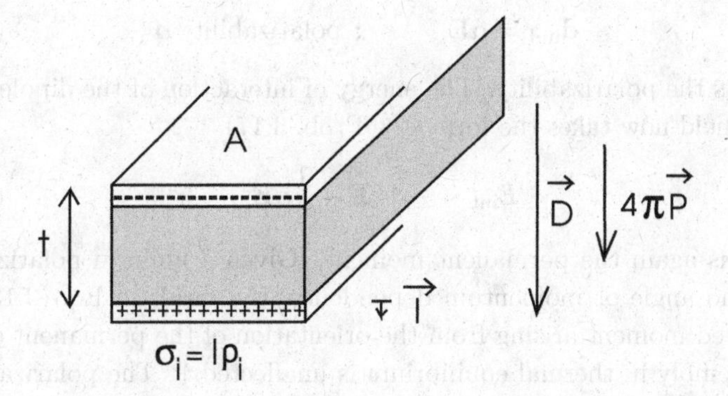

Fig. 3.24 Polarization of a dielectric slab of thickness t and area A in a capacitor. The charge is displaced by $\mathbf{l}$. The volume density of the displaced charge is ρ_i, so that the surface charge density is $\sigma_i = l\rho_i$. The dipole moment per unit volume in the slab is given by $\mathbf{P} = \sigma_i\hat{\mathbf{n}}$, where $\hat{\mathbf{n}}$ is a unit vector normal to the surface in the direction of $\mathbf{D}$.

The electric field in Eq. (3.154) can therefore be written

$$
\mathbf{E} = \mathbf{D} - 4\pi\mathbf{P} \tag{3.156}
$$

where we have restored the directions to the vectors. Note that the polarization $\mathbf{P}$ serves to *decrease* the actual electric field felt by a system in the dielectric. Thus

$$\begin{aligned}\mathbf{D} &= \mathbf{E} + 4\pi\mathbf{P} \\ &= (1 + 4\pi\kappa_e)\mathbf{E} \\ &\equiv \epsilon\mathbf{E}\end{aligned} \tag{3.157}$$

Here we have defined the *dielectric constant*

$$\epsilon = 1 + 4\pi\kappa_e \qquad ; \text{ dielectric constant} \tag{3.158}$$

Equation (3.151) can then be re-written as

$$\mathbf{P} = \kappa_e\mathbf{E} = \left(\frac{\epsilon - 1}{4\pi}\right)\mathbf{E} \tag{3.159}$$

Consider the induced dipole moment in the material. The external field can not only *orient* the individual permanent dipoles, as analyzed in the last section, but it may also *induce* an extra dipole moment in the individual system by distorting its electronic configuration. This induced dipole moment will again be proportional to the actual field in the material, and one can write

$$\mathbf{d}_{\text{ind}} = \alpha\mathbf{E} \qquad ; \text{ polarizability } \alpha \tag{3.160}$$

where α is the polarizability. The energy of interaction of the dipole in the external field now takes the form (see Prob. 3.17)

$$E_{\text{int}} = -\mathbf{d}\cdot\mathbf{E} - \frac{1}{2}\alpha\mathbf{E}^2 \tag{3.161}$$

where $\mathbf{d}$ is again the permanent moment. Given a uniform polarizability α, with no angle or momentum dependence, the result in Eq. (3.131) for the induced moment arising from the orientation of the permanent dipoles in an assembly in thermal equilibrium is unaffected.[36] The polarization $\mathbf{P}$ then simply receives an additional contribution from the polarizability of each system

$$\mathbf{P} = \frac{N}{V}\left(\alpha + \frac{\mu^2}{3k_{\text{B}}T}\right)\mathbf{E} = \left(\frac{\epsilon - 1}{4\pi}\right)\mathbf{E} \tag{3.162}$$

[36]See Prob. 3.16; in this case, the dipole moment μ is the electric dipole moment.

Hence

$$\kappa_e = \frac{\epsilon - 1}{4\pi} = \frac{N}{V}\left(\alpha + \frac{\mu^2}{3k_{\rm B}T}\right) \tag{3.163}$$

Three comments:

- The susceptibility κ_e is directly related to the dielectric constant ϵ;
- The dielectric constant of a material has now been expressed in terms of the properties of the systems of which it is composed;
- This relation gives an explicit temperature dependence to $\epsilon(T)$, which can then be used to measure (α, μ).

Let us go one step further and ask, "What is the actual field acting on a dipole in the medium? Is it just the electric field $\mathbf{E}$ determined from Eq. (3.154)?" The answer is no. The field will be *locally distorted* by the neighboring dipoles. To estimate this effect, imagine that the system sits in a small spherical cavity in the medium, as illustrated in Fig. 3.25(a).

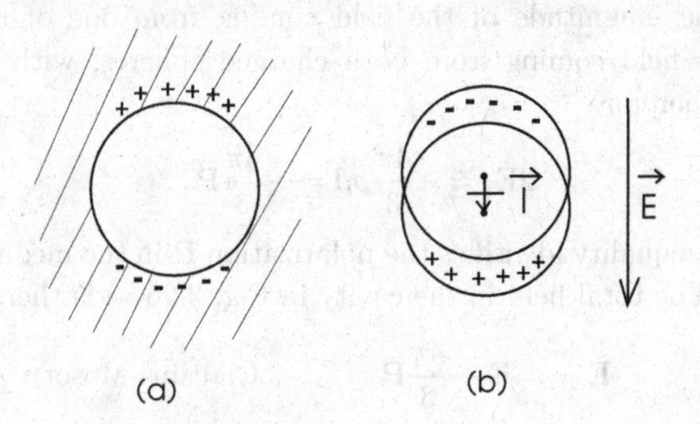

Fig. 3.25 Field in hollow sphere in polarized dielectric (a); is *opposite* to that inside two charged spheres displaced by $\mathbf{l}$ (b).

Now compute the actual field at the center of the cavity. The additional field arising from the polarization charge will be just the *opposite* of that in a dielectric sphere placed in a field $\mathbf{E}$ as illustrated in Fig. 3.25(b), since if that sphere is placed back in the cavity, there is no effect at all. The additional induced field $\mathbf{E}_i$ at the center of this polarized dielectric sphere can be computed with the aid of Gauss' law. One seeks the field a distance $l/2$ away from the centers of two uniformly charged spheres. If the charge density is ρ_i, then Gauss' law gives for the single charged sphere in Fig. 3.26

$$4\pi\left(\frac{l}{2}\right)^2 E_i = 4\pi\left[\frac{4\pi}{3}\left(\frac{l}{2}\right)^3\right]\rho_i$$

$$E_i = \frac{1}{2}\frac{4\pi}{3}l\rho_i \tag{3.164}$$

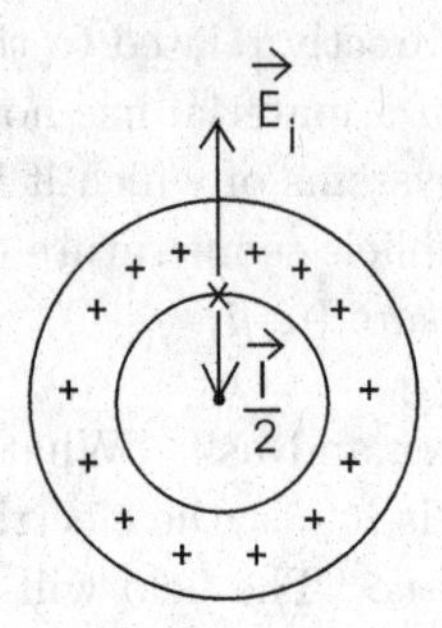

Fig. 3.26 Compute electric field $\mathbf{E}_i$ a distance $\mathbf{l}/2$ away from center of a uniformly charged sphere.

This is the magnitude of the field coming from one of the charged spheres. The field coming from both charged spheres, with the correct direction, is therefore

$$2\mathbf{E}_i = -\frac{4\pi}{3}\rho_i\mathbf{l} = -\frac{4\pi}{3}\mathbf{P} \tag{3.165}$$

where the last equality identifies the polarization $\mathbf{P}$ in the medium through Eq. (3.155). The total field in the cavity in Fig. 3.25(a) is therefore

$$\mathbf{E}_{\text{eff}} = \mathbf{E} + \frac{4\pi}{3}\mathbf{P} \qquad ; \text{Clausius-Mosotti} \tag{3.166}$$

This is the Clausius-Mosotti relation.[37]

The expression for the induced polarization in the medium in Eq. (3.162) should now read

$$\begin{aligned}\mathbf{P} &= \frac{N}{V}\left(\alpha + \frac{\mu^2}{3k_{\text{B}}T}\right)\mathbf{E}_{\text{eff}} \\ &= \frac{N}{V}\left(\alpha + \frac{\mu^2}{3k_{\text{B}}T}\right)\left(\mathbf{E} + \frac{4\pi}{3}\mathbf{P}\right)\end{aligned} \tag{3.167}$$

[37]Remember the additional sign change illustrated in Fig. 3.25. The modification of the field to an effective field is sometimes referred to as the Lorentz-Lorenz effect.

With the insertion of the definition $\mathbf{P} = \kappa_e \mathbf{E}$ from Eq. (3.151), this can be solved for the susceptibility κ_e

$$\begin{aligned}\kappa_e &= \left[1 - \frac{4\pi}{3}\frac{N}{V}\left(\alpha + \frac{\mu^2}{3k_\mathrm{B}T}\right)\right]^{-1} \frac{N}{V}\left(\alpha + \frac{\mu^2}{3k_\mathrm{B}T}\right) \\ &= \frac{\epsilon - 1}{4\pi}\end{aligned} \tag{3.168}$$

This, in turn, can be written as[38]

$$\frac{\epsilon - 1}{\epsilon + 2} = \frac{4\pi}{3}\frac{N}{V}\left(\alpha + \frac{\mu^2}{3k_\mathrm{B}T}\right) \quad ; \text{Debye eqn} \tag{3.169}$$

This is an improved relation for the dielectric constant known as the *Debye equation.*[39]

3.3.4 *Paramagnetic Susceptibility*

Recall some magnetostatics. Let $\mathbf{H}$ be the magnetic field produced by the current i in a solenoid (Fig. 3.27).

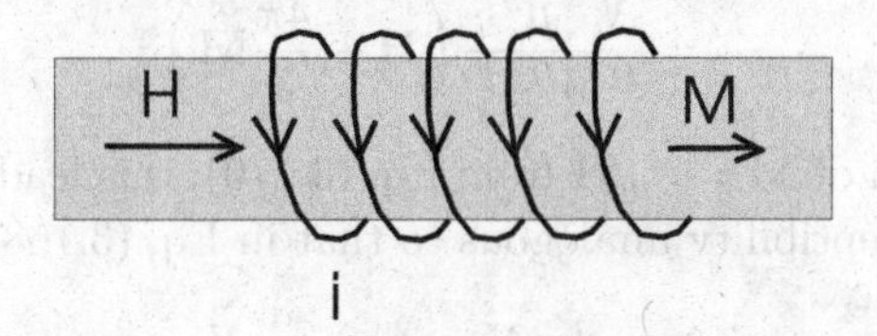

Fig. 3.27 Magnetic material in a solenoid. $\mathbf{H}$ is the magnetic field produced in a solenoid with current i. $\mathbf{M}$ is the induced magnetic moment per unit volume in the material.

If there is a material in the solenoid, then the additional induced magnetic moment per unit volume $\mathbf{M}$ in the material is related to $\mathbf{H}$ by

$$\mathbf{M} = \kappa_m \mathbf{H} \qquad ; \text{ magnetic susceptibility } \kappa_m \tag{3.170}$$

where κ_m is the magnetic susceptibility The result for κ_m arising from a classical gas of permanent magnetic dipoles was obtained in Eq. (3.131)

$$\kappa_m = \frac{N}{V}\frac{\mu^2}{3k_\mathrm{B}T} \tag{3.171}$$

---

[38] Write $(\epsilon - 1)/3 = u/(1-u)$, then $u = (\epsilon - 1)/(\epsilon + 2)$. Once again, here the generic moment μ is the permanent electric dipole moment d.

[39] See [Debye (1988)], and for both this and the next section, see [Van Vleck (1965)].

where μ is now the magnetic dipole moment. The temperature dependence can be displayed as

$$\kappa_m = \frac{C}{T} \qquad ; \text{ Curie's constant } C \tag{3.172}$$

where C is *Curie's constant.* Quantum mechanics refines this result to the expression in Eq. (3.150)

$$\kappa_m = \frac{N}{V}\frac{1}{3}\left(1+\frac{1}{J}\right)\frac{\mu^2}{k_B T} \qquad ; \text{ quantum mechanics} \tag{3.173}$$

These results can be improved by using an *effective field* $\mathbf{H}_{\text{eff}}$ in a spherical cavity in the material. An argument based on "magnetic charges" leads to a result strictly analogous to Eq. (3.166)[40]

$$\mathbf{H}_{\text{eff}} = \mathbf{H} + \frac{4\pi}{3}\mathbf{M} \qquad ; \text{ effective field} \tag{3.174}$$

Equations (3.170) and (3.171) then get generalized to

$$\begin{aligned}\mathbf{M} &= \frac{N}{V}\frac{\mu^2}{3k_B T}\mathbf{H}_{\text{eff}} \\ &= \frac{N}{V}\frac{\mu^2}{3k_B T}\left(\mathbf{H} + \frac{4\pi}{3}\mathbf{M}\right)\end{aligned} \tag{3.175}$$

With the insertion of $\mathbf{M} = \kappa_m \mathbf{H}$ from Eq. (3.170), this leads to a result for the magnetic susceptibility analogous to that in Eq. (3.168)

$$\kappa_m = \left(1 - \frac{4\pi}{3}\frac{N}{V}\frac{\mu^2}{3k_B T}\right)^{-1}\frac{N}{V}\frac{\mu^2}{3k_B T} \tag{3.176}$$

---

[40]For a *dielectric* material in a capacitor, and a *paramagnetic* material in a solenoid,

$$\mathbf{D} = \mathbf{E} + 4\pi\mathbf{P} \qquad ; \mathbf{B} = \mathbf{H} + 4\pi\mathbf{M} \qquad ; \text{ in medium}$$

Here $\mathbf{D}$ is determined from the free charge on the plates, $\mathbf{H}$ is determined from the free current in the wire, and $(\mathbf{E}, \mathbf{B})$ are the true fields that produce the Lorentz force. The polarization $\mathbf{P}$ enhances the electric field $\mathbf{D}$, while the magnetization $\mathbf{M}$ similarly enhances the magnetic field $\mathbf{B}$. Thus one has the strict analogy $(\mathbf{D} \rightleftharpoons \mathbf{B})$, $(\mathbf{E} \rightleftharpoons \mathbf{H})$, and $(\mathbf{P} \rightleftharpoons \mathbf{M})$ [compare Eqs. (3.151) and (3.170)]. This leads to the following analogous relations for spherical *cavities* in these materials (see Prob. 3.18)

$$\mathbf{E}_{\text{eff}} = \mathbf{E} + \frac{4\pi}{3}\mathbf{P} \qquad ; \mathbf{H}_{\text{eff}} = \mathbf{H} + \frac{4\pi}{3}\mathbf{M} \qquad ; \text{ in cavity}$$

This result can be re-written as

$$\kappa_m = \frac{N}{V}\frac{\mu^2}{3k_{\rm B}(T-\Theta)} \qquad ; \text{Curie-Weiss law}$$
$$\Theta = \frac{N}{V}\frac{4\pi\mu^2}{9k_{\rm B}} \tag{3.177}$$

This is the *Curie-Weiss law.* Several comments:

- As $T \to \Theta$ from above, the magnetic susceptibility diverges, $\kappa_m \to \infty$. This happens for *ferromagnetic* materials, which then retain a permanent magnetization $\mathbf{M}$ below Θ;
- Θ is called the *Curie temperature*;
- For high values of the field, one must go back to the full Langevin function $L\,(\mu H_{\rm eff}/k_{\rm B}T)$ in Eqs. (3.128);
- The magnetic polarizability α_m has been neglected in Eq. (3.176) [compare Eq. (3.168)];
- There is also an electronic *diamagnetism* (see later), which must be included;
- In the previous discussion of dielectric materials, one also has the possibility of *ferroelectrics* below a critical temperature T_C; these retain a permanent polarization $\mathbf{P}$ below T_C.

3.3.5 *Thermodynamics*

So far we have computed the Helmholtz free energy

$$\begin{aligned} A(T,V,N,F) &= -k_{\rm B}T \ln \frac{(\text{p.f.})^N}{N!} \\ &= -k_{\rm B}T \ln \frac{(\text{p.f.})_0^N}{N!}(\text{p.f.})_F^N \end{aligned} \tag{3.178}$$

where F is a specified external field. The induced dipole moment of the assembly in the direction of the field is given by Eq. (3.126) as

$$\mathcal{M} = -\left(\frac{\partial A}{\partial F}\right)_{T,V,N} \tag{3.179}$$

Thus one can write the total differential of A in the presence of the external field F as

$$dA = -SdT - PdV + \tilde{\mu}dN - \mathcal{M}dF \tag{3.180}$$

where, to distinguish it, we here use the symbol $\tilde{\mu}$ to denote the chemical potential. This now yields all the thermodynamic quantities through partial differentiation of $A(T, V, N, F)$.[41]

3.4 Chemical Equilibria

We turn next to one of the most interesting and successful applications of the microcanonical ensemble and that is the analysis of *chemical equilibria.* We start with some preliminaries.

3.4.1 *Some Preliminaries*

There are important results that follow immediately from the discussion of the method of steepest descent. Suppose the assembly is composed of several distinct systems with energy levels

$$\begin{aligned} &\varepsilon_1^{(1)}, \varepsilon_2^{(1)}, \varepsilon_3^{(1)}, \cdots && ; \text{ system 1} \\ &\varepsilon_1^{(2)}, \varepsilon_2^{(2)}, \varepsilon_3^{(2)}, \cdots && ; \text{ system 2} \qquad ; \text{ etc.} \end{aligned} \tag{3.181}$$

Define a generating function for each system

$$\begin{aligned} f^{(1)}(z) &= \sum_i z^{\varepsilon_i^{(1)}} \\ f^{(2)}(z) &= \sum_i z^{\varepsilon_i^{(2)}} \qquad ; \text{ etc.} \end{aligned} \tag{3.182}$$

Consider the following product of these generating functions

$$\begin{aligned} \frac{[f^{(1)}(z)]^{N_1}}{N_1!} \frac{[f^{(2)}(z)]^{N_2}}{N_2!} \cdots &= \frac{1}{N_1!} \frac{1}{N_2!} \cdots \\ \times \sum_{\{n^{(1)}\}} \frac{N_1!}{\prod_i n_i^{(1)}!} z^{\sum_i n_i^{(1)} \varepsilon_i^{(1)}} &\sum_{\{n^{(2)}\}} \frac{N_2!}{\prod_i n_i^{(2)}!} z^{\sum_i n_i^{(2)} \varepsilon_i^{(2)}} \times \cdots \end{aligned} \tag{3.183}$$

---

[41]In [Walecka (2000)] there is a very nice argument by Bloch which shows that the work done *by* the magnetized sample in Fig. 3.27, when the external field H is changed slightly by a small movement along the axis of an external magnetic, is $dW = \mathcal{M}dH$. The resulting change in the Helmholtz free energy of the sample is $dA = -\mathcal{M}dH$ [compare Eq. (3.180)]. Note that one could just as well have used $\mathbf{H}$ for the external field in free space in Eq. (3.115) [see also Prob. 3.18(d)].

Here the sums are carried out subject to the constraints

$$\sum_i n_i^{(1)} = N_1 \qquad ; \sum_i n_i^{(2)} = N_2 \qquad ; \text{ etc.} \tag{3.184}$$

Identify the total energy as

$$E = \sum_i n_i^{(1)}\varepsilon_i^{(1)} + \sum_i n_i^{(2)}\varepsilon_i^{(2)} + \cdots \tag{3.185}$$

The goal is then to find the coefficient of z^E in Eq. (3.183), for this coefficient provides the appropriate number of complexions $\Omega(N_1, N_2, \cdots; E)$ for an assembly composed of a fixed number of the distinct systems.[42] The method of steepest descent solves for this coefficient as

$$\Omega(N_1, N_2, \cdots; E) = \frac{1}{N_1!}\frac{1}{N_2!}\cdots\frac{1}{2\pi i}\oint \frac{dz}{z^{E+1}}[f^{(1)}(z)]^{N_1}[f^{(2)}(z)]^{N_2}\cdots \tag{3.186}$$

For large N, saddle-point integration gives the result in Eq. (2.91)[43]

$$\ln\Omega(N_1, N_2, \cdots; E) = \ln\left\{\frac{[f^{(1)}(x_0)]^{N_1}[f^{(2)}(x_0)]^{N_2}\cdots}{N_1!N_2!\cdots}\right\} - E\beta$$

$$; \; x_0 \equiv e^{\beta} \tag{3.187}$$

where $x_0 \equiv e^{\beta}$, or equivalently, $\beta = \ln x_0$. The saddle point is located by

$$\frac{\partial}{\partial x_0}\ln\Omega(N_1, N_2, \cdots; E) = 0 \tag{3.188}$$

which leads to

$$N_1\frac{\sum_i \varepsilon_i^{(1)} x_0^{[\varepsilon_i^{(1)}-1]}}{\sum_i x_0^{\varepsilon_i^{(1)}}} + N_2\frac{\sum_i \varepsilon_i^{(2)} x_0^{[\varepsilon_i^{(2)}-1]}}{\sum_i x_0^{\varepsilon_i^{(2)}}} + \cdots = \frac{E}{x_0} \tag{3.189}$$

The location of the saddle point is therefore determined by

$$N_1\frac{\sum_i \varepsilon_i^{(1)} e^{\beta\varepsilon_i^{(1)}}}{\sum_i e^{\beta\varepsilon_i^{(1)}}} + N_2\frac{\sum_i \varepsilon_i^{(2)} e^{\beta\varepsilon_i^{(2)}}}{\sum_i e^{\beta\varepsilon_i^{(2)}}} + \cdots = E \tag{3.190}$$

This relation implicitly determines $\beta(N_1, N_2, \cdots, E)$.

[42]The systems are non-localized, or localized, according to the factors of $1/N_i!$ retained in Eq. (3.183). For clarity, we temporarily suppress the explicit V dependence in Ω.

[43]See Prob. 3.23; write $N_i = \xi_i N$ where ξ_i is a finite fraction, and then let $N \to \infty$.

In *summary*, Eqs. (3.187) and (3.190) give the logarithm of the constrained sum

$$\Omega(N_1, N_2, \cdots; E) = \frac{1}{N_1!N_2!\cdots} \sum_{\{n^{(1)}\}} \frac{N_1!}{\prod_i n_i^{(1)}!} \sum_{\{n^{(2)}\}} \frac{N_2!}{\prod_i n_i^{(2)}!} \times \cdots \quad (3.191)$$

evaluated under the conditions of fixed $(N_1, N_2, \cdots, N_s, E)$ where

$$\sum_i n_i^{(k)} = N_k \qquad ; k = 1, 2, \cdots, s$$

$$\sum_i n_i^{(1)}\varepsilon_i^{(1)} + \sum_i n_i^{(2)}\varepsilon_i^{(2)} + \cdots + \sum_i n_i^{(s)}\varepsilon_i^{(s)} = E \qquad ; (N_k, E) \text{ fixed} \quad (3.192)$$

Here $k = 1, 2, \cdots, s$ runs over the distinct systems ("species") in the assembly. These results are obtained using the method of steepest descent, and if $N_k = \xi_k N$ where ξ_k is a finite fraction of the total N, they hold when $N \to \infty$.

3.4.2 *Chemical Reactions and the Law of Mass Action*

So far the analysis has assumed fixed values for $(N_1, N_2, \cdots)$. Now include the possibility of a *chemical reaction* that allows an inter-conversion of the types of systems in the assembly. Consider, for example, a reaction

$$A + B \rightleftharpoons AB \qquad ; \text{perfect gases} \quad (3.193)$$

where A and B denote atoms, AB a molecule, and all three species are perfect gases. These species have the individual energy levels and distribution numbers

$$\begin{array}{llll} \varepsilon_1^a, \varepsilon_2^a, \cdots & ; \varepsilon_1^b, \varepsilon_2^b, \cdots & ; \varepsilon_1^{ab}, \varepsilon_2^{ab}, \cdots & ; \text{energy levels} \\ n_1^a, n_2^a, \cdots & ; n_1^b, n_2^b, \cdots & ; n_1^{ab}, n_2^{ab}, \cdots & ; \text{distribution numbers} \end{array} \quad (3.194)$$

Call the total number of the individual species

$$N_1 \equiv \sum_i n_i^a \qquad ; N_2 \equiv \sum_i n_i^b \qquad ; N_{12} \equiv \sum_i n_i^{ab} \quad (3.195)$$

Whether they appear free or in the molecules, the total number N_A of A atoms in the assembly, and the total number N_B of B atoms, are *fixed*, as is the total energy E. Thus one has the constraints

$$\begin{aligned} N_1 + N_{12} &= N_A \qquad ; \ (N_A, N_B, E) \text{ all fixed} \\ N_2 + N_{12} &= N_B \\ \sum_i n_i^a \varepsilon_i^a + \sum_i n_i^b \varepsilon_i^b + \sum_i n_i^{ab} \varepsilon_i^{ab} &= E \end{aligned} \tag{3.196}$$

The total number of complexions for this assembly is given by

$$\begin{aligned} \Omega &= \sum_{N_1,N_2,N_{12}} \frac{1}{N_1!N_2!N_{12}!} \sum_{\{n\}} \frac{N_1!N_2!N_{12}!}{\prod_i n_i^a! \prod_i n_i^b! \prod_i n_i^{ab}!} \\ &= \sum_{N_1,N_2,N_{12}} \Omega(N_1, N_2, N_{12}; E) \qquad ; \text{ three independent perfect gases} \end{aligned} \tag{3.197}$$

where the sums are to be carried out subject to the constraints in Eqs. (3.196), and $\Omega(N_1, N_2, N_{12}; E)$ is identified from the previous section.

We are *now* justified in picking out the largest term in the sum and using Stirling's formula on $\ln N_1!$, $\ln N_2!$, and $\ln N_{12}!$, since these numbers are all very large! To find the maximum of the logarithm of $\Omega(N_1, N_2, N_{12}; E)$, use Eq. (3.187), and set its variation equal to zero[44]

$$\begin{aligned} \delta \ln \Omega(N_1, N_2, N_{12}; E) &= \delta N_1 \ln (\text{p.f.})_{\text{A}} + \delta N_2 \ln (\text{p.f.})_{\text{B}} + \delta N_{12} \ln (\text{p.f.})_{\text{AB}} \\ &\quad - \delta N_1 \ln N_1 - \delta N_2 \ln N_2 - \delta N_{12} \ln N_{12} \\ &= 0 \end{aligned} \tag{3.198}$$

The variation is to be carried out at constant E subject to the constraints

$$\begin{aligned} \delta N_1 + \delta N_{12} &= 0 \\ \delta N_2 + \delta N_{12} &= 0 \end{aligned} \tag{3.199}$$

We again use Lagrange's method of undetermined multipliers. Multiply the first constraint equation by α_A, the second equation by α_B, and add to Eq. (3.198). The coefficients of all the variations must then vanish at the maximum [compare Eqs. (2.24)]

$$\begin{aligned} \ln (\text{p.f.})_{\text{A}} - \ln N_1 + \alpha_A &= 0 \\ \ln (\text{p.f.})_{\text{B}} - \ln N_2 + \alpha_B &= 0 \\ \ln (\text{p.f.})_{\text{AB}} - \ln N_{12} + (\alpha_A + \alpha_B) &= 0 \end{aligned} \tag{3.200}$$

[44]Note that $\partial \ln \Omega(N_1, N_2, \cdots; E)/\partial \beta = [\partial \ln \Omega(N_1, N_2, \cdots; E)/\partial x_0]\, \partial x_0/\partial \beta = 0$ since we are at the saddle point. Thus β can be held fixed when computing this variation. Recall also the frequently used relation $\delta \ln N! = \delta(N \ln N - N) = \delta N \ln N$, which follows from Stirling's formula.

Here and henceforth $N_i \equiv N_i^\star$ denotes the *mean number* of systems of species i, and we leave off the star to save writing. Define

$$e^{\alpha_A} \equiv \lambda_A \qquad ; \; e^{\alpha_B} \equiv \lambda_B \tag{3.201}$$

The exponentiation of Eqs. (3.200) then leads to

$$\begin{aligned} N_1 &= \lambda_A (\text{p.f.})_\text{A} \\ N_2 &= \lambda_B (\text{p.f.})_\text{B} \\ N_{12} &= \lambda_A \lambda_B (\text{p.f.})_\text{AB} \end{aligned} \tag{3.202}$$

The first equation determines α_A, the second equation determines α_B, and the third equation is then not an independent relation.

Form the following ratio

$$\frac{N_1 N_2}{N_{12}} = \frac{(\text{p.f.})_\text{A}(\text{p.f.})_\text{B}}{(\text{p.f.})_\text{AB}} \equiv K_\text{eq}(T,V) \qquad ; \text{ law of mass action} \tag{3.203}$$

Some comments:

- This is the *law of mass action*, which lies at the heart of every introductory chemistry course;
- It is a relation concerning the equilibrium number of systems of each of the species participating in the chemical reaction, in this case the reaction in Eq. (3.193);
- $K_\text{eq}(T,V)$ is known as the *equilibrium constant*;[45]
- *It is a truly remarkable consequence of our two basic statistical assumptions that the equilibrium constant $K_\text{eq}(T,V)$ for a chemical reaction is completely determined in terms of the partition functions for the species involved in it!*[46]

It remains to identify the quantity β. The entropy of the assembly is now given through Eq. (3.187) as

$$\begin{aligned} S &= k_\text{B} \ln \Omega(N_1, N_2, N_{12}; E)_\text{max} \\ &= k_\text{B} \left[\ln \left\{ \frac{[(\text{p.f.})_\text{A}]^{\text{N}_1} [(\text{p.f.})_\text{B}]^{\text{N}_2} [(\text{p.f.})_\text{AB}]^{\text{N}_{12}}}{N_1! N_2! N_{12}!} \right\} - \beta E \right] \end{aligned} \tag{3.204}$$

[45] We have here restored the explicit V dependence to $K_\text{eq}(T,V)$, and we note that the law of mass action can always be re-written as a relation between the particle *densities* of the various participating species (see Prob. 3.21).

[46] See Probs. 3.19–3.21. Note that the equilibrium constant is often defined through the *inverse* of the ratio in Eq. (3.203).

The first and second laws of thermodynamics for an open assembly read[47]

$$dE = TdS - PdV + \sum_{i=1}^{s} \mu_i dN_i \qquad ; \text{ first and second laws} \qquad (3.205)$$

Here the constituents are considered as *species*, of which there are s types, and the sum goes over all of them. The absolute temperature is then introduced into the statistical analysis through the relation

$$\frac{1}{T} = \left(\frac{\partial S}{\partial E}\right)_{V,N_i} \qquad (3.206)$$

If we recall that β is an implicit function of E, then

$$\frac{\partial \ln \Omega_{\max}}{\partial E} = -\beta + \frac{\partial}{\partial \beta}\left[\ln\left\{\frac{[(\text{p.f.})_{\text{A}}]^{N_1}[(\text{p.f.})_{\text{B}}]^{N_2}[(\text{p.f.})_{\text{AB}}]^{N_{12}}}{N_1!N_2!N_{12}!}\right\} - \beta E\right]\frac{\partial \beta}{\partial E}$$
$$= \frac{1}{k_{\text{B}}T} \qquad (3.207)$$

The derivative of the expression in square brackets vanishes by the definition of the saddle point in Eq. (3.190), and Eqs. (3.207) reduce to the familiar result

$$\beta = -\frac{1}{k_{\text{B}}T} \qquad (3.208)$$

The Helmholtz free energy of the assembly is then identified from Eq. (3.204) as

$$A(T, V, N_1, N_2, N_{12}) = -k_{\text{B}}T \ln\left\{\frac{[(\text{p.f.})_{\text{A}}]^{N_1}[(\text{p.f.})_{\text{B}}]^{N_2}[(\text{p.f.})_{\text{AB}}]^{N_{12}}}{N_1!N_2!N_{12}!}\right\}$$
$$; \text{ Helmholtz free energy} \qquad (3.209)$$

The first and second laws give its differential for an open assembly as

$$dA = -SdT - PdV + \sum_{i=1}^{s} \mu_i dN_i \qquad (3.210)$$

In the partition functions for the individual species, all the energies of a system are measured from the same zero (see Fig. 3.28). If the AB molecule is bound as illustrated in this figure, then a term that dominates $K_{\text{eq}}(T, V)$ can be extracted as

$$(\text{p.f.})_{\text{AB}} = e^{\omega/k_{\text{B}}T}(\text{p.f.})^0_{\text{AB}} \qquad (3.211)$$

where the energies in $(\text{p.f.})^0_{\text{AB}}$ are now measured from the ground state of AB.

[47] Here and henceforth, we return to denoting the chemical potential by μ.

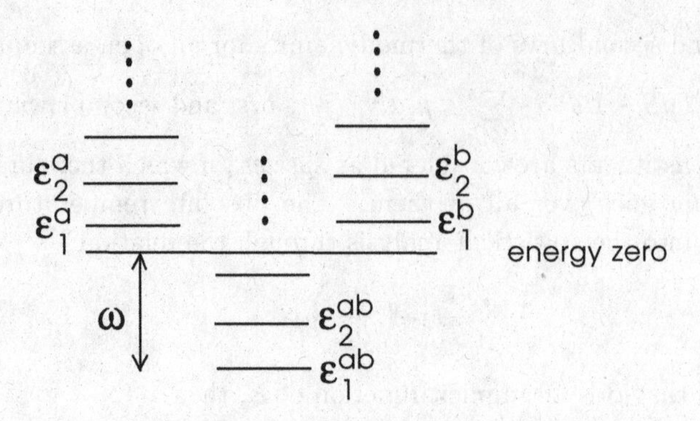

Fig. 3.28 In the partition functions, all the energies of a system are measured from the same zero. Here the AB molecule is bound by an energy ω.

3.4.3 *Chemical Potentials*

Let us use Eq. (3.210) to compute the chemical potential of each species from the free energy in Eq. (3.209)

$$
\begin{aligned}
\mu_1 &= \left(\frac{\partial A}{\partial N_1}\right)_{T,V,N_2,N_{12}} = -k_{\mathrm{B}}T[\ln(\text{p.f.})_A - \ln N_1] = \alpha_A k_{\mathrm{B}}T \\
\mu_2 &= \left(\frac{\partial A}{\partial N_2}\right)_{T,V,N_1,N_{12}} = -k_{\mathrm{B}}T[\ln(\text{p.f.})_B - \ln N_2] = \alpha_B k_{\mathrm{B}}T \\
\mu_{12} &= \left(\frac{\partial A}{\partial N_{12}}\right)_{T,V,N_1,N_2} = -k_{\mathrm{B}}T[\ln(\text{p.f.})_{AB} - \ln N_{12}] = (\alpha_A + \alpha_B)k_{\mathrm{B}}T
\end{aligned}
\tag{3.212}
$$

where Eqs. (3.200) have been used to obtain the final equalities. It follows immediately that the chemical potentials as species satisfy

$$
\mu_1 + \mu_2 = \mu_{12} \quad ; \text{ chemical potentials as species} \tag{3.213}
$$

These chemical potentials are *additively equal* for the chemical reaction in Eq. (3.193). Furthermore, the quantities appearing in Eqs. (3.202) take the form

$$
\begin{aligned}
\lambda_A &= e^{\alpha_A} = e^{\mu_1/k_{\mathrm{B}}T} \\
\lambda_B &= e^{\alpha_B} = e^{\mu_2/k_{\mathrm{B}}T} \\
\lambda_A\lambda_B &= e^{(\mu_1+\mu_2)/k_{\mathrm{B}}T} = e^{\mu_{12}/k_{\mathrm{B}}T}
\end{aligned}
\tag{3.214}
$$

These quantities are known as the *absolute activities.*

Components are those chemical constituents whose numbers can be varied independently.[48] In the present case, the components are the atoms of A and B. We claim that

> *The chemical potential for a substance treated as a component is identical to the chemical potential of that substance treated as a species.*

The proof goes as follows: If the component numbers (N_A, N_B) are changed, then the first two of Eqs. (3.196) and Eq. (3.203) imply that the species numbers (N_1, N_2, N_{12}) change in a corresponding way. This implies that the species numbers are implicit functions of the component numbers. The chemical potential of the component A is the partial derivative of the free energy *at fixed* N_B. The rules for differentiating an implicit function then give

$$\begin{aligned} \mu_A &= \left[\frac{\partial A(T,V,N_1,N_2,N_{12})}{\partial N_A}\right]_{T,V,N_B} \\ &= \frac{\partial A}{\partial N_1}\frac{\partial N_1}{\partial N_A} + \frac{\partial A}{\partial N_2}\frac{\partial N_2}{\partial N_A} + \frac{\partial A}{\partial N_{12}}\frac{\partial N_{12}}{\partial N_A} \end{aligned} \tag{3.215}$$

This expression is to be evaluated subject to the constraints derived from Eqs. (3.196)

$$\begin{aligned} \frac{\partial N_1}{\partial N_A} + \frac{\partial N_{12}}{\partial N_A} &= 1 \qquad ; \text{ constraints} \\ \frac{\partial N_2}{\partial N_A} + \frac{\partial N_{12}}{\partial N_A} &= 0 \end{aligned} \tag{3.216}$$

Since it is here clear from the context what is to be held fixed in each partial derivative in the last three expressions, we suppress the explicit subscripts on them.[49] Substitution of these relations into Eq. (3.215), and the use of Eq. (3.213), then gives

$$\mu_A = \mu_1 + (-\mu_1 - \mu_2 + \mu_{12})\frac{\partial N_{12}}{\partial N_A} = \mu_1 \tag{3.217}$$

A similar calculation gives

$$\mu_B = \mu_2 \tag{3.218}$$

---

[48] The law of mass action implies that one species number cannot be changed experimentally without affecting the others.

[49] See Prob. 3.25.

It follows that

$$
\begin{aligned}
\mu_1 &= \mu_A \\
\mu_2 &= \mu_B \\
\mu_{12} &= \mu_A + \mu_B = \mu_{AB}
\end{aligned}
\tag{3.219}
$$

where the last relation serves to identify $\mu_{\rm AB}$, and ensures that the chemical potentials are again additively conserved. This establishes the result that the chemical potential for a substance treated as a component is identical to the chemical potential of that substance treated as a species.

Equations (3.219) represent general thermodynamic relations. The additional contribution to the energy, or free energy, for an open sample can be written in terms of either species or components since it follows from the above that

$$
\begin{aligned}
\mu_1 dN_1 + \mu_2 dN_2 + \mu_{12} dN_{12} &= \mu_1 dN_1 + \mu_2 dN_2 + (\mu_1 + \mu_2) dN_{12} \\
&= \mu_1 (dN_1 + dN_{12}) + \mu_2 (dN_2 + dN_{12}) \\
&= \mu_A dN_A + \mu_B dN_B \qquad ; \text{ open sample}
\end{aligned}
\tag{3.220}
$$

This verifies the claims made following Eq. (1.24).

3.4.4 *Solid in Equilibrium with Its Vapor*

As another example of chemical equilibria in the microcanonical ensemble, consider a solid in equilibrium with its vapor. This is an assembly with two phases and one component. The argument closely parallels that just given for a chemical reaction, and hence the analysis can be presented rather quickly.

Let

$$
\begin{aligned}
N_1 &= \text{number of systems in the gas phase} \\
N_2 &= \text{number of systems in the solid phase} \\
N_1 + N_2 &= N
\end{aligned}
\tag{3.221}
$$

Here N is the total number of systems in the assembly, which contains two species.

The total number of complexions is

$$\Omega(E,V,N) = \sum_{N_1}\sum_{N_2}\frac{1}{N_1!}\sum_{\{n^{(1)}\}}\sum_{\{n^{(2)}\}}\frac{N_1!}{\prod_i n_i^{(1)}!}\frac{N_2!}{\prod_i n_i^{(2)}!} \tag{3.222}$$

Here

- The sums are constrained by the relations

$$\sum_i n_i^{(1)} = N_1 \qquad ; \sum_i n_i^{(2)} = N_2 \qquad ; N_1 + N_2 = N$$

$$\sum_i n_i^{(1)}\varepsilon_i^{(1)} + \sum_i n_i^{(2)}\varepsilon_i^{(2)} = E \tag{3.223}$$

- In the solid phase, the systems are localized;
- In the gas phase, the systems are non-localized;
- All systems move independently.

A return to Eq. (3.191) shows this is just[50]

$$\Omega(E,V,N) = \sum_{N_1}\sum_{N_2}\Omega(N_1,N_2;E,V)N_2!$$

$$N_1 + N_2 = N \tag{3.224}$$

Once again we are justified in picking out the largest term in the sum, and using Stirling's formula, since the numbers (N_1, N_2) are very large. With the use of the result in Eq. (3.187), the variation of the logarithm of this largest term at fixed E (and fixed β) gives

$$\delta\ln[\Omega(N_1,N_2;E,V)N_2!] = \delta N_1\ln(\text{p.f.})_{(1)} - \delta N_1\ln N_1 + \delta N_2\ln(\text{p.f.})_{(2)} \tag{3.225}$$

The constraint equation gives

$$\delta N_1 + \delta N_2 = 0 \tag{3.226}$$

Multiply this equation by α, add it to the one above, and then set the coefficient of each variation equal to zero at the maximum. This gives

$$\ln(\text{p.f.})_{(1)} - \ln N_1 + \alpha = 0$$

$$\ln(\text{p.f.})_{(2)} + \alpha = 0 \tag{3.227}$$

[50]We restore an explicit dependence on the volume to $\Omega(N_1, N_2; E)$, writing it now as $\Omega(N_1, N_2; E, V)$; however, V is held fixed in the subsequent variations.

Define

$$e^{\alpha} \equiv \lambda \tag{3.228}$$

The exponentiation of Eqs. (3.227) then yields

$$\begin{aligned} \lambda(\text{p.f.})_{(1)} &= N_1 \\ \lambda(\text{p.f.})_{(2)} &= 1 \end{aligned} \tag{3.229}$$

The ratio of these relations gives

$$N_1 = \frac{(\text{p.f.})_{(1)}}{(\text{p.f.})_{(2)}} \tag{3.230}$$

Recall that for independent, non-localized systems one has the equation of state

$$PV = N_1 k_B T \tag{3.231}$$

Hence Eq. (3.230) can be re-written as

$$P = \frac{k_B T}{V} \frac{(\text{p.f.})_{(1)}}{(\text{p.f.})_{(2)}} \equiv g(T)^{-1} \qquad ; \text{ vapor pressure} \tag{3.232}$$

Two comments:

- This is the *vapor pressure* of the solid;
- This vapor pressure is *completely calculable in terms of the partition functions in the gas and in the solid!*[51]

As before, $\beta = -1/k_B T$, and the Helmholtz free energy is identified as[52]

$$A(T, V, N_1, N_2) = -k_B T \ln \frac{(\text{p.f.})_{(1)}^{N_1}}{N_1!} (\text{p.f.})_{(2)}^{N_2} \tag{3.233}$$

where, once again, (N_1, N_2) now represent mean values.

[51] See Prob. 3.22. The result in Eq. (3.232) assumes $V_1 \approx V$, and the factors of V now cancel in this relation. The reason for the notation $g(T)^{-1}$ will become clear in the next section.

[52] Note that the extra factor of $N_2!$ in the first of Eqs. (3.224) does not affect the arguments in Eqs. (3.204)–(3.208).

The chemical potentials are given by

$$\mu_1 = \left(\frac{\partial A}{\partial N_1}\right)_{T,V,N_2} = -k_{\rm B}T\left[\ln{(\text{p.f.})_{(1)}} - \ln N_1\right] = \alpha k_{\rm B}T$$

$$\mu_2 = \left(\frac{\partial A}{\partial N_2}\right)_{T,V,N_1} = -k_{\rm B}T\ln{(\text{p.f.})_{(2)}} = \alpha k_{\rm B}T \tag{3.234}$$

where Eqs. (3.227) have been used to obtain the final equalities. Hence one finds the same chemical potential in both phases

$$\mu_1 = \mu_2 \equiv \mu \qquad ; \text{chemical potential} \tag{3.235}$$

This confirms, through an explicit calculation, the thermodynamic relation in Eq. (1.37). The absolute activity λ is now given by

$$\lambda = e^{\mu/k_{\rm B}T} \qquad ; \text{absolute activity} \tag{3.236}$$

3.4.5 *Surface Adsorption*

Consider a gas composed of a single species with N_1 systems in a box of volume V. Suppose there is a surface in the box composed of another material, where N_S sites on the surface are available to the gas species and N_2 of these sites are occupied (Fig. 3.29).

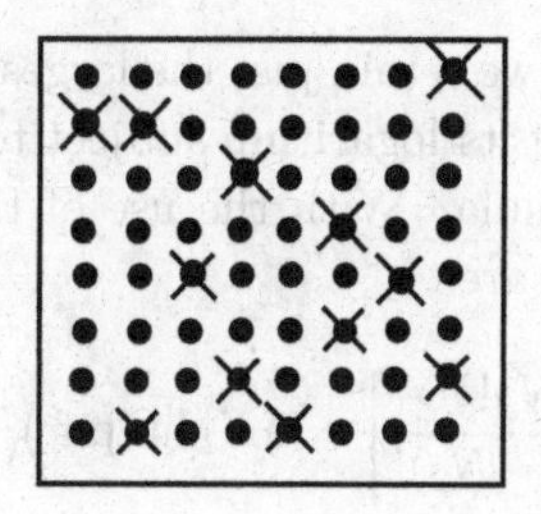

Fig. 3.29 Surface adsorption with N_2 systems on N_S sites.

The problem is to compute the fractional occupation of the sites as a function of the pressure P of the gas, a quantity of interest, for example, in the study of surface catalysis. This is now an assembly with one component and two species. The analysis is a simple extension of the one just carried out in the discussion of vapor pressure.

From our basic counting argument, the number of ways of putting N_2 systems on N_S localized sites is

$$\Omega(N_2, N_S) = \frac{N_S!}{N_2!(N_S - N_2)!} \qquad ; \ N_2 \text{ systems on } N_S \text{ sites} \tag{3.237}$$

The total number of complexions of the assembly is therefore given by Eq. (3.222), augmented with this factor in the summand

$$\Omega(E, V, N) = \sum_{N_1}\sum_{N_2} \frac{1}{N_1!} \sum_{\{n^{(1)}\}} \sum_{\{n^{(2)}\}} \frac{N_1!}{\prod_i n_i^{(1)}!} \frac{N_2!}{\prod_i n_i^{(2)}!} \left[\frac{N_S!}{N_2!(N_S - N_2)!}\right] \tag{3.238}$$

where the sums are subject to the following constraints

$$\sum_i n_i^{(1)} = N_1 \qquad ; \sum_i n_i^{(2)} = N_2 \qquad ; \ N_1 + N_2 = N$$
$$\sum_i n_i^{(1)}\varepsilon_i^{(1)} + \sum_i n_i^{(2)}\varepsilon_i^{(2)} = E \tag{3.239}$$

If we recall Eq. (3.191), this expression can be written as

$$\Omega(E, V, N) = \sum_{N_1}\sum_{N_2} \Omega(N_1, N_2; E, V) \frac{N_S!}{(N_S - N_2)!} \tag{3.240}$$

As in Eqs. (3.225)–(3.227) we retain just the largest term in the sum, which is identified by maximizing its logarithm subject to the number constraint, and employ Stirling's formula. With the use of the result in Eq. (3.187), the appropriate variations are

$$\begin{aligned} \delta \ln \left[\Omega(N_1, N_2; E, V) \frac{N_S!}{(N_S - N_2)!}\right] &= \delta N_1 \ln(\text{p.f.})_{(1)} + \delta N_2 \ln(\text{p.f.})_{(2)} \\ &\quad - \delta N_1 \ln N_1 - \delta N_2 \ln N_2 + \delta N_2 \ln(N_S - N_2) \\ \delta N_1 + \delta N_2 &= 0 \end{aligned} \tag{3.241}$$

Multiply the second equation by α again, add it to the first, and set the coefficients of both variations equal to zero

$$\begin{aligned} \ln(\text{p.f.})_{(1)} + \alpha - \ln N_1 &= 0 \\ \ln(\text{p.f.})_{(2)} + \alpha - \ln N_2 + \ln(N_S - N_2) &= 0 \end{aligned} \tag{3.242}$$

Exponentiate these equations, and define λ as in Eqs. (3.228) and (3.236). The results are

$$N_1 = e^{\mu/k_{\rm B}T}({\rm p.f.})_{(1)}$$
$$\frac{N_2}{N_S - N_2} = e^{\mu/k_{\rm B}T}({\rm p.f.})_{(2)} \tag{3.243}$$

The equation of state for a perfect gas again gives

$$PV = N_1 k_{\rm B}T \tag{3.244}$$

Introduce the fractional occupation of the sites defined by

$$x \equiv \frac{N_2}{N_S} \quad ; \text{ fraction of sites occupied} \tag{3.245}$$

As before, call

$$\frac{k_{\rm B}T}{V}\frac{({\rm p.f.})_{(1)}}{({\rm p.f.})_{(2)}} \equiv \frac{1}{g(T)} \tag{3.246}$$

Three comments:

- This is a known quantity, calculated from the partition functions;[53]
- $({\rm p.f.})_{(2)}$ is now the partition function for a system localized on a site in the surface;
- Since $({\rm p.f.})_{(1)}$ is proportional to the volume V, the ratio $g(T)$ is indeed only a function of temperature.

Take the ratio of the relations in Eqs. (3.243) and then substitute Eqs. (3.244)–(3.246). This yields

$$P = \frac{1}{g(T)}\frac{x}{1-x} \tag{3.247}$$

which implies[54]

$$x = \frac{Pg(T)}{1+Pg(T)} \quad ; \text{ Langmuir adsorption isotherm} \tag{3.248}$$

This is known as the *Langmuir adsorption isotherm* [Langmuir (1916)]. It gives the fractional occupation of the sites on the surface as a function of the pressure of the gas. It is sketched in Fig. 3.30, where the possible onset of multilayer adsorption is also indicated.

---

[53] See Prob. 3.24.

[54] We now see the reason for the notation in Eq. (3.246).

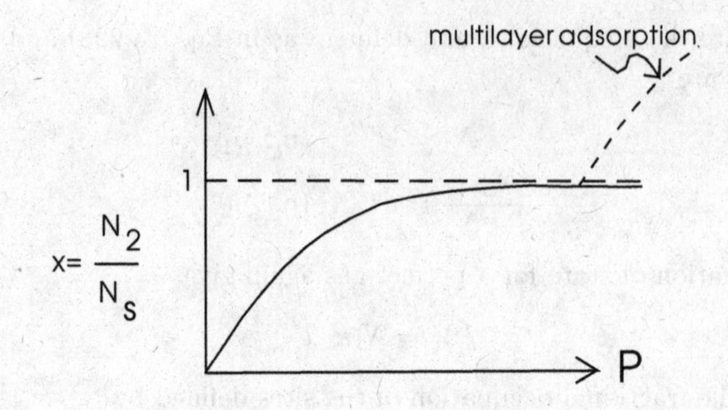

Fig. 3.30 Sketch of the Langmuir adsorption isotherm in Eq. (3.248), where $x = N_2/N_S$ is the fractional occupation of sites on the surface, and P is the gas pressure. Also indicated is the possible onset of multilayer adsorption.

Chapter 4

The Canonical Ensemble

So far we have seen many successful applications of the *microcanonical ensemble*, a collection of identical independent systems, either localized or non-localized, in thermal equilibrium and distributed in energy according to the Boltzmann distribution. The full power of statistical mechanics becomes evident when the analysis is extended to include *interactions* between the systems, and one deals with the *canonical ensemble.* Although many analyses will start here,[1] the present approach works up to this through the microcanonical ensemble. To reach this goal, we turn to an examination of the constant-temperature partition function.

4.1 Constant-Temperature Partition Function

Consider the equilibrium ensemble at a temperature T formed by a *collection of assemblies*, each of which, in turn, is formed from a collection of systems. Imagine that all the assemblies are in *thermal contact*, and hence free to exchange energy. One can picture this, for example, as a 10-ton block of Pb divided into 10^7 little cubes, all in thermal contact (Fig. 4.1).[2] Recall the previous discussion of systems in thermal equilibrium in an assembly. The mean energy of a system is given by

$$\varepsilon = \frac{E}{N} = \frac{\sum_i \varepsilon_i e^{-\varepsilon_i/k_{\rm B}T}}{\sum_i e^{-\varepsilon_i/k_{\rm B}T}} = k_{\rm B}T^2 \frac{\partial}{\partial T} \ln{(\text{p.f.})} \tag{4.1}$$

[1] See, for example, [Walecka (2000)].
[2] See [Rushbrooke (1949)].

where

$$(\text{p.f.}) = \sum_i e^{-\varepsilon_i/k_B T}$$
$$= \sum_j \omega_j e^{-\varepsilon_j/k_B T} \qquad ; \text{ degeneracy explicit} \qquad (4.2)$$

The second line explicitly exhibits the degeneracy ω_j of the energy level ε_j.[3]

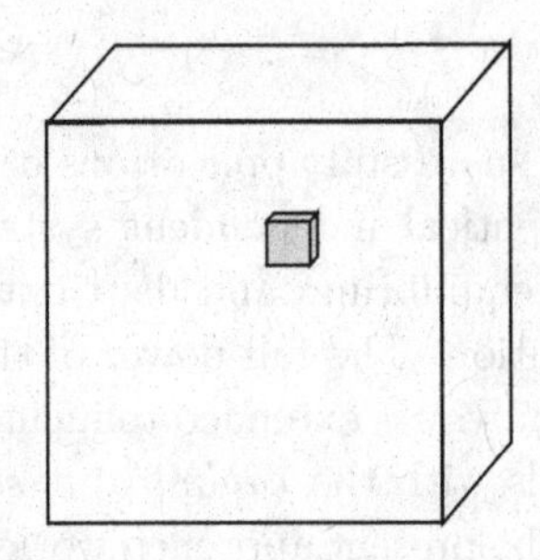

Fig. 4.1 A 10-ton block of Pb (the "ensemble") divided into 10^7 little cubes (the "assemblies"), all in thermal contact.

Let us now make the re-identification

$$\text{System} \rightarrow \text{Assembly}$$
$$\text{Assembly} \rightarrow \text{Ensemble} \qquad (4.3)$$

All our previous results can be taken over! We now have a set of energy levels and degeneracies of an *assembly*

$$E_1, E_2, E_3, \cdots \qquad ; \text{ energies of the assembly}$$
$$\Omega_1, \Omega_2, \Omega_3, \cdots \qquad ; \text{ degeneracies of the assembly} \qquad (4.4)$$

Here the Ω_j represent the number of states of the assembly with a given E_j; they are evidently the same quantities Ω introduced previously in the discussion of the microcanonical ensemble. The above analogy indicates that the mean energy of an *assembly* (the little cubes in Fig. 4.1) in this

[3]The first sum goes over all the states, and the second goes over the energy levels.

ensemble (the big cube in Fig. 4.1) is

$$E = k_{\rm B}T^2 \frac{\partial}{\partial T} \ln (\text{P.F.}) \qquad ; \text{ mean energy of assembly}$$
$$(\text{P.F.}) = \sum_j \Omega_j e^{-E_j/k_{\rm B}T} \tag{4.5}$$

The corresponding Helmholtz free energy for the assembly can be identified from the previous thermodynamic relation in Eq. (2.48)

$$E = -T^2 \frac{\partial}{\partial T} \left(\frac{A}{T} \right)_{N,V} \tag{4.6}$$

Hence we arrive at the following expression for the Helmholtz free energy of an assembly[4]

$$A(T, V, N) = -k_{\rm B}T \ln (\text{P.F.}) \qquad ; \text{ free energy of assembly}$$
$$(\text{P.F.}) = \sum_j \Omega_j e^{-E_j/k_{\rm B}T} \qquad ; \text{ canonical partition function} \tag{4.7}$$

A set of assemblies in thermal equilibrium at the temperature T, and distributed in energy according to the Boltzmann distribution, is said to form the *canonical ensemble*. The quantity (P.F.) is known as the *canonical partition function*; it involves a sum over all possible values of the energies and degeneracies of the assembly.

A few comments:

- Equations (4.7) form the *second fundamental relation* of statistical mechanics, the first being Boltzmann's equation in Eq. (1.46);[5]
- Equations (4.7) provide the Helmholtz free energy $A(T, V, N)$ of an assembly in terms of the energies and degeneracies of the assembly;
- All the thermodynamic properties of the assembly follow from $A(T, V, N)$;
- On physical grounds, the thermodynamic properties of the assembly at temperature T should be the same whether or not it is in actual thermal contact with the other members of the ensemble;
- We shall, in fact, subsequently show that in the limit where the assembly consists of very many systems so that $N \to \infty$, all assemblies in the canonical ensemble will effectively have an energy equal to their mean value E;

[4] In the microcanonical case, this is the Helmholtz free energy *per system*.

[5] Equations (4.7) may be considered as an alternative to $S = k_{\rm B} \ln \Omega$ as the starting point of statistical mechanics.

- The energies E_j are the eigenvalues of the total hamiltonian H, and thus interactions between the systems in the assembly are now immediately included in the analysis.

In the previous discussion of the microcanonical ensemble we had

$$\begin{aligned} S &= k_{\rm B} \ln \Omega \\ A &= E - TS \\ A &= -k_{\rm B} T \ln \left(\Omega e^{-E/k_{\rm B}T}\right) \end{aligned} \tag{4.8}$$

This amounts to keeping only *one term* in the sum in Eq. (4.7). Can we understand why the two approaches are equivalent? Again, the basic idea is that for very large N, the largest term dominates the sum. This is explicitly illustrated below.

4.1.1 *Independent Localized Systems*

Let us examine what the present approach gives for an assembly composed of independent localized systems. Here one has the following energy levels, degeneracies, and distribution numbers for the systems in the assembly

$$\begin{aligned} &\varepsilon_1, \varepsilon_2, \varepsilon_3, \cdots && ; \text{ energies} \\ &\omega_1, \omega_2, \omega_3, \cdots && ; \text{ degeneracies} \\ &n_1, n_2, n_3, \cdots && ; \text{ distribution numbers} \end{aligned} \tag{4.9}$$

The distribution number n_k is now the number of systems with energy ε_k.

Since each of the n_k systems in the energy level ε_k can go into any one of the ω_k degenerate states, the total number of complexions Ω_j with a total energy E_j follows from our previous analysis as

$$\begin{aligned} \Omega_j &= \sum_{\{n\}} \frac{N!}{\prod_i n_i!} \prod_k \omega_k^{n_k} \\ E_j &= \sum_k n_k \varepsilon_k \qquad ; N = \sum_k n_k \end{aligned} \tag{4.10}$$

where the degeneracy factor $\omega_1^{n_1} \omega_2^{n_2} \omega_3^{n_3} \cdots$ has now been explicitly exhibited in Ω_j (Fig. 4.2).[6]

[6] We could, of course, handle the degeneracies as before by simply letting the ε_i become equal at the end [compare Eq. (4.12)], but here we prefer to include the degeneracies from the outset.

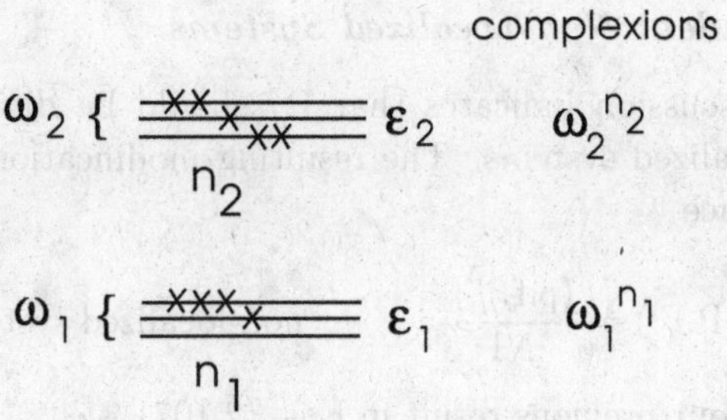

complexions

$\omega_2 \{$ ε_2 $\omega_2^{n_2}$

n_2

$\omega_1 \{$ ε_1 $\omega_1^{n_1}$

n_1

Fig. 4.2 Each of the given n_k systems in the energy level ε_k in Fig. 2.1 can go into one of the ω_k degenerate states, increasing the number of complexions for a given n_k by a factor of $\omega_k^{n_k}$ (compare Prob. 4.3).

The canonical partition function is given by

$$
\begin{aligned}
(\text{P.F.}) &= \sum_j \Omega_j e^{-E_j/k_\text{B}T} \\
&= \sum_{\{n\}} \frac{N!}{\prod_i n_i!} \prod_k \omega_k^{n_k} e^{-\sum_l n_l \varepsilon_l/k_\text{B}T} \qquad ; \sum_k n_k = N \qquad (4.11)
\end{aligned}
$$

This sum can now be done *exactly*, since for fixed $\sum_k n_k = N$,

$$
\begin{aligned}
(\text{P.F.}) &= \sum_{\{n\}} N! \frac{(\omega_1 e^{-\varepsilon_1/k_\text{B}T})^{n_1}}{n_1!} \frac{(\omega_2 e^{-\varepsilon_2/k_\text{B}T})^{n_2}}{n_2!} \frac{(\omega_3 e^{-\varepsilon_3/k_\text{B}T})^{n_3}}{n_3!} \cdots \\
&= \left(\sum_j \omega_j e^{-\varepsilon_j/k_\text{B}T} \right)^N = (\text{p.f.})^N \qquad (4.12)
\end{aligned}
$$

Here the second line follows from the use of the multinomial theorem. Hence

$$
\begin{aligned}
(\text{P.F.}) &= (\text{p.f.})^N \\
A(T, V, N) &= -N k_\text{B} T \ln (\text{p.f.}) \qquad (4.13)
\end{aligned}
$$

Three observations:

- This exactly reproduces our previous result for independent localized systems in Eqs. (2.103)!
- There is no "method of steepest descent" involved, nor use of Stirling's formula!
- Removal of the constraint of fixed $E = \sum_i n_i \varepsilon_i$ greatly simplifies the sums!

4.1.2 *Independent Non-Localized Systems*

Our previous discussion indicates that Ω_j should be divided by $N!$ for identical non-localized systems. The resulting modification of Eqs. (4.13) is to simply replace

$$(\text{P.F.}) = \frac{(\text{p.f.})^N}{N!} \qquad ; \text{ non-localized systems} \tag{4.14}$$

This reproduces our previous result in Eqs. (2.107).

Since we previously only evaluated $S = k_B \ln \Omega$ to $O(\ln N)$ through the method of steepest descent, we conclude that the microcanonical and canonical approaches, as presented here, differ at most by terms of this order. Let us see how this works.

4.2 Classical Limit

Consider the classical limit of the canonical partition function for non-localized systems

$$(\text{P.F.}) = \frac{1}{N!\,h^{3N}} \int \cdots \int e^{-H(p,q)/k_B T} dp_1 \cdots dp_{3N} dq_1 \cdots dq_{3N} \tag{4.15}$$

Here the full hamiltonian $H(p,q)$ is used for the energy $E(p,q)$ in the exponential, which allows us to easily include interactions. It follows from Eqs. (4.13) and thermodynamics that

$$(\text{P.F.}) = e^{-A/k_B T} = e^{-E/k_B T} e^{TS/k_B T} \tag{4.16}$$

where E is the mean energy of an assembly. The mean entropy S of an assembly in the canonical ensemble is therefore

$$S = k_B \ln \left\{ \frac{e^{E/k_B T}}{N! h^{3N}} \int \cdots \int e^{-H(p,q)/k_B T} dp_1 \cdots dp_{3N} dq_1 \cdots dq_{3N} \right\}$$
$$; \text{ canonical ensemble} \tag{4.17}$$

This is to be compared with our previous expression for the entropy of the assembly with the microcanonical ensemble in Eqs. (2.170)

$$S = k_B \ln \left\{ \frac{1}{N!\,h^{3N}} \int \cdots \int dp_1 \cdots dp_{3N} dq_1 \cdots dq_{3N} \right\}$$
$$E \leq H(p,q) \leq E + \Delta E$$
$$; \text{ microcanonical ensemble} \tag{4.18}$$

How do we understand that Eqs. (4.17) and (4.18) give the same result for the entropy? Let us again consider the perfect gas with hamiltonian

$$H(p,q) = \sum_{i=1}^{N} \frac{1}{2m}\mathbf{p}_i^2 \tag{4.19}$$

In this case, we can obtain analytic expressions for all the quantities involved.

For the *microcanonical ensemble*, Eqs. (2.192) provide the equivalent expression

$$\begin{aligned} S &= k_B \ln \Omega = k_B \ln \omega(E) \\ \omega(E) &= \frac{1}{N!h^{3N}} \int \cdots \int dp_1 \cdots dp_{3N} dq_1 \cdots dq_{3N} \\ & 0 \leq H(p,q) \leq E \end{aligned} \tag{4.20}$$

where $\omega(E)$ is the result arising from the total phase space volume contained inside the surface of constant E. This is evaluated in Eqs. (2.185) and (2.199) as

$$\begin{aligned} \omega(E) &= \frac{V^N}{N!\,h^{3N}} \frac{(2\pi m E)^{3N/2}}{\Gamma(1+3N/2)} \quad ; \; E = \frac{3}{2}Nk_BT \\ &= \frac{1}{N!}\left[V\left(\frac{2\pi m k_B T}{h^2}\right)^{3/2}\right]^N \frac{(3N/2)^{3N/2}}{(3N/2)!} \end{aligned} \tag{4.21}$$

where the second line makes use of the equation of state $E = (3/2)Nk_BT$.[7]

For the *canonical ensemble*, Eq. (4.17) provides an effective Ω

$$\begin{aligned} S &= k_B \ln \Omega_{\text{eff}} \\ \Omega_{\text{eff}} &= e^{E/k_BT}\,(\text{P.F.}) \end{aligned} \tag{4.22}$$

Here

$$\begin{aligned} (\text{P.F.}) &= \frac{1}{N!\,h^{3N}} \int \cdots \int dp_1 \cdots dp_{3N} dq_1 \cdots dq_{3N} \exp\left\{-\frac{1}{2mk_BT}\sum_{i=1}^{3N} p_i^2\right\} \\ &= \frac{V^N}{N!\,h^{3N}} \left[\int_{-\infty}^{\infty} dp\, e^{-p^2/2mk_BT}\right]^{3N} \end{aligned} \tag{4.23}$$

[7] We also assume $3N/2$ is an integer in the second line.

The standard gaussian integral then gives[8]

$$(\text{P.F.}) = \frac{1}{N!}\left[V\left(\frac{2\pi m k_{\text{B}}T}{h^2}\right)^{3/2}\right]^N \tag{4.24}$$

Hence in the canonical ensemble

$$\Omega_{\text{eff}} = e^{E/k_{\text{B}}T}\,(\text{P.F.}) = \frac{e^{E/k_{\text{B}}T}}{N!}\left[V\left(\frac{2\pi m k_{\text{B}}T}{h^2}\right)^{3/2}\right]^N \tag{4.25}$$

The equation of state can again be employed to give

$$\ln\left(e^{E/k_{\text{B}}T}\right) = \ln\left(e^{3N/2}\right) = \frac{3N}{2} \qquad ;\ E = \frac{3}{2}Nk_{\text{B}}T \tag{4.26}$$

Thus

$$\ln\Omega_{\text{eff}} = \ln(\text{P.F.}) + \frac{3N}{2} \tag{4.27}$$

When the logarithm is taken of Eq. (4.21), one has with the use of Stirling's formula (certainly valid here)

$$\begin{aligned}\ln\left\{\frac{(3N/2)^{3N/2}}{(3N/2)!}\right\} &= \frac{3N}{2}\ln\frac{3N}{2} - \frac{3N}{2}\ln\frac{3N}{2} + \frac{3N}{2} + O(\ln N) \\ &= \frac{3N}{2} + O(\ln N)\end{aligned} \tag{4.28}$$

With the neglect of terms of $O(\ln N)$, one therefore has from Eqs. (4.21) and (4.24)

$$\ln\omega(E) = \ln(\text{P.F.}) + \frac{3N}{2} \tag{4.29}$$

Equations (4.27) and (4.29) imply that the expressions in Eqs. (4.20) and (4.22) for the entropy of the assembly in the microcanonical and canonical ensembles are *identical*

$$S = k_{\text{B}}\ln\omega(E) = k_{\text{B}}\ln\Omega_{\text{eff}} \tag{4.30}$$

[8] Note that by explicit calculation we obtain $(\text{P.F.}) = (\text{p.f.})^N/N!$, in accordance with Eq. (4.14).

4.3 Energy Distribution

To obtain insight into the above result, consider the integral in the canonical ensemble in more detail. Write it as[9]

$$\Omega_{\text{eff}} = e^{E/k_{\text{B}}T}(\text{P.F.}) = e^{E/k_{\text{B}}T}\int_0^\infty \frac{\partial\omega(E')}{\partial E'}e^{-E'/k_{\text{B}}T}\,dE'$$

$$\omega(E) = K(N,V)E^{3N/2} \tag{4.31}$$

where, as before, the functional form of $\omega(E)$ follows from Eq. (4.21). Thus

$$\Omega_{\text{eff}} = \int_0^\infty \frac{\partial\omega(E')}{\partial E'}e^{(E-E')/k_{\text{B}}T}\,dE' \tag{4.32}$$

With the use of the last of Eqs. (4.31), this expression can be re-written as

$$\Omega_{\text{eff}} = \frac{\partial\omega(E)}{\partial E}E\int_0^\infty \left(\frac{E'}{E}\right)^{(3N/2-1)}\frac{dE'}{E}e^{(E-E')/k_{\text{B}}T} \tag{4.33}$$

From Eqs. (2.192), the factor in front of the integral reproduces the result for the entropy in the microcanonical ensemble (there is a negligible contribution to the logarithm from the extra E). Consider, then, the factor arising from the remaining integral. Introduce

$$E' = Ex \qquad ; \; E = (3/2)Nk_{\text{B}}T \tag{4.34}$$

This gives

$$\begin{aligned}\int_0^\infty \left(\frac{E'}{E}\right)^{(3N/2-1)}\frac{dE'}{E}e^{(E-E')/k_{\text{B}}T} &= \int_0^\infty x^{n-1}\,e^{n(1-x)}\,dx \\ &= \frac{e^n}{n^n}\Gamma(n) \qquad ; \; n \equiv \frac{3N}{2}\end{aligned} \tag{4.35}$$

The second equality follows from the integral representation of the Γ-function in Eq. (2.176). We are interested in the limit of this expression as $n = 3N/2 \to \infty$. Write

$$\int_0^\infty x^{n-1}\,e^{n(1-x)}\,dx \equiv \int_0^\infty I(x)dx \tag{4.36}$$

and consider the integrand $I(x)$. Observe that:

[9] If the exponential is omitted, and the integral is carried out only to E, one must obtain $\omega(E)$ [see Eq. (2.172)].

(1) For large n, the logarithm of the integrand takes the form

$$\ln I(x) = n[\ln x + 1 - x] \leq 0 \qquad ; \; n \to \infty \tag{4.37}$$

(2) The maximum of this logarithm occurs at

$$\begin{aligned} \frac{\partial}{\partial x} \ln I(x) &= \frac{n}{x} - n = 0 \\ \Longrightarrow \qquad x &= 1 \qquad ; \; I(1) = 1 \end{aligned} \tag{4.38}$$

(3) The curvature of the logarithm there is

$$\begin{aligned} \left[\frac{\partial^2}{\partial x^2} \ln I(x)\right]_{x=1} &= -n \\ &\to -\infty \end{aligned} \tag{4.39}$$

The integrand $I(x)$ is strongly peaked at $x = 1$ as $n \to \infty$ (Fig. 4.3).

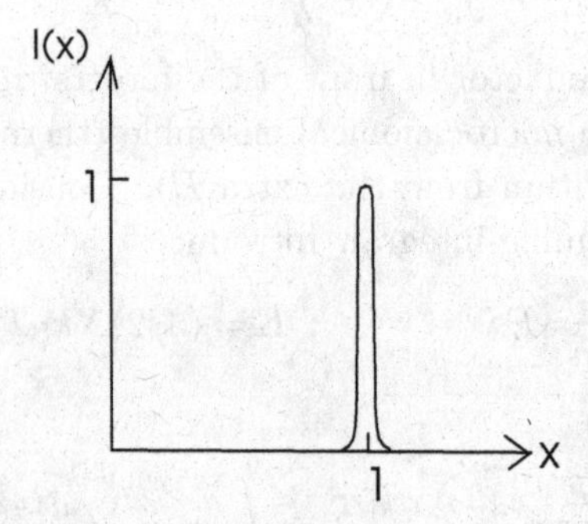

Fig. 4.3 Sketch of the integrand $I(x)$ in Eq. (4.36) as $n \to \infty$. It is strongly peaked at $x = 1$, which implies $E' = E$. The width of this curve is $\sim 1/\sqrt{n}$ (see Prob. 4.2).

This is exactly the situation we had in the method of steepest descent, and the same technique can be used to evaluate the integral. Expand the logarithm of the integrand as a Taylor series in $(x - 1)$, and keep up through quadratic terms

$$\begin{aligned} \int_0^\infty I(x)dx &= \int_0^\infty e^{\ln I(x)}\, dx \\ &\approx \int_0^\infty e^{-(x-1)^2 n/2}\, dx \approx \int_{-\infty}^\infty e^{-nu^2/2}\, du \\ &= \left(\frac{2\pi}{n}\right)^{1/2} \left[1 + O\left(\frac{1}{n}\right)\right] \qquad ; \; n \to \infty \end{aligned} \tag{4.40}$$

As before, this result holds to $O(1/n)$. Hence, from Eqs. (4.35)

$$\frac{e^n}{n^n}\Gamma(n) = \left(\frac{2\pi}{n}\right)^{1/2}\left[1 + O\left(\frac{1}{n}\right)\right] \quad ; \; n \to \infty \tag{4.41}$$

Notice that we have just derived Stirling's formula![10]

The remaining integral in Eq. (4.33) thus contributes only a term of $O(\ln N)$ to the entropy, whereas the leading term is of $O(N)$. Therefore we have the result that as $N \to \infty$

$$\ln\left[\int_0^\infty \frac{\partial\omega(E')}{\partial E'} e^{(E-E')/k_BT}\, dE'\right] = \ln\frac{\partial\omega(E)}{\partial E} + O(\ln N)$$

$$\text{canonical ensemble} \qquad\qquad \text{microcanonical ensemble} \tag{4.42}$$

In this limit, the sum (integral) in the logarithm on the l.h.s. can be replaced by its largest term (maximum value of the integrand), which occurs at $E' = E$. This is why the canonical and microcanonical ensembles yield the same result for the entropy of an assembly [compare Eqs. (4.8)].

Let us examine this result in more detail:

- For very large N, the integrand in Eq. (4.33) is sharply peaked at $E' = E$. The width of the energy distribution is of $O(1/\sqrt{N})$ [see Prob. 4.2]. Each assembly effectively contributes to the mean entropy S of an assembly at the mean energy E;
- One component of this peaking is the very rapid rise in the phase space volume $\partial\omega(E')/\partial E'$;
- A second component is the very rapid fall in the Boltzmann factor $e^{(E-E')/k_BT}$ for $E' > E$;
- This argument that the energy distribution is strongly peaked in the limit $N \to \infty$, and each assembly can actually be assigned the mean energy E, is ultimately the justification for the use of the canonical ensemble in statistical mechanics.[11]

4.4 Summary of Results So Far

It is useful to provide a summary of the working results obtained so far.

---

[10]See Prob. 4.1.

[11]Compare [Walecka (2000)]. See also the later discussion in Chapter 6.

4.4.1 *Microcanonical Ensemble*

The microcanonical ensemble is appropriate for an assembly of identical, independent systems. The Helmholtz free energy of the assembly is obtained as

$$\begin{aligned} A(T, V, N) &= -k_B T \ln (\text{p.f.})^N && ; \text{localized} \\ &= -k_B T \ln \frac{(\text{p.f.})^N}{N!} && ; \text{non-localized} \end{aligned} \tag{4.43}$$

where the first expression holds for localized systems, and the second for non-localized systems. The partition function is given by

$$(\text{p.f.}) = \sum_i e^{-\varepsilon_i / k_B T} \equiv \sum_j \omega_j e^{-\varepsilon_j / k_B T} \tag{4.44}$$

Here $\varepsilon_1, \varepsilon_2, \varepsilon_3, \cdots$ are the energy levels available to a system and the last relation explicitly exhibits their degeneracies $\omega_1, \omega_2, \omega_3, \cdots$.[12]

The classical limit of the partition function, valid when many states contribute to it, is obtained as

$$(\text{p.f.}) = \frac{1}{h^s} \int \cdots \int e^{-\varepsilon(p,q)/k_B T} \, dp_1 \cdots dp_s dq_1 \cdots dq_s \qquad ; \text{classical limit} \tag{4.45}$$

where $\varepsilon(p.q) \equiv h(p, q)$ is the energy of one system, and s is the number of its degrees of freedom.

The thermodynamic properties of the assembly are then obtained through the use of the first and second laws of thermodynamics

$$dA = -SdT - PdV + \mu dN \tag{4.46}$$

4.4.2 *Canonical Ensemble*

The canonical ensemble is appropriate for assemblies of identical systems, both *with and without interactions*. The Helmholtz free energy of an assembly is obtained as

$$A(T, V, N) = -k_B T \ln (\text{P.F.}) \tag{4.47}$$

The canonical partition function is given by

$$(\text{P.F.}) = \sum_j \Omega_j e^{-E_j / k_B T} \tag{4.48}$$

[12]The first sum goes over all the states, and the second goes over the energy levels.

Here $E_1, E_2, E_3, \cdots$ are the energy levels available to the *assembly*, and $\Omega_1, \Omega_2, \Omega_3, \cdots$ are their degeneracies.

For a very large number N of systems in the assemblies, the assembly energies at the temperature T are sharply-peaked about the mean value

$$E = k_B T^2 \frac{\partial}{\partial T} \ln(\text{P.F.}) \tag{4.49}$$

The classical limit of the canonical partition function, here valid for non-localized systems with s degrees of freedom, is obtained as

$$(\text{P.F.}) = \frac{1}{N!\, h^{sN}} \int \cdots \int e^{-H(p,q)/k_B T}\, dp_1 \cdots dp_{sN} dq_1 \cdots dq_{sN}$$
$$\text{; classical limit} \qquad \text{; non-localized} \tag{4.50}$$

where $E(p,q) = H(p,q)$ is the classical energy of the assembly.

All thermodynamic properties of the assembly can now be obtained through the use of the first and second laws of thermodynamics

$$dA = -SdT - PdV + \mu dN \tag{4.51}$$

Chapter 5

Applications of the Canonical Ensemble

5.1 Solids

As a first application of the canonical ensemble, consider the thermodynamic properties of solids. Given a mechanical model of the solid, the goal is to calculate the canonical partition function and corresponding Helmholtz free energy

$$A(T,V,N) = -k_{\rm B}T\ln{\rm (P.F.)}$$
$$({\rm P.F.}) = \sum_j e^{-E_j/k_{\rm B}T} \tag{5.1}$$

where the sum now goes over all the states of an assembly. As an introduction, recall the Einstein model of a solid.

5.1.1 *Einstein Model*

Einstein modeled a solid as a collection of $3N$ one-dimensional simple harmonic oscillators with a single oscillator frequency ν_0 (see Fig. 2.1). The partition function for the one-dimensional oscillator was calculated in Eq. (2.46), and thus from Eqs. (4.13)

$$({\rm P.F.}) = ({\rm p.f.})^{3N}$$
$$({\rm p.f.}) = \frac{e^{-h\nu_0/2k_{\rm B}T}}{1-e^{-h\nu_0/k_{\rm B}T}} \tag{5.2}$$

Equations (5.1) then reproduce the previous microcanonical results in Eqs. (2.51)–(2.55). Although a significant result of the Einstein model is that the heat capacity C_V goes to zero at low temperature, it goes to zero *too fast*, vanishing exponentially while experiment indicates a power-series behavior as $T \to 0$ (see Fig. 5.1).

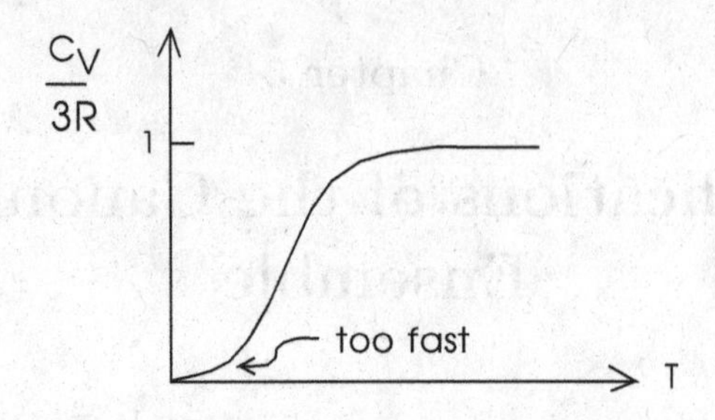

Fig. 5.1 Sketch of the molar heat capacity $C_V/3R$ in the Einstein model of a solid. The exponential decrease to zero at low temperature is too fast (compare Prob. 5.6).

5.1.2 *Normal Modes*

The Einstein model is readily generalized. One can still talk about harmonic restoring forces, but now picture the solid as a huge molecule. There are N systems (atoms) with $3N - 6$ vibrational degrees of freedom. There will be $3N - 6 \approx 3N$ *normal coordinates* q_i describing the *normal modes of oscillations of the solid*, each of which behaves like an uncoupled simple harmonic oscillator. For small oscillations about equilibrium, the hamiltonian for the assembly of interacting systerms can always be reduced to the form[1]

$$H = \sum_{i=1}^{3N} \frac{1}{2} \left(\dot{q}_i^2 + \omega_i^2 q_i^2\right) \qquad ; \text{ normal modes} \tag{5.3}$$

The normal modes consist of $3N$ oscillators of *different* frequencies $\omega_i = 2\pi\nu_i$, with $i = 1, 2, \cdots, 3N$. The oscillators are quantized as in elementary quantum mechanics, with the result that

$$E = \sum_{i=1}^{3N} h\nu_i(p_i + 1/2) \qquad ; \ p_i = 0, 1, 2, \cdots \tag{5.4}$$

The frequencies ν_i constitute the *normal-mode spectrum.* Since the normal modes arise from the coupling between the atoms, we are considering the excitations of the entire many-body solid, and no longer are we talking about the excitations of independent systems.

[1] See [Fetter and Walecka (2003a)].

Since we now have the energies E_j, characterized by a given set of oscillator quantum numbers $(p_1, p_2, p_3, \cdots)$, the canonical partition function follows from Eq. (5.1) as

$$(\text{P.F.}) = \sum_{p_1}\sum_{p_2}\cdots \exp\left\{-\frac{h}{k_{\text{B}}T}\sum_i \nu_i\left(p_i + \frac{1}{2}\right)\right\} \tag{5.5}$$

The canonical partition function *factors*, and the result for each oscillator is just that in Eq. (2.46). Thus

$$\begin{aligned}(\text{P.F.}) &= \sum_{p_1} e^{-h\nu_1(p_1+1/2)/k_{\text{B}}T}\sum_{p_2} e^{-h\nu_2(p_2+1/2)/k_{\text{B}}T} \times \cdots \\ &= \prod_{i=1}^{3N}\frac{e^{-h\nu_i/2k_{\text{B}}T}}{1-e^{-h\nu_i/k_{\text{B}}T}}\end{aligned}$$

$$A(T,V,N) = k_{\text{B}}T\sum_{i=1}^{3N}\ln\left(1-e^{-h\nu_i/k_{\text{B}}T}\right) + \frac{1}{2}\sum_{i=1}^{3N}h\nu_i \tag{5.6}$$

Hence, just as in Eqs. (2.51)–(2.53),

$$E = -T^2\frac{\partial}{\partial T}\left(\frac{A}{T}\right) = k_{\text{B}}T\sum_{i=1}^{3N}\frac{u_i}{e^{u_i}-1} + \frac{1}{2}k_{\text{B}}T\sum_{i=1}^{3N}u_i$$

$$C_V = \frac{\partial E}{\partial T} = k_{\text{B}}\sum_{i=1}^{3N}\frac{u_i^2 e^{u_i}}{(e^{u_i}-1)^2} \qquad ; \; u_i \equiv \frac{h\nu_i}{k_{\text{B}}T} \tag{5.7}$$

These results are completely general, assuming only small oscillations about equilbrium for the solid.

Several comments:

- The normal modes of a macroscopic solid will be closely spaced, and therefore the sum over modes can be replaced by an integral over frequencies

$$\sum_i \to \int g(\nu)d\nu \qquad ; \text{ macroscopic solid} \tag{5.8}$$

Here

$$g(\nu)d\nu \equiv \text{number of modes between } \nu \text{ and } \nu + d\nu \tag{5.9}$$

We refer to $g(\nu)$ as the *spectral weight*.

- The *wavelength* of the normal modes is limited, since it makes no sense to talk about normal-mode oscillations of a crystal with a wavelength smaller that the interatomic spacing.[2] Thus there will be a *maximum normal-mode frequency* ν_m

$$\nu_m \equiv \text{maximum normal-mode frequency} \tag{5.10}$$

- Since the total number of normal modes is the number of degrees of freedom, one has

$$\int_0^{\nu_m} g(\nu)d\nu = 3N \qquad ; \text{number of degrees of freedom} \tag{5.11}$$

- Einstein simply said that (see Fig. 5.2)

$$g(\nu) = 3N\delta(\nu - \nu_0) \qquad ; \text{Einstein model} \tag{5.12}$$

- The whole problem now has been reduced to calculating a realistic spectral weight $g(\nu)$!

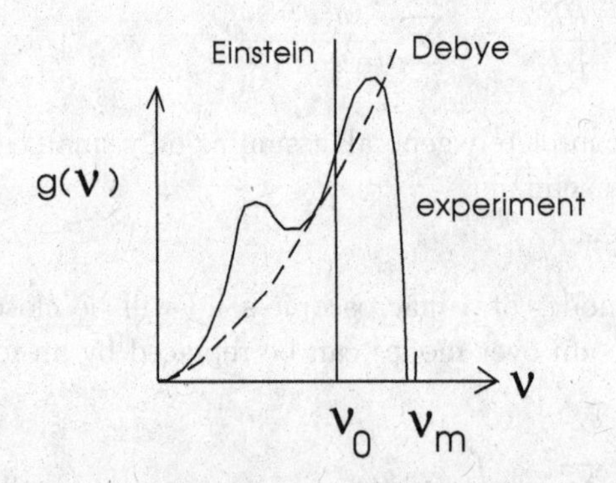

Fig. 5.2 Typical normal-mode distribution $g(\nu)$ in a solid (see [Davidson (2003); Kittel (2004)]). The Einstein model has $g(\nu) = 3N\delta(\nu - \nu_0)$. The quadratic spectrum of the Debye model is discussed below.

[2]See Prob. 5.4.

5.1.3 *Debye Model*

Debye modeled the solid as a continuous, isotropic, elastic medium [Debye (1912)]. The normal modes in this medium are *sound waves*, satisfying

$$c = \nu\lambda \qquad ; \text{ sound velocity}$$

$$\Longrightarrow \qquad k = \frac{\omega}{c} = \frac{2\pi\nu}{c} \tag{5.13}$$

5.1.3.1 *Normal-Mode Spectrum*

Suppose one is interested in the standing waves in a solid with clamped boundaries. The wave equation in the medium reduces to the scalar Helmholtz Eq. (2.110), and we have already looked at the standing waves in a box where the wave function vanishes on the boundaries in Eqs. (2.111)–(2.112). The wave numbers must satisfy (see Fig. 5.3)

$$k_x = \frac{\pi n_x}{L} \quad ; \; k_y = \frac{\pi n_y}{L} \quad ; \; k_z = \frac{\pi n_z}{L} \tag{5.14}$$

where the n_i run over the positive integers

$$n_i = 1, 2, 3, \cdots \qquad ; \; i = x, y, z \tag{5.15}$$

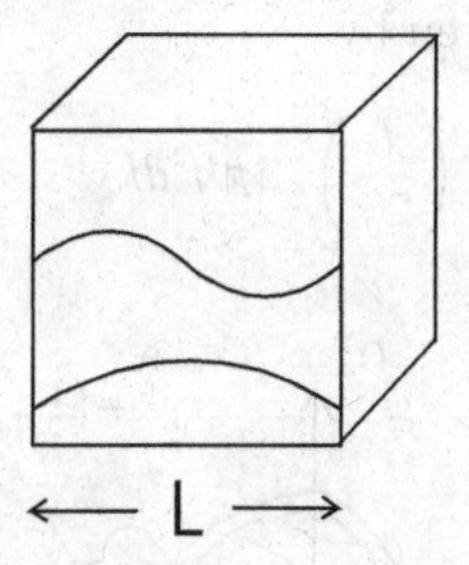

Fig. 5.3 Normal modes in a continuous, isotropic elastic medium with clamped boundaries. We illustrate $kL = 2\pi L/\lambda = n\pi$ for $n = 1, 2$, in the x-direction.

Let us find the number of modes lying between k and $k + dk$ where

$$k^2 = k_x^2 + k_y^2 + k_z^2 \tag{5.16}$$

Consider n-space where the coordinate axes are (n_x, n_y, n_z). Each mode can be associated with a unit volume in this space, by associating that volume with the point (n_x, n_y, n_z) in a one-to-one fashion. This can be done, for example, by using the point in the upper right-hand corner in

Fig. 5.3 for each unit cube. The problem of finding the number of modes is then reduced to finding the volume in this n-space. Now

$$n^2 = n_x^2 + n_y^2 + n_z^2 = \left(\frac{L}{\pi}\right)^2 k^2 \tag{5.17}$$

If the size of the box is large, then there are very many modes between k and $k + dk$, and one can simply compute the corresponding volume in n-space as an *integral.* Thus[3]

$$\sum_{n_x,n_y,n_z} \to \int\int\int_{\text{first octant}} dn_x dn_y dn_z = \left(\frac{L}{\pi}\right)^3 \int\int\int_{\text{first octant}} dk_x dk_y dk_z \quad ; L \to \infty \tag{5.18}$$

Since the n_i are positive integers, the integral is taken over the first octant in n-space, and correspondingly, the first octant in k-space. The integral over the first octant is 1/8 the integral over all k, and thus as $L \to \infty$

$$\sum_{n_x,n_y,n_z} \to \frac{1}{8}\left(\frac{L}{\pi}\right)^3 \int_{\text{all } k} d^3k \qquad ; L \to \infty \tag{5.19}$$

For a given magnitude of k, the volume between the two surfaces in the first octant in Fig. 5.4 is given by

$$\sum_{n_x,n_y,n_z} \to \left(\frac{L}{2\pi}\right)^3 4\pi k^2 dk \qquad ; L \to \infty \tag{5.20}$$

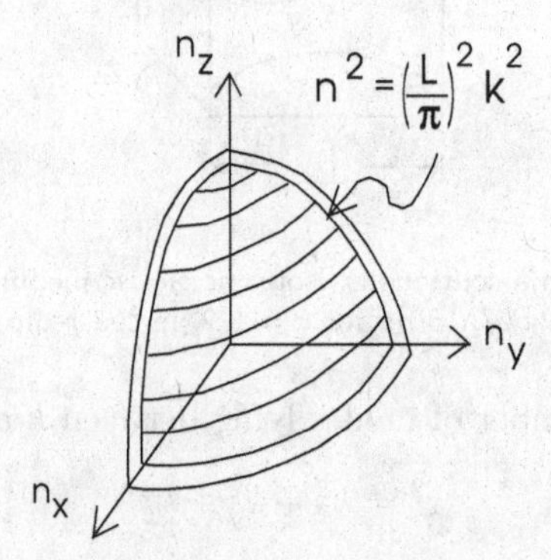

Fig. 5.4 Counting of normal modes for large L. The volume between the two surfaces in the first octant in n-space is $(1/8)4\pi n^2 dn = (1/8)(L/\pi)^3 4\pi k^2 dk$.

[3]For the formal proof of the conversion of the sum to an integral in this case, see Prob. 5.1. We assume here a summand $f(\mathbf{k}^2)$ that makes all of these sums finite.

As this is the number of normal modes between the corresponding ν and $\nu + d\nu$, one has

$$g(\nu)d\nu = \left(\frac{L}{2\pi}\right)^3 4\pi k^2 dk$$
$$k = \frac{\omega}{c} = \frac{2\pi\nu}{c} \tag{5.21}$$

Thus we have the lovely, simple, and very important expression for the normal-mode distribution

$$g(\nu)d\nu = V\frac{4\pi\nu^2 d\nu}{c^3} \tag{5.22}$$

where $L^3 = V$ is the volume of the box, and c is the sound velocity.

Suppose we had imposed *periodic boundary conditions* on the box in Fig. 5.3 instead of clamped boundaries. Take as solutions to the scalar Helmholtz equation

$$\psi_{\mathbf{k}}(\mathbf{x}) = \frac{1}{\sqrt{L^3}}e^{i\mathbf{k}\cdot\mathbf{x}} \tag{5.23}$$

The wave numbers must now satisfy

$$k_x = \frac{2\pi n_x}{L} \quad ; \; k_y = \frac{2\pi n_y}{L} \quad ; \; k_z = \frac{2\pi n_z}{L} \tag{5.24}$$

where the n_i run over *all* the integers

$$n_i = 0, \pm 1, \pm 2, \cdots \quad ; \; i = x, y, z \tag{5.25}$$

The previous arguments on converting the sums to integrals again hold, only now one no longer integrates only over the first octant, but over all n-space, and correspondingly, all k-space. Thus

$$\sum_{n_x, n_y, n_z} \rightarrow \left(\frac{L}{2\pi}\right)^3 \int_{\text{all } k} d^3k \quad ; L \rightarrow \infty \tag{5.26}$$

One arrives at *exactly the same expression as in Eq. (5.19)!* Hence one obtains the important result that

The number of normal modes per unit volume

$$\frac{1}{V}g(\nu)d\nu = \frac{4\pi\nu^2 d\nu}{c^3} \tag{5.27}$$

for the scalar wave equation is an intensive quantity, independent of the boundary conditions, as $L \rightarrow \infty$.

Let us use this ν^2 spectrum of normal modes up to the cut-off ν_m, where ν_m is determined from Eq. (5.11) by matching the correct total number of degrees of freedom

$$\int_0^{\nu_m} g(\nu)d\nu = 3N \tag{5.28}$$

It follows that

$$g(\nu)d\nu = 9N\frac{\nu^2 d\nu}{\nu_m^3} \qquad ; \nu \leq \nu_m \tag{5.29}$$

5.1.3.2 *Thermodynamics*

Introduce the dimensionless variable

$$u = \frac{h\nu}{k_{\rm B}T} \tag{5.30}$$

Define the *Debye temperature* as

$$\theta_D \equiv \frac{h\nu_m}{k_{\rm B}} \qquad ; \text{Debye temperature} \tag{5.31}$$

Then from Eq. (5.29) the spectral weight is given by

$$g(\nu)d\nu = 9N\left(\frac{T}{\theta_D}\right)^3 u^2 du \tag{5.32}$$

Hence from Eqs. (5.7)–(5.8)

$$E = 9Nk_{\rm B}T\left(\frac{T}{\theta_D}\right)^3 \int_0^{\theta_D/T} \frac{u^3du}{e^u - 1} + \frac{9}{8}Nk_{\rm B}\theta_D \qquad ; \text{Debye model}$$

$$C_V = 9Nk_{\rm B}\left(\frac{T}{\theta_D}\right)^3 \int_0^{\theta_D/T} \frac{u^4e^u\, du}{(e^u - 1)^2} \tag{5.33}$$

This is the Debye model for the energy and specific heat of a homogeneous, isotropic solid.

Two limiting cases of this result are of particular interest:

(1) For $T \to \infty$:[4]

$$C_V \to 9Nk_{\rm B}\left(\frac{T}{\theta_D}\right)^3 \int_0^{\theta_D/T} u^2du = 3Nk_{\rm B} \tag{5.34}$$

This reproduces the law of Dulong and Petit.

[4]Which implies $u \to 0$.

The result in Eq. (5.34) follows from the classical equipartition theorem (Prob. 2.14) and the fact that we have incorporated the correct number of vibrational modes; indeed, this result provides a nice check that Eq. (5.28) is, in fact, correct.

(2) For $T \to 0$:[5]

$$
\begin{aligned}
C_V &\to 9Nk_\mathrm{B}\left(\frac{T}{\theta_D}\right)^3 \int_0^\infty \frac{u^4 e^u \, du}{(e^u - 1)^2} \\
&= \frac{12\pi^4}{5} Nk_\mathrm{B}\left(\frac{T}{\theta_D}\right)^3
\end{aligned}
\tag{5.35}
$$

This is the celebrated *Debye T^3-law*, which reproduces the experimental low-temperature behavior of the specific heat of solids.

We proceed to a discussion of these results.

5.1.3.3 *Discussion*

- If the one parameter θ_D left in Eqs. (5.33) is determined empirically, the Debye model provides an excellent description of the specific heat of monatomic solids. Several examples are shown in Fig. 5.5;

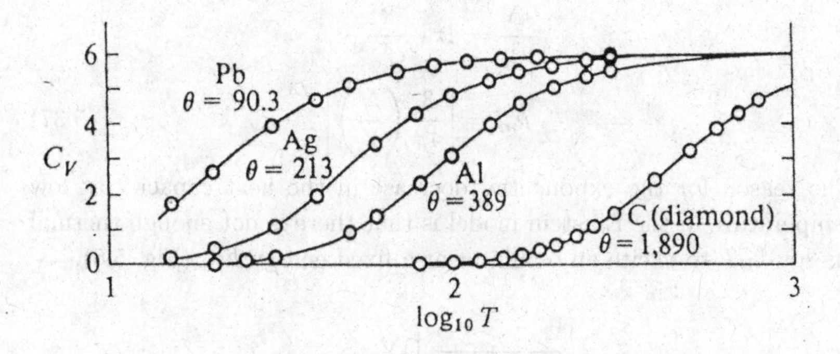

Fig. 5.5 Observed molar heat capacity in cal/mole-°K of some monatomic solids and fits with various values of the Debye temperature $\theta \equiv \theta_D$ in °K. From [Fetter and Walecka (2003)].

- A determination of $\theta_D = h\nu_m/k_\mathrm{B}$ yields the cut-off frequency ν_m;

[5]Note $\int_0^\infty u^4 e^u \, du/(e^u - 1)^2 = 4\pi^4/15$.

- The actual sound waves in an elastic solid are more complicated than the simple scalar sound waves analyzed here. One employs the vector displacement $\mathbf{s}$ of each little element in the solid, and the analysis leads to two transverse modes of propagation, similar to the transverse displacement of a string, and one longitudinal mode, corresponding to a density wave.[6] Each of these types of waves has its own velocity of propagation, and the simple scalar normal-mode spectral weight in Eq. (5.22) should really be replaced by

$$\begin{aligned} g(\nu)d\nu &= 4\pi V\left(\frac{2}{c_t^3}+\frac{1}{c_l^3}\right)\nu^2 d\nu \\ &\equiv 4\pi V\left(\frac{3}{c_{\rm av}^3}\right)\nu^2 d\nu \end{aligned} \tag{5.36}$$

where $c_{\rm av}$ is the average sound velocity in the solid.

- The spectral weight in Eq. (5.36) is again quadratic in ν, and a reproduction of the correct number of degrees of freedom again leads to Eq. (5.29);
- Since Eq. (5.29) is valid, *all the subsequent analysis in terms of θ_D still holds.* In particular, Eqs. (5.33) are still correct. The only new feature is the improved relation between ν_m and the velocity of sound

$$\begin{aligned} \frac{9N}{\nu_m^3} &= 4\pi V\frac{3}{c_{\rm av}^3} \\ \Longrightarrow \qquad \nu_m &= \left[\frac{3}{4\pi}\left(\frac{N}{V}\right)\right]^{1/3} c_{\rm av} \end{aligned} \tag{5.37}$$

- The reason for the exponential decrease in the heat capacity at low temperature in the Einstein model is that there is not enough thermal energy $k_{\rm B}T$ to excite an oscillator at a fixed energy $h\nu_0$ (Fig. 5.6).

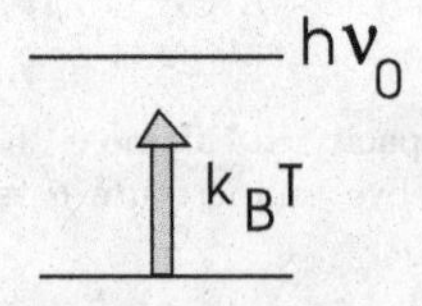

Fig. 5.6 Thermal energy $k_{\rm B}T$ relative to the oscillator energy $h\nu_0$ at low temperature in the Einstein model.

[6] See, for example, [Fetter and Walecka (2003a)].

In the Debye model, there are long-wavelength modes available in the macroscopic solid at arbitrarily low energy $h\nu$;

- In more detailed fits to the data, the parameter $\theta_D(T)$ may deviate from a constant value by the order of 10% (see Fig. 5.7). While appropriate at very low temperature where the long-wavelength normal modes dominate, the model of a continuous, homogeneous, isotropic elastic medium is too simple to describe the detailed behavior of solids at all temperatures. It is therefore appropriate to develop an improved description of their normal modes.

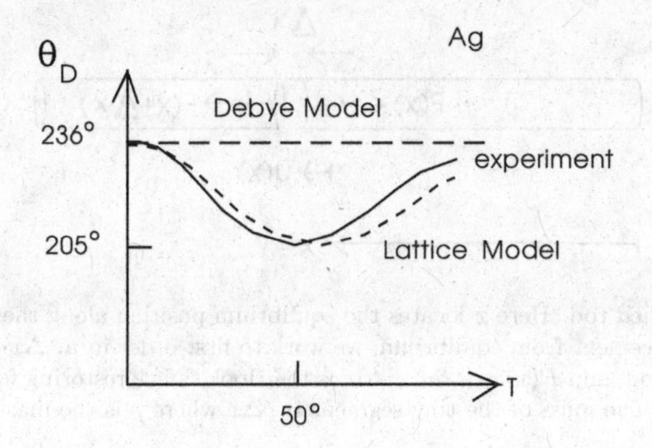

Fig. 5.7 Sketch of a detailed fit to the experimental heat capacity of silver with a temperature-dependent parameter $\theta_D(T)$. Also shown are a constant $\theta_D(0)$ from the Debye model, and the results from a lattice model with a more realistic normal-mode spectrum. (After [Davidson (2003)].)

5.1.4 *Improved Normal-Mode Spectrum*

We start by reviewing the theory of one-dimensional sound waves in a solid rod in continuum mechanics.

5.1.4.1 *Longitudinal Waves in a Rod*

Consider a long uniform rod and let x denote the position along the rod (Fig. 5.8). Suppose the rod is stretched so that each element is displaced by a small distance $u(x)$; we work to lowest order in u. Consider a small element of the rod Δx. There will be a force across each face of this element given by Hooke's law

$$F = \kappa\varepsilon \qquad ; \text{ Hooke's law}$$
$$\varepsilon = \frac{\partial u}{\partial x} \qquad ; \text{ strain} \tag{5.38}$$

where κ is a constant and $\varepsilon = \partial u/\partial x$ is the local strain in the rod.[7] The net force on the small element is then

$$F_{\rm el} = \kappa\left[\varepsilon(x + \Delta x) - \varepsilon(x)\right] = \kappa\frac{\partial^2 u}{\partial x^2}\Delta x \tag{5.39}$$

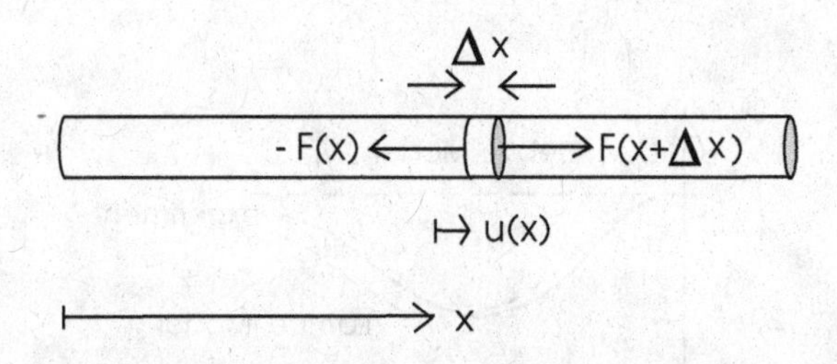

Fig. 5.8 Stretched rod. Here x locates the equilbrium position along the rod, and $u(x)$ is a small displacement from equilbrium; we work to first order in u. Δx denotes a tiny segment of the rod, and $F(x) = \kappa\, \partial u(x)/\partial x$ is the Hooke's law restoring force across any transverse face. The mass of the tiny segment is $\rho\Delta x$ where ρ is the mass density.

Newton's second law for the small segment gives

$$\rho\Delta x\frac{\partial^2 u}{\partial t^2} = \kappa\frac{\partial^2 u}{\partial x^2}\Delta x \tag{5.40}$$

where ρ is the mass density of the rod, assumed here to be constant. Thus the dispacement $u(x,t)$ obeys the one-dimensional wave equation

$$\frac{\partial^2 u}{\partial x^2} = \frac{1}{c^2}\frac{\partial^2 u}{\partial t^2} \qquad ; \text{ wave equation}$$
$$c = \left(\frac{\kappa}{\rho}\right)^{1/2} \qquad ; \text{ sound velocity} \tag{5.41}$$

This equation describes longitudinal sound waves that propagate down the rod with a velocity c.

[7] The partial derivative is appropriate when $u(x,t)$ acquires a time dependence; Newton's law requires the force at any instant in time. See also Prob. 5.7.

The following normal-mode solution satisfies this wave equation[8]

$$u(x,t) = Ae^{i(kx-\omega t)} \qquad ; \; k^2 = \frac{\omega^2}{c^2} \tag{5.42}$$

If the ends of a long rod of total length L are joined to form a circle, then periodic boundary conditions apply. In this case

$$\begin{aligned} & u(x+L) = u(x) && ; \text{ periodic boundary conditions} \\ \Longrightarrow \quad & kL = 2\pi n && ; \; n = 0, \pm 1, \pm 2, \cdots \end{aligned} \tag{5.43}$$

In the limit of large L, the number of normal modes can be counted as before

$$\sum_n \to \int_{-\infty}^{\infty} dn = \frac{L}{2\pi}\int_{-\infty}^{\infty} dk \qquad ; \; L \to \infty \tag{5.44}$$

Here the integral goes over all n, and consequently over all k. It follows that

$$\begin{aligned} \sum_n & \to \frac{L}{2\pi} 2 \int_0^\infty dk && ; \; L \to \infty \\ & = \frac{2L}{c}\int_0^\infty d\nu \end{aligned} \tag{5.45}$$

where we now integrate only over positive k in the first line. The second line then follows from the relation

$$k = \frac{\omega}{c} = \frac{2\pi\nu}{c} \tag{5.46}$$

We are now in a position to identify the normal-mode spectral weight for this problem

$$g(\nu) = \frac{2L}{c} \qquad ; \text{ spectral weight} \tag{5.47}$$

In this one-dimensional continuum problem, the normal-mode spectral weight is independent of ν.

If we again introduce a maximum normal-mode frequency ν_m, and equate the number of normal modes with the number of degrees of freedom

---

[8] Recall that normal modes are solutions where everything oscillates with the *same frequency.* The displacement $u(x,t)$ is real, so one really wants to use the real part of this expression.

N for this one-dimensional problem, then

$$\int_0^{\nu_m} g(\nu)d\nu = N$$
$$\Longrightarrow \qquad \nu_m = \frac{Nc}{2L} \tag{5.48}$$

Hence

$$g(\nu) = \frac{N}{\nu_m} \qquad ; \text{ spectral weight} \tag{5.49}$$

This analysis illustrates the underlying basis of the Debye model, at least in this one-dimensional problem. The goal is now to improve this normal-model analysis by employing a more realistic description of the solid. We thus turn to a *lattice model.*

5.1.4.2 *Lattice Model*

Consider a linear set of mass points m connected with springs, with an equilibrium spacing of a (Fig. 5.9). Let μ_n be the displacement of the nth mass from equilibrium, and $\mathcal{K}$ be the nearest-neighbor spring constant. Newton's second law for the nth mass then reads

$$m\frac{d^2\mu_n}{dt^2} = \mathcal{K}(\mu_{n+1} - \mu_n) - \mathcal{K}(\mu_n - \mu_{n-1}) \tag{5.50}$$

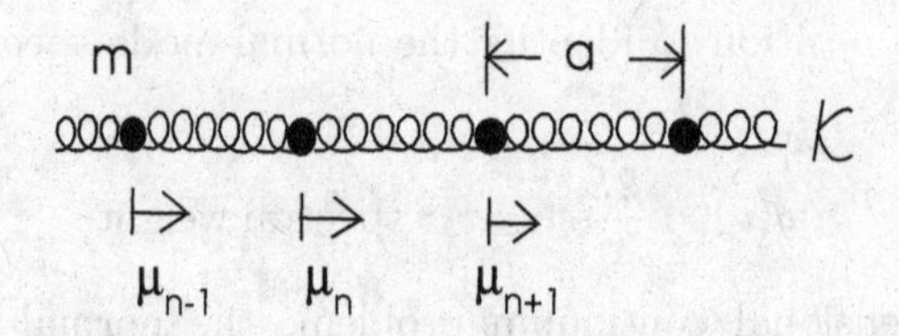

Fig. 5.9 Linear set of mass points m connected by springs with an equlibrium separation of a. μ_n is the displacement of the nth mass from equilibrium. The nearest-neighbor spring constant is $\mathcal{K}$.

If there are N masses, and the two ends are joined around a frictionless cylinder to form a long circular chain of length L, then

$$Na = L \qquad ; \text{ total length} \tag{5.51}$$

Let us look for normal-mode waves going down the chain of the form

$$\begin{aligned}\mu_n &= Ae^{i(kx-\omega t)} \\ &= Ae^{i(kna-\omega t)} \qquad ; \; x = na \end{aligned} \tag{5.52}$$

where $x = na$ locates the position of the nth mass. Periodic boundary conditions are then appropriate

$$\begin{aligned} \mu_{n+N} &= \mu_n \qquad &; \text{ p.b.c.} \\ \Longrightarrow \qquad kNa &= 2\pi p \qquad &; \; p = 0, \pm 1, \pm 2, \cdots \end{aligned} \tag{5.53}$$

A complete set of solutions to this problem is obtained by taking the normal-mode frequency ω to be positive and then using $\pm|k|$ to describe waves moving to the right and left along the chain.

Substitution of Eq. (5.52) into Eq. (5.50) gives

$$\begin{aligned} -m\omega^2 &= \mathcal{K}(e^{ika} + e^{-ika} - 2) \\ &= -2\mathcal{K}(1 - \cos ka) \\ &= -4\mathcal{K}\sin^2 \frac{ka}{2} \\ \text{or;} \qquad \omega^2 &= 4\frac{\mathcal{K}}{m}\sin^2 \frac{ka}{2} \end{aligned} \tag{5.54}$$

Since $\omega = 2\pi\nu$, this yields the following expression for the positive normal-mode frequencies

$$\nu_p = \frac{1}{\pi}\left(\frac{\mathcal{K}}{m}\right)^{1/2} \sin\left(\frac{\pi|p|}{N}\right) \qquad ; \; p = 0, \pm 1, \pm 2, \cdots \tag{5.55}$$

normal-mode frequencies

This result is sketched in Fig. 5.10.

We make several comments on these lattice-model results:

(1) Let $p \to p \pm N$. Then from Eq. (5.53)

$$ka \to ka \pm 2\pi \qquad ; \; p \to p \pm N \tag{5.56}$$

This does not produce a new solution in Eq. (5.52). Thus one can limit the values of p to the finite set[9]

$$p = 0, \pm 1, \pm 2, \cdots, \pm\frac{N}{2} \tag{5.57}$$

[9] One of the end values $\pm N/2$ should actually be omitted, leading to precisely N normal modes; here we assume N is even.

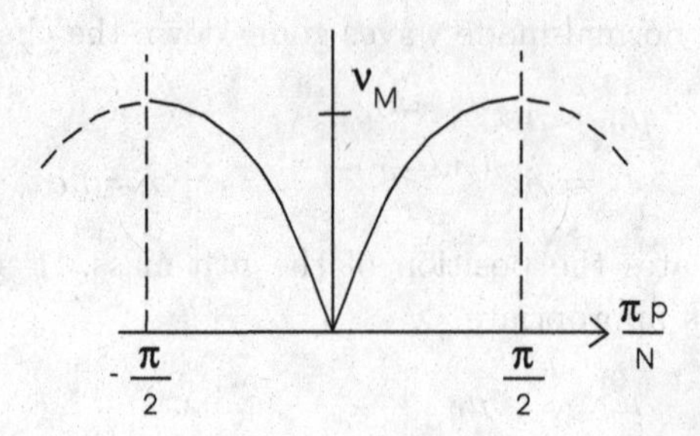

Fig. 5.10 Normal-mode spectrum in Eq. (5.55). The maximum value of the normal-mode frequency $\nu_M = (1/\pi)(\mathcal{K}/m)^{1/2}$ occurs at $p = \pm N/2$.

(2) The maximum value of the normal-mode frequency, which occurs at $p = \pm N/2$, is

$$\nu_M = \frac{1}{\pi}\left(\frac{\mathcal{K}}{m}\right)^{1/2} \qquad \text{; maximum value} \tag{5.58}$$

This value can be compared with the cut-off frequency ν_m in the Debye model in Eq. (5.48)

$$\nu_m = \frac{Nc}{2L} = \frac{N}{2L}\left(\frac{\kappa}{\rho}\right)^{1/2} \text{; Debye} \tag{5.59}$$

Make the identifications

$$F = \kappa\varepsilon \to \kappa\left(\frac{\mu_{n+1} - \mu_n}{a}\right) \qquad ; \kappa \to \mathcal{K}a \qquad ; \rho \to \frac{m}{a} \tag{5.60}$$

This yields

$$\nu_m = \frac{1}{2}\left(\frac{\mathcal{K}}{m}\right)^{1/2} \tag{5.61}$$

Thus

$$\nu_m = \frac{\pi}{2}\nu_M \qquad \text{; Debye} \tag{5.62}$$

(3) The *spectral weight* is obtained in the following fashion

$$\sum_p \to 2\int_0^{N/2} dp = 2\int_0^{\nu_M} \left(\frac{dp}{d\nu}\right) d\nu$$
$$= 2\int_0^{\nu_M} \left(\frac{d\nu}{dp}\right)^{-1} d\nu \qquad (5.63)$$

The derivative is evaluated from Eq. (5.55), with the result

$$\sum_p \to 2\int_0^{\nu_M} \left[\frac{\pi}{N}\nu_M \cos\frac{\pi p}{N}\right]^{-1} d\nu \qquad (5.64)$$

The normal-mode spectral weight is thus identified as

$$g(\nu) = \frac{2N}{\pi\nu_M}\frac{1}{\cos(\pi p/N)} = \frac{2N}{\pi(\nu_M^2 - \nu^2)^{1/2}} \quad ; \text{ spectral weight} \qquad (5.65)$$

This result is sketched in Fig. 5.11, where it is compared with the Debye result in Eq. (5.49).

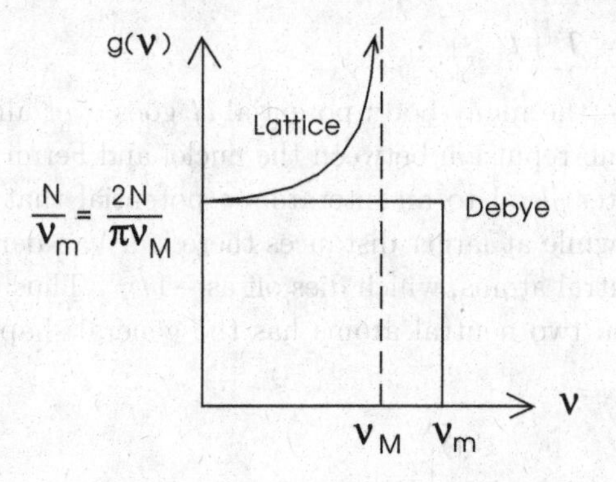

Fig. 5.11 Normal-mode spectral weight in Eq. (5.65), compared with the Debye result in Eq. (5.49). Here $\nu_m = (\pi/2)\nu_M$.

A few observations:

- The lattice and Debye spectral weights coincide at $\nu = 0$, that is, for long wavelengths;
- There is an actual maximum normal-mode frequency ν_M in the lattice model, which provides a more realistic theory of elasticity;
- The number of normal modes is N in both cases (see Prob. 5.5);

- 3-D lattice calculations have been carried out that include nearest-neighbor couplings (see [Davidson (2003)]);
- The observed spectral weight of a monatomic solid has the character shown in Fig. 5.2. There are two peaks, and the *Nernst-Lindemann approximation* places a pair of peaks in the spectral weight $g(\nu)$, with 1/2 the strength at ν_m, and 1/2 at $\nu_m/2$;
- Additional development of the lattice model in statistical mechanics can be found in [Walecka (2000)].

5.2 Imperfect Gases

As a second application of the canonical ensemble, we turn to the classical theory of imperfect gases.[10] Consider N identical non-localized point systems of mass m in a volume V, but now allow *interactions* between the systems. Assume the many-body hamiltonian has the form

$$\begin{aligned} H(p,q) &= \sum_{i=1}^{N} \frac{\mathbf{p}_i^2}{2m} + \sum_{i<j\leq N} U(r_{ij}) \qquad ; \; r_{ij} = |\mathbf{r}_i - \mathbf{r}_j| \\ &= \mathcal{T} + \mathcal{U} \end{aligned} \tag{5.66}$$

where the sum in the many-body potential $\mathcal{U}$ goes over all *pairs* of particles. The Coulomb repulsion between the nuclei and Fermi pressure of the electrons (see later) lead to an interatomic potential that is repulsive at short distances, while at larger distances there is a Van der Waal's attraction between neutral atoms, which dies off as $-1/r^6$. Thus the interatomic potential between two neutral atoms has the general shape illustrated in Fig. 5.12.

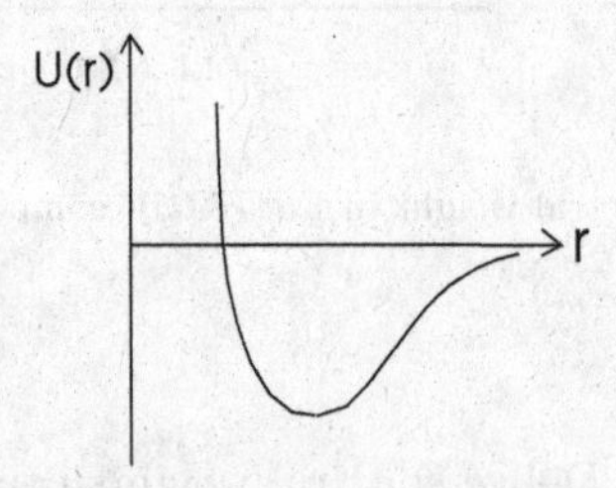

Fig. 5.12 General shape of the interatomic potential between two neutral atoms.

[10] See, for example, [Mayer and Mayer (1977)].

5.2.1 *Configuration Integral*

The canonical partition function in the classical limit is given by

$$(\text{P.F.}) = \frac{1}{N!\,h^{3N}} \int \cdots \int dq_1 \cdots dq_{3N} dp_1 \cdots dp_{3N}\, e^{-H(p,q)/k_B T} \tag{5.67}$$

The momentum integrals are easily performed using our previous results, leading to

$$(\text{P.F.}) = \frac{1}{N!}\left(\frac{2\pi m k_B T}{h^2}\right)^{3N/2} Q_N$$

$$Q_N \equiv \int \cdots \int d\tau_1 \cdots d\tau_N \exp\left[-\frac{1}{k_B T}\sum_{i<j\le N} U(r_{ij})\right] \quad ; \text{ configuration integral} \tag{5.68}$$

The problem has been reduced to computing the *configuration integral* Q_N, an integral over $3N$-dimensional configuration space with an integrand $\exp\left[-(1/k_B T)\sum_{i<j\le N} U(r_{ij})\right]$.[11] The Helmholtz free energy of the assembly, and corresponding pressure, then follow from Eqs. (5.68) as

$$A(T,V,N) = -k_B T \ln(\text{P.F.})$$

$$P = -\left(\frac{\partial A}{\partial V}\right)_{T,N} = k_B T \frac{\partial}{\partial V} \ln Q_N \tag{5.69}$$

For a *perfect* gas, each spatial integral over $d\tau$ simply yields the volume V, and therefore

$$Q_N = V^N \quad ; \text{ perfect gas} \tag{5.70}$$

One thus recovers the equation of state of a perfect gas

$$\frac{P}{k_B T} = \frac{N}{V} \equiv \frac{1}{v} \quad ; \text{ perfect gas} \tag{5.71}$$

where we have defined the volume per particle by $v \equiv V/N$.[12]

To express deviations from this relation for an *imperfect* gas, one uses a power series in $1/v$

$$\frac{P}{k_B T} = \sum_{n=1}^{\infty} \frac{c_n(T)}{v^n} = \frac{1}{v} + \frac{B(T)}{v^2} + \frac{C(T)}{v^3} + \cdots \quad ; \text{ virial expansion} \tag{5.72}$$

[11] Here $d\tau \equiv d^3x$.

[12] Note that low density is *large* v. Note also that the virial expansion in Eq. (5.72) reduces to the perfect gas result at very low density.

This is known as the *virial expansion*, with $B(T)$ the second virial coefficient, $C(T)$ the third virial coeffcient, and so on.

As an example, consider the semi-empirical Van der Waal's equation of state, seen in most introductory courses

$$\left(P + \frac{a}{v^2}\right)(v - b) = k_{\rm B}T \qquad ; \text{ Van der Waal's} \tag{5.73}$$

where a and b are constants. This expression can be re-arranged to read

$$P = \frac{k_{\rm B}T}{v-b} - \frac{a}{v^2} = k_{\rm B}T\left[\frac{1}{v} + \frac{1}{v^2}\left(b - \frac{a}{k_{\rm B}T}\right) + \cdots\right] \tag{5.74}$$

The second virial coefficient with the Van der Waal's equation of state is then identified as

$$B(T)_{\rm VW} = b - \frac{a}{k_{\rm B}T} \qquad ; \text{ Van der Waal's} \tag{5.75}$$

A measurement of the temperature dependence of $B(T)_{\rm VW}$ can be used determine the two parameters (a, b) for any substance.

5.2.2 *Second Virial Coefficient*

To get a feel for the configuration integral in the canonical ensemble, we present an introductory discussion of the second virial coefficient. First, regroup the terms in the many-body potential $\mathcal{U}$, isolating the interactions with the Nth system.[13] Thus

$$\begin{aligned}\mathcal{U} &= U_{1N} + U_{2N} + \cdots + U_{N-1,N} + \sum_{i<j<N} U_{ij} \\ &\equiv U_{1N} + U_{2N} + \cdots + U_{N-1,N} + {\sum}' U_{ij} \\ \exp\left(-\frac{\mathcal{U}}{k_{\rm B}T}\right) &= \exp\left(-\frac{1}{k_{\rm B}T}\sum_{i<N} U_{iN}\right)\exp\left(-\frac{1}{k_{\rm B}T}{\sum}' U_{ij}\right)\end{aligned} \tag{5.76}$$

The coordinates of the Nth system no longer appear in the additional term $\sum'$, defined in the second line. Write the first factor, where the coordinates of the Nth system *do* appear, as

$$\exp\left(-\frac{1}{k_{\rm B}T}\sum_{i<N} U_{iN}\right) = 1 + \left[\exp\left(-\frac{1}{k_{\rm B}T}\sum_{i<N} U_{iN}\right) - 1\right] \tag{5.77}$$

[13] Remember the systems are now all labeled.

Now carry out $\int d\tau_N$. The first term on the r.h.s. of Eq. (5.77) again gives the volume V, and thus

$$\int d\tau_N \, e^{-\mathcal{U}/k_{\mathrm{B}}T} = e^{[-(1/k_{\mathrm{B}}T)\sum' U_{ij}]}\,[V+W]$$

$$W \equiv \int d\tau_N \left[\exp\left(-\frac{1}{k_{\mathrm{B}}T}\sum_{i<N} U_{iN}\right) - 1\right] \qquad (5.78)$$

We now assume that W, the remaining integral over $d\tau_N$, is independent of the coordinates of the ith system. What does this assumption entail?

- This assumes *two-body collisions*, and is essentially a *low-density approximation*;
- It relies on the fact that $e^{[-U(r)/k_{\mathrm{B}}T]} - 1$, where $U(r)$ is the two-body potential, goes to zero outside the range of the interatomic force, typically a distance of several angstroms (see Fig. 5.13);

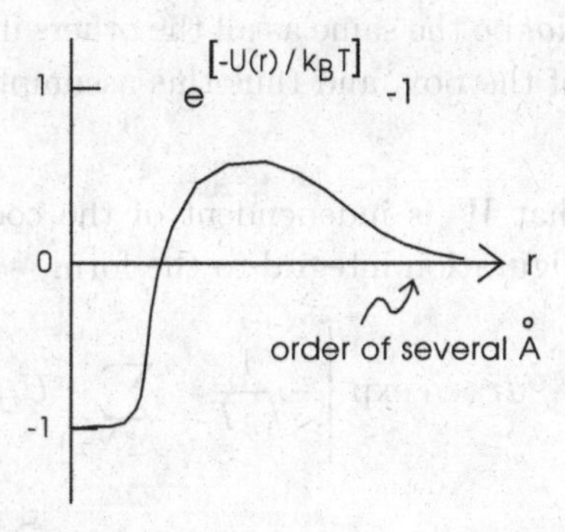

Fig. 5.13 Sketch of the quantity $e^{[-U(r)/k_{\mathrm{B}}T]} - 1$, where $U(r)$ is the interatomic two-body potential, with a typical range of several Å.

- Thus the integrand in Eq. (5.78) will only be non-zero when the coordinate of the Nth atom is close to one of the other atoms;
- If the other atoms are far enough away when one is performing $\int d\tau_N$ over the region around a given atom in W, say atom 1 in Fig. 5.14, then atom N is outside the range of the potentials of these other atoms, for example atoms 2 and 3 in Fig. 5.14, and they do not affect the non-zero contribution to the integral coming from the region around atom 1;
- Exactly the same argument holds when the particle N is near one of these other atoms. There are $N-1$ of these other atoms all now giving

identical contributions, and thus[14]

$$W = (N-1)\mathcal{B}$$
$$\mathcal{B} = 4\pi \int_0^\infty \left[e^{-U(r)/k_{\rm B}T} - 1\right] r^2 dr \tag{5.79}$$

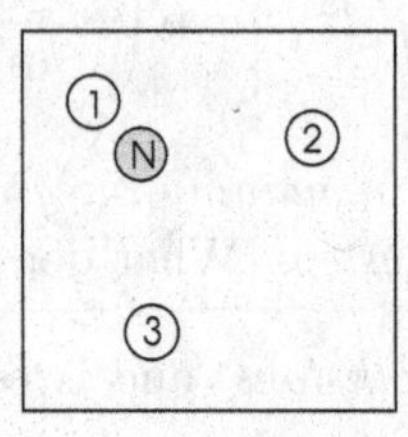

Fig. 5.14 Two-body collisions. Here particle N is within the range of the potential of particle 1, but outside the range of the potentials of particles 2 and 3. Thus those potentials vanish when one is evaluating the non-zero contribution to the configuration integral in Eq. (5.78) coming from the region around particle 1.

- The integral will *not* be the same as all the others if the given particle is near the surface of the box, and thus this assumption neglects surface effects.

The assumption that W is independent of the coordinates of the ith atom reduces the configuration integral to the form

$$Q_N = \left\{\int \cdots \int d\tau_1 \cdots d\tau_{N-1} \exp\left[-\frac{1}{k_{\rm B}T} \sum_{i<j\leq N-1} U_{ij}\right]\right\}[V + (N-1)\mathcal{B}] \tag{5.80}$$

Now *repeat the argument*. The configuration integral therefore becomes

$$Q_N = [V + (N-1)\mathcal{B}][V + (N-2)\mathcal{B}] \cdots [V + \mathcal{B}]V \tag{5.81}$$

where the last integral sees the entire volume V. Hence

$$Q_N = \prod_{p=0}^{N-1} (V + p\mathcal{B}) = V^N \prod_{p=0}^{N-1} \left(1 + \frac{p\mathcal{B}}{V}\right) \tag{5.82}$$

[14]The positions of atoms $(1, 2, \cdots, N-1)$ are held fixed while performing $\int d\tau_N$; however, they are themselves eventually integrated over in Q_N. Although not important at low density, there will be contributions to Q_N where three or more atoms overlap. We will return shortly to an exact linked-cluster analysis of the configuration integral.

Now assume *low density*, so that

$$\frac{N\mathcal{B}}{V} = \frac{\mathcal{B}}{v} \ll 1 \qquad ; \text{ low density} \tag{5.83}$$

Then

$$\begin{aligned}\ln Q_N &= N \ln V + \sum_{p=0}^{N-1} \ln\left(1 + \frac{p\mathcal{B}}{V}\right) \\ &\approx N \ln V + \frac{\mathcal{B}}{V}\sum_{p=0}^{N-1} p\end{aligned} \tag{5.84}$$

Here the expansion $\ln(1 + p\mathcal{B}/V) \approx p\mathcal{B}/V$ is justified by Eq. (5.83). For the large N of the canonical ensemble

$$\sum_{p=0}^{N-1} p \to \int_0^{N-1} p\, dp = \frac{1}{2}(N-1)^2 \approx \frac{1}{2}N^2 \tag{5.85}$$

The logarithm of the configuration integral therefore takes the following form at large N and low density

$$\ln Q_N = N\left[\ln V + \frac{N\mathcal{B}}{2V}\right] \qquad ; v \to \infty \tag{5.86}$$

The expression for the pressure in Eq. (5.69) then lets us identify the second virial coefficient

$$\begin{aligned}\frac{P}{k_{\mathrm{B}}T} &= \frac{1}{v} - \frac{\mathcal{B}}{2}\frac{1}{v^2} + \cdots \\ B(T) &= -\frac{\mathcal{B}}{2}\end{aligned} \tag{5.87}$$

Therefore, from Eq. (5.79),

$$B(T) = 2\pi \int_0^\infty \left[1 - e^{-U(r)/k_{\mathrm{B}}T}\right] r^2 dr \qquad ; \text{ 2nd virial coefficient} \tag{5.88}$$

This is an extremely useful relation, derived in the canonical ensemble, that expresses the second virial coefficient for an imperfect gas as an integral over the interatomic two-body potential.[15]

[15] Recall Fig. 5.13.

For example, take an interatomic potential of the form (see Fig. 5.15)

$$\begin{aligned} U(r) &= \infty \qquad &; r < \sigma \\ &= -\frac{c}{r^6} \qquad &; r > \sigma \end{aligned} \tag{5.89}$$

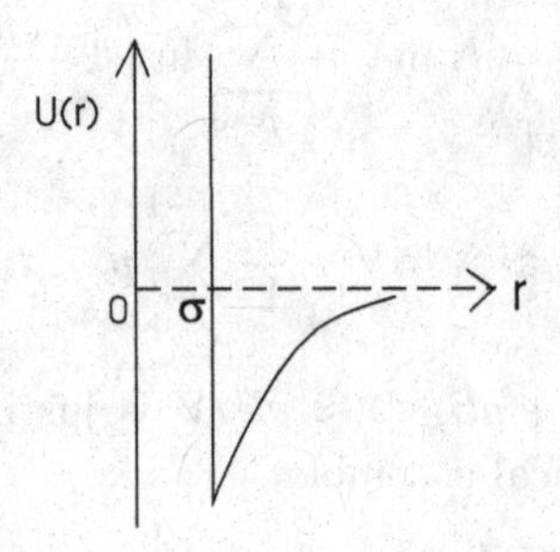

Fig. 5.15 Model interatomic two-body potential in Eq. (5.89) used for the calculation of the second virial coefficient $B(T)$. We refer to this as the Van der Waal's potential.

Then

$$\frac{B(T)}{2\pi} = \int_0^\sigma r^2\,dr + \int_\sigma^\infty \left[1 - \exp\left(\frac{c}{r^6 k_\mathrm{B} T}\right)\right] r^2\,dr \tag{5.90}$$

Consider this expression at high temperature where

$$\begin{aligned} \frac{B(T)}{2\pi} &= \frac{\sigma^3}{3} - \int_\sigma^\infty \frac{c}{r^6 k_\mathrm{B} T} r^2\,dr \qquad ; T \to \infty \\ &= \frac{\sigma^3}{3} - \frac{c}{3\sigma^3 k_\mathrm{B} T} \end{aligned} \tag{5.91}$$

Hence the high-temperature second virial coefficient calculated from the interatomic potential in Eq. (5.89) is

$$B(T) = \frac{2\pi\sigma^3}{3}\left[1 - \frac{c}{\sigma^6 k_\mathrm{B} T}\right] \qquad ; T \to \infty \tag{5.92}$$

This has precisely the same form as the second virial coefficient in Eq. (5.75), obtained from the Van der Waal's equation of state[16]

$$B(T)_\mathrm{VW} = b - \frac{a}{T} \tag{5.93}$$

[16] See Prob. 5.12.

5.2.3 *General Analysis of Configuration Integral*

With this introduction, we turn to the general analysis of the configuration integral for an imperfect gas. The typical form of the interatomic two-body potential is sketched in Fig. 5.12, and the corresponding quantity $e^{-U(r)/k_BT} - 1$ is sketched in Fig. 5.13. The latter goes to zero outside the range of the two-body potential. It is then useful to introduce a pair interaction function f_{ij} through the relation

$$e^{-U(r_{ij})/k_BT} \equiv 1 + f_{ij} \qquad ; \text{ pair interaction} \tag{5.94}$$

The quantity $f_{ij}(r_{ij})$ has the following properties:

(1) f_{ij} corresponds to collisions between systems (i, j);
(2) It is symmetric in i and j

$$f_{ij} = f_{ji} \qquad ; \text{ symmetric} \tag{5.95}$$

(3) It goes to zero outside of the range of the two-body force

$$f_{ij} = 0 \qquad ; \ r_{ij} \text{ outside range of force} \tag{5.96}$$

(4) It depends on the temperature T.

Since the exponentials *factor*, the configuration integral in Eq. (5.68) can be written

$$Q_N = \int \cdots \int d\tau_1 \cdots d\tau_N \prod_{i<j\leq N} (1 + f_{ij}) \tag{5.97}$$

The particles are now all labeled, and the product goes over all pairs of particles.[17]

5.2.3.1 *Linked-Cluster Expansion*

Multiply out the repeated product. What does the result look like? Suppose, for example, that $N = 3$ and there are three systems. Then

$$\begin{aligned}(1 + f_{12})(1 + f_{13})(1 + f_{23}) = 1 &+ (f_{12} + f_{13} + f_{23}) \\ &+ (f_{12}f_{13} + f_{13}f_{23} + f_{12}f_{23} + f_{12}f_{13}f_{23})\end{aligned} \tag{5.98}$$

The result can be analyzed in terms of *clusters*:

- A term f_{ij} by itself represents a cluster of the two systems i and j;

[17] The factor of $1/N!$ in (P.F.) takes care of the counting for non-localized particles. Remember $f_{ij} = f_{ji}$ is symmetric.

- $f_{ij}f_{kl}$ where the indices are distinct would represent two clusters of two systems (there aren't any here);
- $f_{ij}f_{jk}$ and $f_{ij}f_{jk}f_{ik}$, where the indices are *linked*, represent clusters of three systems. The possible linked clusters of three labeled systems are illustrated pictorially in Fig. 5.16. All these terms appear in Eq. (5.98).

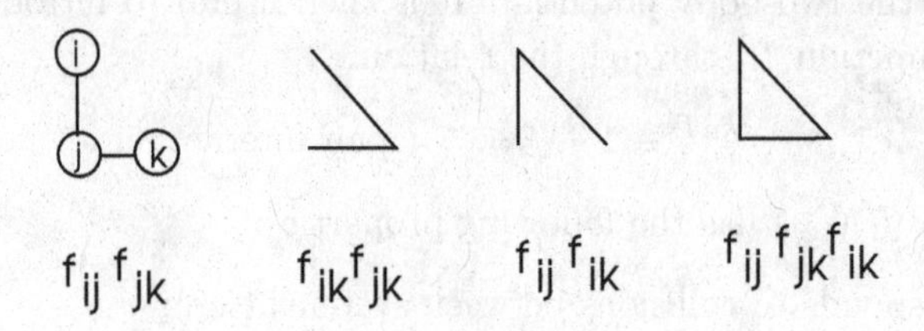

Fig. 5.16 Linked clusters of three systems (i, j, k).

- *Since the coordinates are now all linked in a cluster, the more complicated the cluster, the more complicated the multiple integral that must be done in evaluating* Q_N.

A *general term* in the repeated product in Eq. (5.97) can be characterized as having

$$\begin{aligned}
&n_1 \quad \text{isolated systems} \\
&n_2 \quad \text{clusters of two systems} \\
&n_3 \quad \text{clusters of three systems} \\
&\quad\vdots \\
&n_l \quad \text{clusters of } l \text{ systems}
\end{aligned} \tag{5.99}$$

Since a given system must lie in one and only one of these clusters[18]

$$\sum_{l=1}^{\infty} l n_l = N \qquad ; \text{each system in a cluster} \tag{5.100}$$

Suppose the (labeled) systems (i, j, k) form one of the clusters of 3 systems. This can happen several ways, as illustrated in Fig. 5.16. *Add these contributions together, and take out a factor*

$$(f_{ij}f_{jk} + f_{ik}f_{jk} + f_{ij}f_{ik} + f_{ij}f_{jk}f_{ik}) \quad ; \text{cluster of the 3 systems } (i, j, k) \tag{5.101}$$

[18]Note that n_l must vanish for $l > N$.

As a *shorthand* for such a term, write[19]

$$\sum \left(\prod_{p<q} f_{pq} \right) \qquad \text{; shorthand for cluster} \quad (5.102)$$

Since we have used systems (i, j, k) in this cluster of 3, they will no longer appear in the general term we are constructing. Define

$$\int\int\int \sum \left(\prod_{p<q} f_{pq} \right) d\tau_i d\tau_j d\tau_k \equiv VB_3 \qquad \text{; cluster of 3 systems} \quad (5.103)$$

Note that B_3 is independent of V, if wall effects are neglected.

More generally, define

$$B_l(T) \equiv \frac{1}{V} \int \cdots \int d\tau_1 \cdots d\tau_l \sum \left(\prod_{p<q} f_{pq} \right) \qquad \text{; cluster of } l \text{ systems}$$

$$B_1 \equiv 1 \qquad (5.104)$$

It follows that

(1) B_l is independent of V;
(2) B_l is independent of *which* l systems lie in the cluster.

The distribution of clusters in Eq. (5.99) then contributes the following general term to the configuration integral

$$(VB_1)^{n_1}(VB_2)^{n_2} \cdots (VB_l)^{n_l} \cdots \equiv \prod_l (VB_l)^{n_l} \qquad \text{; general term} \quad (5.105)$$

The configuration integral can thus be written as a sum over all possible sets of cluster distribution numbers $(n_1, n_2, \cdots, n_l, \cdots) \equiv \{n_l\}$ as follows

$$Q_N = \sum_{\{n_l\}} g(n_l) \prod_l (VB_l)^{n_l} \qquad (5.106)$$

$$g(n_l) = \text{number of ways } N \text{ labeled objects can form } \{n_l\} \text{ clusters}$$

The counting problem is one we are already familiar with (see Fig. 5.17). The only new wrinkle is that the rearrangement of labels *within* a cluster

[19]See [Rushbrooke (1949)].

does not give a new contribution. The answer is[20]

$$g(n_l) = \frac{N!}{n_1!n_2!n_3!\cdots(1!)^{n_1}(2!)^{n_2}(3!)^{n_3}\cdots} = \frac{N!}{\prod_l n_l! \prod_l (l!)^{n_l}}$$

$$N = \sum_{l=1}^{\infty} l n_l \tag{5.107}$$

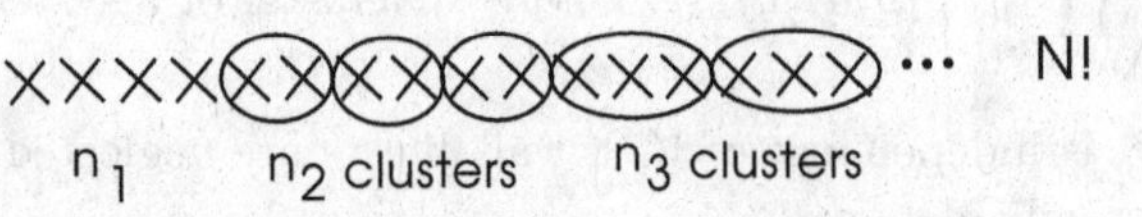

Fig. 5.17 N labeled objects divided up into n_1 isolated systems, n_2 clusters of 2, n_3 clusters of 3, and so on. The number of possibilites is given in Eq. (5.107).

Thus we are led to the following relations

$$\frac{Q_N}{N!} = \sum_{\{n_l\}} \frac{1}{\prod_l n_l!} \prod_l \left(\frac{VB_l}{l!}\right)^{n_l} \quad ; \text{ linked-cluster expansion}$$

$$N = \sum_{l=1}^{\infty} l n_l \quad ; \text{ number of systems} \tag{5.108}$$

The sum in Q_N goes over all sets of cluster distribution numbers $\{n_l\} = (n_1, n_2, \cdots, n_l, \cdots)$ satisfying the second relation, and the cluster integrals are given in Eq. (5.104). *Equations (5.108) form an exact linked-cluster decomposition of the configuration integral.*

Recall from Eq. (5.68) that

$$(\text{P.F.}) = \left(\frac{2\pi m k_B T}{h^2}\right)^{3N/2} \frac{Q_N}{N!} \tag{5.109}$$

Define

$$g_l(T) \equiv \left(\frac{2\pi m k_B T}{h^2}\right)^{3l/2} \frac{B_l}{l!} \tag{5.110}$$

[20] In the example in Eq. (5.98), one has $(n_1, n_2, n_3) = (3,0,0), (1,1,0), (0,0,1)$, with a corresponding $g(n_l) = 1, 3, 1$, respectively. See also Prob. 5.13.

Then Eq. (5.108) can be recast as[21]

$$(\text{P.F.}) = \sum_{\{n_l\}} \prod_l \frac{(Vg_l)^{n_l}}{n_l!} = \sum_{\{n_l\}} \frac{(Vg_1)^{n_1}}{n_1!} \frac{(Vg_2)^{n_2}}{n_2!} \frac{(Vg_3)^{n_3}}{n_3!} \cdots$$

$$N = \sum_{l=1}^{\infty} l n_l \tag{5.111}$$

5.2.3.2 *Summation of Series*

The series in Eq. (5.111) can be summed for large N using our old friend, the method of steepest descent. Define a generating function as

$$\exp\left\{V\sum_{l=1}^{\infty} g_l z^l\right\} = \sum_{n=0}^{\infty} \frac{1}{n!}\left(V\sum_{l=1}^{\infty} g_l z^l\right)^n \quad ; \text{ generating function} \tag{5.112}$$

Then, with the aid of the multinomial theorem on the r.h.s.,

$$\exp\left\{V\sum_{l=1}^{\infty} g_l z^l\right\} = \sum_{n=0}^{\infty} \sum_{\{n\}} \frac{(Vg_1 z)^{n_1}}{n_1!} \frac{(Vg_2 z^2)^{n_2}}{n_2!} \frac{(Vg_3 z^3)^{n_3}}{n_3!} \cdots$$

$$n = \sum_{i=0}^{\infty} n_i \tag{5.113}$$

where $\{n\}$ stands for the set of distribution numbers $(n_1, n_2, n_3, \cdots)$, which must sum to n.

The canonical partition function in Eq. (5.111) is just the coefficient of z^N in this series, where

$$(z)^{n_1}(z^2)^{n_2}(z^3)^{n_3} \cdots = z^{\sum_l l n_l} = z^N \tag{5.114}$$

As in Eq. (2.70), this coefficient is extracted with a contour integral

$$(\text{P.F.}) = \frac{1}{2\pi i} \oint_C \frac{\exp\left\{V\sum_{l=1}^{\infty} g_l z^l\right\}}{z^N} \frac{dz}{z} \tag{5.115}$$

The analysis then parallels that given previously:[22]

[21] Remember that n_l must vanish for $l > N$.

[22] We will later carry out an exact summation of the series in Eq. (5.111) when we talk about the grand partition function, which reproduces Eqs. (5.120).

(1) The saddle point at $z = \lambda$ is located by setting the derivative of the logarithm of the integrand equal to zero

$$\left[\frac{d}{dz}\ln I(z)\right]_{z=\lambda} = 0$$

$$\Longrightarrow \qquad V\sum_{l=1}^{\infty} lg_l\lambda^{l-1} - \frac{N+1}{\lambda} = 0 \tag{5.116}$$

Since $N \gg 1$, this gives[23]

$$\frac{N}{V} = \frac{1}{v} = \sum_{l=1}^{\infty} lg_l\lambda^l \qquad ; \text{ saddle point} \tag{5.117}$$

(2) At the saddle point, the logarithm of (P.F.) is given by

$$\ln(\text{P.F.}) = V\sum_{l=1}^{\infty} g_l\lambda^l - N\ln\lambda \tag{5.118}$$

where we have again used $N \gg 1$, and neglected terms of $O(\ln N)$.

Now compute the pressure from Eq. (5.69). Note that $g_l(T)$ is independent of V, and by the definition of the saddle point

$$\frac{\partial}{\partial\lambda}\ln(\text{P.F.}) = 0 \qquad ; \text{ definition of saddle point} \tag{5.119}$$

Therefore

$$\frac{P}{k_BT} = \sum_{l=1}^{\infty} g_l(T)\lambda^l$$

$$\frac{1}{v} = \sum_{l=1}^{\infty} l\, g_l(T)\lambda^l \qquad ; \text{ parametric equation of state} \tag{5.120}$$

Here the coefficients $g_l(T)$ are given by Eq. (5.110); they involve the cluster integrals over the potential in Eq. (5.104). The quantity λ is now a *parameter* which must be chosen to fit the density $1/v = N/V$.

Equations (5.120) present an exact parametric equation of state for an imperfect gas.

[23] We assume that for real positive λ this relation has one solution where λ vanishes as $1/v \to 0$.

5.2.3.3 *Interpretation*

A nice interpretation of these results is obtained by defining

$$Vg_l\lambda^l \equiv n_l^\star \tag{5.121}$$

Then Eqs. (5.120) take the form

$$\frac{PV}{k_{\rm B}T} = \sum_{l=1}^{\infty} n_l^\star$$
$$N = \sum_{l=1}^{\infty} l n_l^\star \tag{5.122}$$

A comparison with the equation of state of a perfect gas leads to the interpretation of $n_l^\star$ as the mean number of clusters of l systems.

What is λ? From Eq. (5.118) one has

$$A = -k_{\rm B}T\ln({\rm P.F.}) = -k_{\rm B}T\left[V\sum_{l=1}^{\infty} g_l\lambda^l - N\ln\lambda\right]$$
$$= -PV + Nk_{\rm B}T\ln\lambda \tag{5.123}$$

Now it is a result from thermodynamics that the Gibbs free energy can be written

$$G = E - TS + PV = A + PV = N\mu \quad ; \text{Gibbs free energy} \tag{5.124}$$

where μ is the chemical potential (see Prob. 5.14). Thus

$$Nk_{\rm B}T\ln\lambda = N\mu$$
$$\text{or ;} \qquad \lambda = e^{\mu/k_{\rm B}T} \qquad ; \text{absolute activity} \tag{5.125}$$

This is the *absolute activity* of the gas.

5.2.3.4 *Virial Expansion*

Let us make the connection of the cluster analysis to the virial expansion in Eq. (5.72). The second of Eqs. (5.120) is an implicit relation for $\lambda(v)$.

$$\frac{1}{v} = g_1\lambda + 2g_2\lambda^2 + 3g_3\lambda^3 + \cdots \tag{5.126}$$

Inversion of this series (carried out below) gives λ as power series in $1/v$.

$$\lambda = \frac{a_1}{v} + \frac{a_2}{v^2} + \cdots \tag{5.127}$$

This first of Eqs. (5.120) then yields the virial expansion through

$$\frac{P}{k_{\rm B}T} = g_1\lambda + g_2\lambda^2 + g_3\lambda^3 + \cdots \tag{5.128}$$

To invert the series in Eq. (5.126) and find the coefficients $(a_1, a_2, \cdots)$ in Eq. (5.127), substitute the latter into the former, and equate powers of $1/v$

$$\frac{1}{v} = g_1\left(\frac{a_1}{v} + \frac{a_2}{v^2} + \cdots\right) + 2g_2\left(\frac{a_1}{v} + \cdots\right)^2 + \cdots$$

$$\Longrightarrow \qquad g_1 a_1 = 1$$

$$g_1 a_2 + 2g_2 a_1^2 = 0$$

$$\vdots \tag{5.129}$$

The solution to these equations is

$$a_1 = \frac{1}{g_1}$$

$$a_2 = -\frac{2g_2}{g_1}a_1^2 = -\frac{2g_2}{g_1^3}$$

$$\vdots \tag{5.130}$$

Thus from Eq. (5.127)

$$\lambda = \frac{1}{g_1 v} - \frac{2g_2}{g_1^3 v^2} + \cdots \tag{5.131}$$

Substitution of this result into Eq. (5.128) then leads to

$$\frac{P}{k_{\rm B}T} = g_1\left(\frac{1}{g_1 v} - \frac{2g_2}{g_1^3 v^2} + \cdots\right) + g_2\left(\frac{1}{g_1 v} + \cdots\right)^2 + \cdots$$

$$= \frac{1}{v} - \frac{g_2}{g_1^2}\frac{1}{v^2} + \cdots \tag{5.132}$$

The second virial coefficient is then identified as[24]

$$B(T) = -\frac{g_2}{g_1^2} \tag{5.133}$$

[24]For the third virial coefficient, see Prob. 5.17.

Substitution of the definitions in Eqs. (5.110) and (5.104) expresses the second virial coefficient as

$$B(T) = -\frac{B_2}{2B_1^2} = -\frac{1}{2}\int_0^\infty 4\pi r^2\, dr\left[e^{-U(r)/k_{\rm B}T} - 1\right] \tag{5.134}$$

or

$$B(T) = 2\pi \int_0^\infty \left[1 - e^{-U(r)/k_{\rm B}T}\right] r^2\, dr \tag{5.135}$$

This reproduces our previous result in Eq. (5.88).

5.2.4 *Law of Corresponding States*

Let us return to Van der Waal's equation of state

$$\left(P + \frac{a}{v^2}\right)(v - b) = k_{\rm B}T$$

$$\text{or ;} \qquad P(v) = \frac{k_{\rm B}T}{v-b} - \frac{a}{v^2} \tag{5.136}$$

Here (a, b) are constants, determined empirically for any gas, and $P(v)$ is an isotherm. Three isotherms of the Van der Waal's equation of state are sketched in Fig. 5.18.

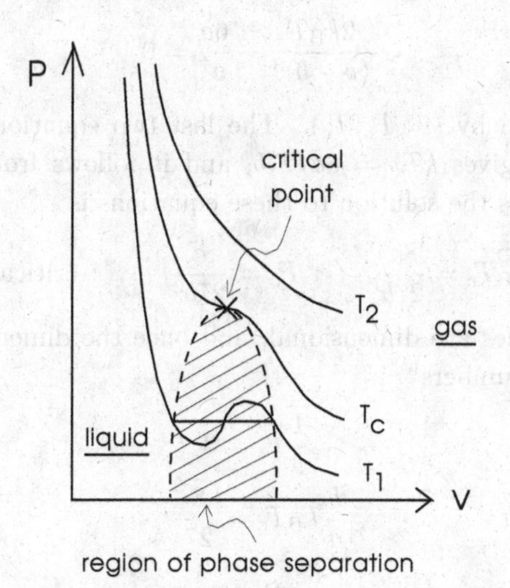

Fig. 5.18 Sketch of the Van der Waal's equation of state. Three isotherms $P(v)$ are shown (see text). All reduce to the perfect gas result $Pv = k_{\rm B}T$ at low density, where $v = V/N \to \infty$.

- At high temperature T_2, and large-enough v, the behavior is that of a perfect gas;
- At low-enough temperature T_1, there is a region of *phase separation* into a gas and liquid;[25]
- There is a *critical isotherm* at temperature T_c, whose slope vanishes at the critical point, which is also a point of inflection.

The *critical point* is located by imposing the following three conditions on the pressure

$$
\begin{aligned}
P &= P(v) \\
\frac{\partial P}{\partial v} &= 0 \\
\frac{\partial^2 P}{\partial v^2} &= 0 \qquad ; \text{ critical point}
\end{aligned}
\tag{5.137}
$$

It follows from Eq. (5.136) that the critical point is given by the solution to the following three equations in the three unkowns (v, P, T):

$$
\begin{aligned}
\frac{k_B T}{v-b} - \frac{a}{v^2} &= P \\
-\frac{k_B T}{(v-b)^2} + \frac{2a}{v^3} &= 0 \\
\frac{2k_B T}{(v-b)^3} - \frac{6a}{v^4} &= 0
\end{aligned}
\tag{5.138}
$$

Label the solution by (v_c, P_c, T_c). The last two equations give $v_c = 3b$, the second then gives $kT_c = 8a/27b$, and it follows from the first that $P_c = a/27b^2$. Thus the solution to these equations is

$$
v_c = 3b \qquad ; \; k_B T_c = \frac{8a}{27b} \qquad ; \; P_c = \frac{a}{27b^2} \qquad ; \text{ critical point} \tag{5.139}
$$

These quantities are dimensional, and once the dimensions are scaled out, we just get numbers

$$
\begin{aligned}
\frac{1}{b}\, v_c &= 3 \\
\frac{b}{a}\, k_B T_c &= \frac{8}{27} \\
\frac{b^2}{a}\, P_c &= \frac{1}{27}
\end{aligned}
\tag{5.140}
$$

[25] We will return later to a discussion of phase separations.

The dimensionless ratio $P_c v_c / k_B T_c$ is also a pure number

$$\frac{P_c v_c}{k_B T_c} = \frac{3}{8} \tag{5.141}$$

It is very instuctive to introduce *reduced* quantities, which are the *ratios* to the critical values

$$v_r \equiv \frac{v}{v_c} \quad ; \; P_r \equiv \frac{P}{P_c} \quad ; \; T_r \equiv \frac{T}{T_c} \quad ; \text{ reduced quantities} \tag{5.142}$$

Written in terms of these quantities, Van der Waal's equation of state becomes (see Prob. 5.19)

$$\left(P_r + \frac{3}{v_r^2}\right)\left(v_r - \frac{1}{3}\right) = \frac{8}{3}T_r$$
$$\frac{P_c v_c}{k_B T_c} = \frac{3}{8} \tag{5.143}$$

We proceed to discuss these results:

- The constants (a, b) referring to a given substance no longer appear in Eqs. (5.143); they are now *universal relations* applicable to any gas! Within this framework, the measurement of the equation of state for any gas, re-expressed in terms of these reduced variables, should lie on a *universal curve*;[26]
- This result is derived within the framework of the Van der Waal's equation of state. Is there an analogous relation that can be derived within the general theory of imperfect gases?
- First, re-express Eq. (5.143) as

$$\frac{P_r v_r}{T_r} = \frac{8v_r}{(3v_r - 1)} - \frac{3}{v_r T_r} \tag{5.144}$$

- The generalization to the *law of corresponding states* says that for *any* imperfect gas

$$\frac{P_r v_r}{T_r} = f(v_r, T_r) \quad ; \text{ law of corresponding states} \tag{5.145}$$

  Again, once the function $f(v_r, T_r)$ has been determined, the measurement of the equation of state for any gas, re-expressed in terms of these reduced variables, should lie on a *universal curve*;

[26] By equation of state, we here refer to the isotherms $P(v)$; the full equation of state is actually a surface, $P = P(v, T)$.

- When applied to a perfect gas, the relation in Eq. (5.145) gives

$$\frac{Pv}{k_{\rm B}T} = f(v_r, T_r)\frac{P_c v_c}{k_{\rm B}T_c}$$

$$\rightarrow 1 \qquad ; v \rightarrow \infty \qquad ; \text{all } T \qquad (5.146)$$

The last relation follows since all imperfect gases become perfect gases at low enough density. Thus

$$\frac{P_c v_c}{k_{\rm B}T_c} = \frac{1}{f(\infty, T_r)} \equiv \frac{1}{f(\infty)} \qquad ; \text{universal constant} \qquad (5.147)$$

Since the l.h.s. is independent of T_r, the r.h.s. must also be; hence it is a universal constant.[27]

We proceed to a demonstration for imperfect gases of the law of corresponding states in Eq. (5.145).

5.2.4.1 *Derivation*

We make two assumptions:

(1) Classical statistics is applicable;
(2) The interatomic two-body potential has the general form

$$U(r) = \epsilon\, \Theta\left(\frac{r}{\sigma}\right) \qquad ; \text{two-body potential} \qquad (5.148)$$

Two examples of such potentials are:

- The Lennard-Jones potential (see Fig. 5.19)[28]

$$U(r) = 4\epsilon\left[\left(\frac{\sigma}{r}\right)^{12} - \left(\frac{\sigma}{r}\right)^{6}\right] \qquad ; \text{Lennard-Jones} \qquad (5.149)$$

- The Van der Waal's potential

$$U(r) = \epsilon \cdot \infty \qquad ; r < \sigma \qquad ; \text{Van der Waal's}$$

$$= -\frac{\epsilon}{(r/\sigma)^6} \qquad ; r > \sigma \qquad (5.150)$$

---

[27] In Eq. (5.144) for example, $f(\infty, T_r) \equiv f(\infty) = 8/3$.
[28] Compare Prob. 5.11.

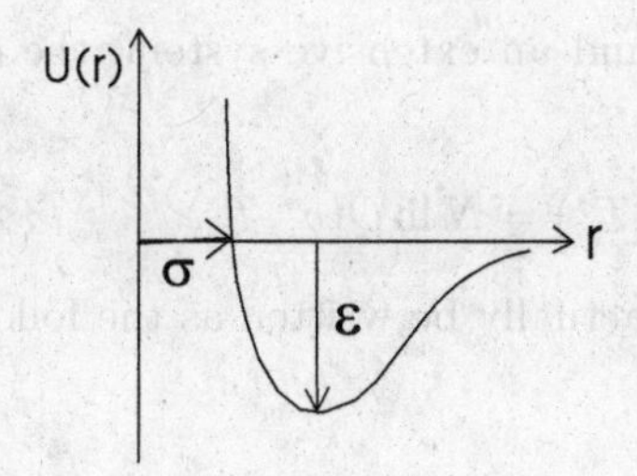

Fig. 5.19 The Lennard-Jones potential.

The equation of state is determined by the configuration integral

$$Q_N(V,T) = \int \cdots \int e^{-U(\mathbf{r}_1,\cdots,\mathbf{r}_N)/k_BT}\, d^3\mathbf{r}_1 \cdots d^3\mathbf{r}_N$$

$$P = k_BT\frac{\partial}{\partial V}\ln Q_N(V,T) \qquad ; \text{ equation of state} \qquad (5.151)$$

Introduce *dimensionless* variables

$$\mathbf{r}_i^* \equiv \frac{\mathbf{r}_i}{\sigma} \qquad ; T^* \equiv \frac{k_BT}{\epsilon} \qquad (5.152)$$

It follows from Eq. (5.148) that

$$U(\mathbf{r}_1, \cdots \mathbf{r}_N) = \epsilon\Theta(\mathbf{r}_1^*, \cdots, \mathbf{r}_N^*) \qquad (5.153)$$

In terms of these new variables, the configuration integral take the form

$$Q_N(V,T) = \sigma^{3N}\int_0^{V^{1/3}/\sigma} \cdots \int_0^{V^{1/3}/\sigma} dq_1^* \cdots dq_{3N}^*\, e^{-\Theta(q_1^*,\cdots,q_{3N}^*)/T^*} \qquad (5.154)$$

Here we have assumed a cubical box with sides of length $L = V^{1/3}$.

Now take the dimensions out of *all* the thermodynamic variables

$$V^* \equiv \frac{V}{\sigma^3} \qquad ; P^* \equiv \frac{P\sigma^3}{\epsilon} \qquad ; T^* \equiv \frac{k_BT}{\epsilon} \qquad (5.155)$$

One then arrives at an equation of state from which the parameters (ϵ, σ) of the potential have disappeared[29]

$$P^* = T^*\frac{\partial}{\partial V^*}\ln Q_N(V^*, T^*) \qquad (5.156)$$

[29] The derivative of the logarithm eliminates the overall factor of σ^{3N} in Eq. (5.154).

For fixed $v^* = V^*/N$ and an extensive system, the quantity $\ln Q_N$ must scale as N [30]

$$\ln Q_N(V^*, T^*) = N \ln Q(v^*, T^*) \qquad ; N \to \infty \tag{5.157}$$

Hence Eq. (5.156) can actually be written as the following *universal equation of state*

$$P^* = T^* \frac{\partial}{\partial v^*} \ln Q(v^*, T^*) \qquad ; \text{ universal equation of state} \tag{5.158}$$

This result has several consequences:

(1) As in Eqs. (5.137), the critical point is located by the following three relations in the three unknowns (v^*, P^*, T^*)

$$\begin{aligned} P^* &= P^*(v^*, T^*) \\ \frac{\partial P^*}{\partial v^*} &= 0 \\ \frac{\partial^2 P^*}{\partial v^{*2}} &= 0 \qquad ; \text{ critical point} \end{aligned} \tag{5.159}$$

If the solution to these equations is labeled by (v_c^*, P_c^*, T_c^*), then these quantities are *universal constants* independent of (ϵ, σ);

(2) Introduce *reduced* variables

$$\begin{aligned} P_r &\equiv \frac{P^*}{P_c^*} = \frac{P}{P_c} \\ T_r &\equiv \frac{T^*}{T_c^*} = \frac{T}{T_c} \\ v_r &\equiv \frac{v^*}{v_c^*} = \frac{v}{v_c} \end{aligned} \tag{5.160}$$

It follows from Eq. (5.158) that

$$\frac{P_r v_r}{T_r} = f(v_r, T_r) \tag{5.161}$$

where $f(v_r, T_r)$ is some function of the indicated variables, which can absorb the constants (v_c^*, P_c^*, T_c^*). The result is the *law of corresponding states*

$$\frac{P_r v_r}{T_r} = f(v_r, T_r) \qquad ; \text{ law of corresponding states} \tag{5.162}$$

[30] (v^*, T^*) are now *intensive* variables; the precise form of $\ln Q_N(V^*, T^*)$ depends on the shape of the potential Θ in Eqs. (5.148) and (5.154).

(3) As in Eq. (5.147), one then concludes that

$$\frac{P_c v_c}{k_{\rm B} T_c} = \text{universal constant} \tag{5.163}$$

For either of the two-parameter potentials described above, the values of (ϵ, σ) can be obtained, for example, from a fit to the temperature dependence of the second virial coefficient $B(T)$.[31] We give two comparisons with experiment, based on the Lennard-Jones potential [Davidson (2003)].

- The measured (P_c^*, T_c^*, v_c^*) are indeed constant for the entire set of gases Ne, A, Xe, N_2, O_2, CH_4

$$\begin{aligned} P_c^* &= \frac{P_c \sigma^3}{\epsilon} = 0.127 \pm 0.015 \qquad ; \text{ experiment} \\ T_c^* &= \frac{k_{\rm B} T_c}{\epsilon} = 1.28 \pm 0.03 \\ v_c^* &= \frac{v_c}{\sigma^3} = 3.0 \pm 0.03 \end{aligned} \tag{5.164}$$

- These numbers give[32]

$$\frac{P_c^* v_c^*}{T_c^*} = \frac{P_c v_c}{k_{\rm B} T_c} = 0.298 \qquad ; \text{ Lennard-Jones} \tag{5.165}$$

The experimental values of this quantity for a variety of gases are shown in Table 5.1.

Table 5.1 Measured values for a variety of gases [Davidson (2003)].

Substance	$P_c v_c / k_{\rm B} T_c$
He	0.300
H_2	0.304
Ne	0.307
O_2	0.292
CH_4	0.290
Ethane	0.267
n-Heptane	0.258
n-Octane	0.258
Benzene	0.265
Acetylene	0.275

[31] See Probs. 5.11–5.12.
[32] Compare with the result for a Van der Waal's gas in Eq. (5.141).

The overall agreement is quite impressive. The law of corresponding states works best, of course, for similar substances. It is useful for predicting (P, V, T) behavior.[33]

These two applications of the canonical ensemble, normal modes in solids and the configuration integral in the classical theory of imperfect gases, serve to demonstrate the power and beauty of statistical mechanics. We turn next to the grand canonical ensemble, which provides the most general framework for the theory.

[33]In a classic work, the universal behavior of various quantities in the vicinity of the critical point has been analyzed by Wilson using renormalization-group techniques [Wilson (1971)]. A useful, but somewhat more advanced, reference here is [Amit (2005)]; see also [Plischke and Bergersen (2006)].

Chapter 6

The Grand Canonical Ensemble

So far we have developed two equivalent approaches to statistical mechanics:

(1) The *microcanonical ensemble* with

$$\begin{aligned} S &= k_{\text{B}} \ln \Omega \\ \Omega &= \Omega(E, V, N) \qquad ; \text{ number of complexions} \end{aligned} \tag{6.1}$$

(2) The *canonical ensemble* with

$$\begin{aligned} A(T, V, N) &= -k_{\text{B}} T \ln (\text{P.F.}) \\ (\text{P.F.}) &= \sum_j \Omega_j e^{-E_j/k_{\text{B}}T} \quad ; \text{ canonical partition function} \end{aligned} \tag{6.2}$$

In addition to allowing us to readily include interactions, an advantage of the second approach is that when one restrictive condition is removed, namely that of fixed energy $\sum_i \varepsilon_i n_i = E$, many of the sums involved can be carried out *analytically*. The question arises as to whether the second restrictive condition, that of fixed number of particles $\sum_i n_i = N$, can also be removed.

To get from the first approach to the second, we looked at an *ensemble of assemblies* which could interchange energy at a constant temperature T. We let

$$e^{\beta E} \Omega \to \sum_E e^{\beta E} \Omega(E, V, N) \tag{6.3}$$

where β was the Lagrange multiplier for the condition $\sum_i \varepsilon_i n_i = E$. We then identified

$$\beta = -\frac{1}{k_{\text{B}}T} \tag{6.4}$$

We observed that for large N, only the mean energy $\langle E\rangle \equiv E^*$ contributes appreciably for any assembly in the ensemble.

This *suggests* that we try letting

$$e^{\alpha N}e^{\beta E}\,\Omega \to \sum_E \sum_N e^{\alpha N}e^{\beta E}\,\Omega(E,V,N) \tag{6.5}$$

where α is the Lagrange multiplier for the condition of fixed $\sum_i n_i = N$, and then identify

$$\alpha = \frac{\mu}{k_{\rm B}T} \tag{6.6}$$

It should again be true that if it is large enough, only the mean number of particles $\langle N\rangle \equiv N^*$ will contribute appreciably for any assembly in the ensemble.

One can think of this as an ensemble of assemblies that are now free to exchange both energy and particles at fixed temperature T and fixed chemical potential μ (Fig. 6.1).

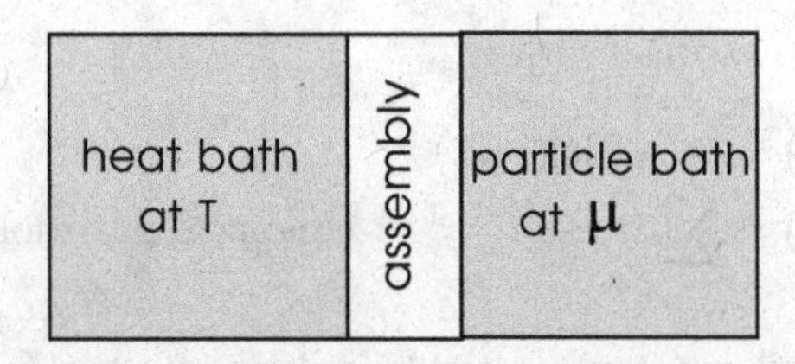

Fig. 6.1 An assembly that is free to interchange both energy and particles at fixed temperature T and chemical potential μ. The particles can be exchanged, for example, through a semi-permeable membrane.

6.1 Grand Partition Function

We define the *grand partition function* as[1]

$$({\rm G.P.F.}) \equiv \sum_N ({\rm P.F.})_N\, e^{N\mu/k_{\rm B}T} \qquad ;\ \text{grand partition function}$$

$$= \sum_N \sum_E \Omega(E,V,N)e^{-E/k_{\rm B}T}e^{N\mu/k_{\rm B}T} \tag{6.7}$$

[1]The notation $({\rm P.F.})_N$ implies that the canonical partition function is computed for given N.

Sometimes it is more convenient to use the absolute activity λ

$$\lambda = e^{\mu/k_BT} \qquad ; \text{ absolute activity} \tag{6.8}$$

In this case Eqs. (6.7) become

$$\begin{aligned}(\text{G.P.F.}) &= \sum_N (\text{P.F.})_N \lambda^N \\ &= \sum_N \sum_E \Omega(E,V,N) e^{-E/k_BT} \lambda^N\end{aligned} \tag{6.9}$$

The thermodynamic variables are now (T, V, μ). To what thermodynamic function does (G.P.F.) correspond? Write

$$\begin{aligned}(\text{G.P.F.}) &= \sum_N t(N) \\ t(N) &= (\text{P.F.})_N \lambda^N\end{aligned} \tag{6.10}$$

Let us, as before, pick out the largest term in this sum. This is obtained by maximizing its logarithm

$$\left[\frac{\partial}{\partial N} \ln t(N)\right]_{T,V,\mu} = \frac{\partial}{\partial N} \ln (\text{P.F.})_N + \ln \lambda = 0 \tag{6.11}$$

From the previous analysis of the canonical ensemble

$$\left[\frac{\partial}{\partial N} \ln (\text{P.F.})_\text{N}\right]_{T.V} = -\frac{1}{k_BT}\left(\frac{\partial A}{\partial N}\right)_{T,V} = -\frac{\mu}{k_BT} \tag{6.12}$$

Hence Eq. (6.11) reproduces the relation

$$\begin{aligned}\ln \lambda &= \frac{\mu}{k_BT} \\ \text{or ;} \qquad \lambda &= e^{\mu/k_BT}\end{aligned} \tag{6.13}$$

Consider the logarithm of the grand partition function. The replacement of the sum in Eq. (6.10) by its largest term, obtained at the mean value N^*, is certainly valid if N^* is large enough. *Now denote this mean value, used in the subsequent thermodynamic arguments, by* $N^\star \equiv N$.[2] Then

$$\begin{aligned}k_BT \ln (\text{G.P.F.}) &= k_BT \ln (\text{P.F.})_N + N\mu \\ &= -A + N\mu = -A + G = PV\end{aligned} \tag{6.14}$$

Here $G = N\mu$ is the Gibbs free energy (recall Prob. 5.14).

[2] As before, for simplicity, we suppress the star in the thermodynamic arguments.

The quantity $-PV \equiv \Phi(T, V, \mu)$ is known as the *thermodynamic potential*[3]

$$-PV \equiv \Phi(T, V, \mu) \qquad ; \text{ thermodynamic potential} \qquad (6.15)$$

This leads to the *third fundamental relation* of statistical mechanics, which expresses the thermodynamic potential of an assembly in the grand canonical ensemble in terms of the grand partition function

$$\Phi(T, V, \mu) = -k_{\rm B}T \ln(\text{G.P.F.}) \qquad ; \text{ grand canonical ensemble}$$
$$(\text{G.P.F.}) = \sum_N (\text{P.F.})_N \, e^{N\mu/k_{\rm B}T} \qquad (6.16)$$

Alternatively

$$(\text{G.P.F.}) = \sum_N (\text{P.F.})_N \, \lambda^N \qquad (6.17)$$

Let us do some thermodynamics. We have[4]

$$\begin{aligned}
\Phi &= -PV = A - G \\
&= A - N\mu \\
d\Phi &= -SdT - PdV + \mu dN - Nd\mu - \mu dN \\
&= -SdT - PdV - Nd\mu \qquad (6.18)
\end{aligned}$$

Hence the entropy, pressure, and particle number can be obtained from the thermodynamic potential $\Phi(T, V, \mu)$ by partial differentiation

$$S = -\left(\frac{\partial \Phi}{\partial T}\right)_{V,\mu} \quad ; P = -\left(\frac{\partial \Phi}{\partial V}\right)_{T,\mu} \quad ; N = -\left(\frac{\partial \Phi}{\partial \mu}\right)_{T,V} \qquad (6.19)$$

An even simpler expression for the pressure is obtained directly from the definition of the thermodynamic potential

$$P = \frac{k_{\rm B}T}{V} \ln(\text{G.P.F.}) \qquad (6.20)$$

It follows from the second line of Eqs. (6.18), and the first and second law of thermodynamics, that an allowable transition for an assembly at

---

[3] Although the conventional notation for the thermodynamic potential is $\Omega(T, V, \mu)$ (see, for example, [Fetter and Walecka (2003)]), we will use $\Phi(T, V, \mu)$ to avoid confusion with the number of complexions $\Omega(E, V, T)$.

[4] Recall that now $N \equiv N^*$ is the mean number of particles in an assembly in the grand canonical ensemble; recall also Eq. (4.51).

fixed (T, V, μ) is characterized by the condition

$$\delta\Phi\Big|_{T,V,\mu} = đQ - T\delta S \leq 0 \tag{6.21}$$

Hence *an assembly in equilbrium at fixed* (T, V, μ) *will minimize its thermodynamic potential*[5]

$$\delta\Phi\Big|_{T,V,\mu} \geq 0 \qquad ; \text{ equilbrium} \tag{6.22}$$

It is often more convenient to do the thermodynamics directly in terms of the absolute activity $\lambda(\mu, T)$ rather than the chemical potential μ. This requires some simple manipulations

$$\begin{aligned}
\mu &= k_{\rm B}T \ln \lambda \\
d\mu &= k_{\rm B}\, dT \ln \lambda + k_{\rm B}T \frac{d\lambda}{\lambda} \\
&= \mu \frac{dT}{T} + k_{\rm B}T \frac{d\lambda}{\lambda}
\end{aligned} \tag{6.23}$$

Use

$$N\mu + TS = E + PV \tag{6.24}$$

Then from Eq. (6.18)

$$d\Phi = -\left(\frac{E + PV}{T}\right) dT - PdV - Nk_{\rm B}T \frac{d\lambda}{\lambda} \tag{6.25}$$

It follows that

$$\begin{aligned}
\frac{E + PV}{T} &= -\left(\frac{\partial\Phi}{\partial T}\right)_{V,\lambda} \\
&= k_{\rm B} \ln (\text{G.P.F.}) + k_{\rm B}T \frac{\partial}{\partial T} \ln (\text{G.P.F.})
\end{aligned} \tag{6.26}$$

The use of Eq. (6.20) to cancel equal terms reduces this to

$$E = k_{\rm B}T^2 \frac{\partial}{\partial T} \ln (\text{G.P.F.}) \tag{6.27}$$

[5] See Chapter 1, and Prob. 6.1.

A combination of Eqs. (6.25)–(6.27) and (6.16) implies that with the variable set (T, V, λ)

$$
\begin{aligned}
E &= k_{\mathrm{B}}T^2 \frac{\partial}{\partial T} \ln \mathrm{(G.P.F.)} &&; \text{variable set } (T, V, \lambda) \\
N &= \lambda \frac{\partial}{\partial \lambda} \ln \mathrm{(G.P.F.)} && \lambda = e^{\mu/k_{\mathrm{B}}T} \\
P &= k_{\mathrm{B}}T \frac{\partial}{\partial V} \ln \mathrm{(G.P.F.)}
\end{aligned} \tag{6.28}
$$

6.2 Relation to Previous Results

We have *assumed* that the relation in Eqs. (6.16) is valid. Let us show that this assumption reproduces all of our previous results.

6.2.1 *Independent Non-Localized Systems*

It is an exact result that the canonical partition function for independent, non-localized systems is given by Eq. (4.14)

$$
\mathrm{(P.F.)}_N = \frac{\mathrm{(p.f.)}^N}{N!} \tag{6.29}
$$

The grand partition function is then calculated analytically as

$$
\mathrm{(G.P.F.)} = \sum_{N=0}^{\infty} \frac{\mathrm{(p.f.)}^N}{N!} \lambda^N = \exp\left[\,\lambda \mathrm{(p.f.)}\,\right] \tag{6.30}
$$

It follows from the second of Eqs. (6.28) and Eq. (6.20) that[6]

$$
\begin{aligned}
N &= \lambda \mathrm{(p.f.)} \\
PV &= k_{\mathrm{B}}T\lambda \mathrm{(p.f.)} = Nk_{\mathrm{B}}T
\end{aligned} \tag{6.31}
$$

Thus we recover the equation of state of a perfect gas. Futhermore, it follows from these relations and Eq. (6.13) that

$$
\begin{aligned}
\mu N - PV &= Nk_{\mathrm{B}}T \ln \lambda - PV \\
&= Nk_{\mathrm{B}}T[\ln N - \ln \mathrm{(p.f.)}] - Nk_{\mathrm{B}}T \\
G - PV &= -k_{\mathrm{B}}T \ln \frac{\mathrm{(p.f.)}^N}{N!} + O(\ln N)
\end{aligned} \tag{6.32}
$$

[6]Remember that now $N = N^*$ is the mean number of particles in an assembly in the grand canonical ensemble; note that the first of Eqs. (6.31) allows us to invert and find $\lambda(T, V, N)$.

where the last line follows from $G = \mu N$ and the use of Stirling's formula. Hence, with the neglect of terms of $O(\ln N)$, we recover the previous result that for an assembly of independent, non-localized systems in the grand canonical ensemble

$$A(T,V,N) = -k_{\rm B}T \ln \frac{({\rm p.f.})^N}{N!} \tag{6.33}$$

6.2.2 *Independent Localized Systems*

The canonical partition function for independent, localized systems is given by Eq. (4.13) as

$$({\rm P.F.})_N = ({\rm p.f.})^N \tag{6.34}$$

The grand partition function is then evaluated analytically as

$$({\rm G.P.F.}) = \sum_{N=0}^{\infty} ({\rm p.f.})^N \lambda^N = \frac{1}{1-\lambda({\rm p.f.})} \tag{6.35}$$

It follows from Eq. (6.20) and the second of Eqs. (6.28) that

$$\begin{aligned} PV &= -k_{\rm B}T \ln\left[1-\lambda({\rm p.f.})\right] \\ N &= \frac{\lambda({\rm p.f.})}{1-\lambda({\rm p.f.})} \end{aligned} \tag{6.36}$$

The inversion of the second relation gives

$$\lambda({\rm p.f.}) = \frac{N}{N+1} \tag{6.37}$$

Then, as above,[7]

$$\begin{aligned} \mu N - PV &= Nk_{\rm B}T\left[\ln\frac{N}{N+1} - \ln({\rm p.f.})\right] + k_{\rm B}T\ln\frac{1}{N+1} \\ &= -k_{\rm B}T\ln({\rm p.f.})^N + O(1) \end{aligned} \tag{6.38}$$

Thus, with the neglect of terms of relative order $1/N$, we have for the Helmholtz free energy of an assembly of independent localized systems in the grand canonical ensemble

$$A(T,V,N) = -k_{\rm B}T\ln({\rm p.f.})^N \tag{6.39}$$

This is our previous result.

[7]Recall $\mu = k_{\rm B}T\ln\lambda$, and use $1-\lambda({\rm p.f.}) = (N+1)^{-1}$. Note that in this derivation, one has the strict inequality $\lambda({\rm p.f.}) < 1$.

6.2.3 *Imperfect Gases*

The linked-cluster expansion of the canonical partition function for imperfect gases is given in Eqs. (5.111) as

$$\begin{aligned}(\text{P.F.})_N &= \sum_{\{n_l\}} \prod_l \frac{(Vg_l)^{n_l}}{n_l!} \\ \sum_{l=1}^{\infty} ln_l &= N\end{aligned} \tag{6.40}$$

The grand partition function is again obtained as

$$\begin{aligned}(\text{G.P.F.}) &= \sum_N (\text{P.F.})_N \lambda^N = \sum_N (\text{P.F.})_N \lambda^{\sum_l ln_l} \\ &= \sum_{\{n_l\}} \prod_l \frac{(Vg_l\lambda^l)^{n_l}}{n_l!} \\ &= \sum_{n_1}\sum_{n_2}\sum_{n_3} \cdots \frac{(Vg_1\lambda)^{n_1}}{n_1!} \frac{(Vg_2\lambda^2)^{n_2}}{n_2!} \frac{(Vg_3\lambda^3)^{n_3}}{n_3!} \cdots\end{aligned} \tag{6.41}$$

The sum over all N produces an *unconstrained* sum over all distributions $\{n_l\} = (n_1, n_2, n_3, \cdots)$, leading to the product of exponentials

$$\begin{aligned}(\text{G.P.F.}) &= e^{Vg_1\lambda} e^{Vg_2\lambda^2} e^{Vg_3\lambda^3} \cdots \\ &= \prod_l \exp(Vg_l\lambda^l)\end{aligned} \tag{6.42}$$

Thus we have an exact evaluation of the grand partition function for imperfect gases as

$$(\text{G.P.F.}) = \exp\left(V \sum_{l=1}^{\infty} g_l \lambda^l\right) \tag{6.43}$$

It follows from Eq. (6.20) and the second of Eqs. (6.28) that

$$\begin{aligned}\frac{P}{k_{\text{B}}T} &= \sum_{l=1}^{\infty} g_l \lambda^l \\ \frac{N}{V} &= \sum_{l=1}^{\infty} l g_l \lambda^l\end{aligned} \tag{6.44}$$

This exactly reproduces our parametric equation of state for imperfect gases in Eqs. (5.120), and nowhere did we have to invoke the method of steepest descent!

We thus see that the basic assumption of the grand canonical ensemble, summarized in Eqs. (6.16), reproduces our previous results with no further approximations. Equations (6.16) therefore constitute the most general starting point for statistical mechanics.

In the canonical ensemble, the assemblies are free to interchange energy at fixed temperature, and we argued that for large N, the energies of the assemblies are sufficiently peaked at the mean value $\langle E \rangle$ that the mean value can be used for the energy E of an assembly in thermodynamic arguments. In the grand canonical ensemble the assemblies are further free to exchange particles at fixed chemical potential μ, and we anticipated that the particle number in the assemblies are sufficiently peaked at the mean value $\langle N \rangle$ that the mean value can be again be used for the particle number N of an assembly in thermodynamic arguments. It is time to investigate these assumptions in more detail.

6.3 Fluctuations

Recall the definition of the *mean-square deviation* of a quantity x from its mean value $\langle x \rangle$ in some normalized distribution

$$\begin{aligned}\langle (\Delta x)^2 \rangle &= \langle (x - \langle x \rangle)^2 \rangle \\ &= \langle x^2 - 2x\langle x \rangle + \langle x \rangle^2 \rangle \\ &= \langle x^2 \rangle - 2\langle x \rangle^2 + \langle x \rangle^2 \\ \langle (\Delta x)^2 \rangle &= \langle x^2 \rangle - \langle x \rangle^2 \qquad ; \text{ mean-square deviation} \end{aligned} \tag{6.45}$$

Let us apply this result.

6.3.1 *Distribution of Energies in the Canonical Ensemble*

In the canonical ensemble, the thermodynamic quantities (T, V, N) are specified, the assemblies are distributed in energy according to a Boltzmann distribution, and the canonical partition function is given by

$$(\text{P.F.}) = \sum_i e^{-E_i/k_B T} \tag{6.46}$$

where the sum here goes over all the states of an assembly. The mean energy of an assembly is given by

$$\langle E\rangle = k_{\rm B}T^2\frac{\partial}{\partial T}\ln(\text{P.F.})$$
$$= \frac{\sum_i E_i e^{-E_i/k_{\rm B}T}}{\sum_i e^{-E_i/k_{\rm B}T}} \tag{6.47}$$

Identify this mean energy with the thermodynamic variable E, so that $\langle E\rangle \equiv E$. The constant-volume heat capacity is given by

$$C_V = \left(\frac{\partial E}{\partial T}\right)_{V,N}$$
$$= \frac{1}{k_{\rm B}T^2}\left[\frac{\sum_i E_i^2 e^{-E_i/k_{\rm B}T}}{\sum_i e^{-E_i/k_{\rm B}T}} - \left(\frac{\sum_i E_i e^{-E_i/k_{\rm B}T}}{\sum_i e^{-E_i/k_{\rm B}T}}\right)^2\right] \tag{6.48}$$

Thus the mean-square deviation of the energy of an assembly in the canonical ensemble is given by the constant-volume heat capacity!

$$k_{\rm B}T^2C_V = \langle E^2\rangle - \langle E\rangle^2 = \langle(\Delta E)^2\rangle \tag{6.49}$$

For a *perfect gas*

$$E = \frac{3}{2}Nk_{\rm B}T$$
$$C_V = \frac{3}{2}Nk_{\rm B}$$
$$\langle(\Delta E)^2\rangle = \frac{3}{2}N(k_{\rm B}T)^2 \tag{6.50}$$

The quantity of physical interest is the ratio of the root-mean-square deviation to the mean value itself, and for a perfect gas this is

$$\frac{\sqrt{\langle(\Delta E)^2\rangle}}{E} = \sqrt{\frac{2}{3N}} \qquad ;\ \text{perfect gas} \tag{6.51}$$

This ratio vanishes as $1/\sqrt{N}$ for large N, as previously claimed.[8]

6.3.2 *Distribution of Particle Numbers in the Grand Canonical Ensemble*

In the grand canonical ensemble, the thermodynamic quantities (T, V, μ) are specified, the assemblies are distributed in energy and particle number

[8]See Prob. 4.2.

according to Boltzmann distributions, and the grand partition function is given by

$$(\text{G.P.F.}) = \sum_N \left[\sum_i e^{-E_i/k_B T} \right]_N e^{\mu N/k_B T} = \sum_N (\text{P.F.})_N \lambda^N \tag{6.52}$$

The mean particle number is given by the second of Eqs. (6.28)

$$\langle N \rangle = \lambda \frac{\partial}{\partial \lambda} \ln (\text{G.P.F.}) = \frac{\sum_N N\, (\text{P.F.})_N \lambda^N}{\sum_N (\text{P.F.})_N \lambda^N} \tag{6.53}$$

Consider a further derivative of this quantity

$$\lambda \frac{\partial}{\partial \lambda} \langle N \rangle = \frac{\sum_N N^2 (\text{P.F.})_N \lambda^N}{\sum_N (\text{P.F.})_N \lambda^N} - \left(\frac{\sum_N N\, (\text{P.F.})_N \lambda^N}{\sum_N (\text{P.F.})_N \lambda^N} \right)^2 \tag{6.54}$$

Thus the mean-square deviation of the particle number in the grand canonical ensemble is given by

$$\lambda \frac{\partial}{\partial \lambda} \langle N \rangle = \langle (\Delta N)^2 \rangle \tag{6.55}$$

Now

$$\begin{aligned} \lambda \frac{\partial}{\partial \lambda} \langle N \rangle &= \frac{\partial}{\partial \ln \lambda} \langle N \rangle = k_B T \left(\frac{\partial \langle N \rangle}{\partial \mu} \right)_{T,V} \\ &= \frac{k_B T}{(\partial \mu / \partial N)_{T,V}} \end{aligned} \tag{6.56}$$

where we again write the thermodynamic variable $\langle N \rangle \equiv N$ in the last line. We claim it now follows from thermodynamics that

$$\begin{aligned} \left(\frac{\partial \mu}{\partial N} \right)_{T,V} &= \frac{V}{N^2 \kappa_c} \\ \kappa_c &\equiv -\frac{1}{V} \left(\frac{\partial V}{\partial P} \right)_{T,N} \qquad \text{; isothermal compressibility} \end{aligned} \tag{6.57}$$

Here κ_c is the *isothermal compressibility.* The fractional mean-square deviation in particle number is therefore given by Eqs. (6.55)–(6.57) as

$$\frac{\langle (\Delta N)^2 \rangle}{N^2} = \frac{k_B T\, \kappa_c}{V} \tag{6.58}$$

To establish Eqs. (6.57), start from

$$\mu = \left[\frac{\partial A(V,T,N)}{\partial N} \right]_{V,T} \tag{6.59}$$

The Helmholtz free energy is extensive, so for large N we can write[9]

$$A(V,T,N) = Na(v,T) \qquad ; \; v = \frac{V}{N} \tag{6.60}$$

Now differentiate this expression with respect to N at fixed (V,T)

$$\begin{aligned}\mu &= a(v,T) + N\left[\frac{\partial a(v,T)}{\partial v}\right]\left(-\frac{V}{N^2}\right) \\ &= a(v,T) - v\left[\frac{\partial a(v,T)}{\partial v}\right]\end{aligned} \tag{6.61}$$

Differentiate once more

$$\begin{aligned}\left(\frac{\partial \mu}{\partial N}\right)_{T,V} &= \frac{\partial}{\partial v}\left[a - v\frac{\partial a}{\partial v}\right]\left[\frac{\partial v}{\partial N}\right] \\ &= \left[\frac{\partial a}{\partial v} - \frac{\partial a}{\partial v} - v\left(\frac{\partial^2 a}{\partial v^2}\right)\right]\left[-\frac{V}{N^2}\right] \\ &= \frac{v^2}{N}\left(\frac{\partial^2 a}{\partial v^2}\right)_T\end{aligned} \tag{6.62}$$

Now re-insert Eq. (6.60)

$$\left(\frac{\partial \mu}{\partial N}\right)_{T,V} = \frac{V^2}{N^2}\left[\frac{\partial^2 A(V,T,N)}{\partial V^2}\right]_{T,N} \tag{6.63}$$

Thus

$$\left(\frac{\partial \mu}{\partial N}\right)_{T,V} = -\frac{V^2}{N^2}\left(\frac{\partial P}{\partial V}\right)_{T,N} \tag{6.64}$$

as claimed in Eqs. (6.57).[10]

The isothermal compressibility is readily calculated for a *perfect gas*

$$\begin{aligned}V &= \frac{Nk_{\mathrm{B}}T}{P} \\ \kappa_c &= \frac{1}{P} \qquad ; \text{ perfect gas}\end{aligned} \tag{6.65}$$

---

[9]Recall the argument leading to Eq. (5.157).
[10]Remember $(\partial P/\partial V)_{T,N} = [(\partial V/\partial P)_{T,N}]^{-1}$.

Hence from Eq. (6.58)

$$\frac{\langle(\Delta N)^2\rangle}{N^2} = \frac{k_B T}{PV} = \frac{1}{N}$$
$$\frac{\sqrt{\langle(\Delta N)^2\rangle}}{N} = \frac{1}{\sqrt{N}} \qquad ; \text{ perfect gas} \tag{6.66}$$

The ratio of the root-mean-square deviation of the particle number to the mean value itself vanishes as $1/\sqrt{N}$ for large N, and as with the energy, the particle number of an assembly in the grand canonical ensemble is sharply-defined.

At the critical point in an *imperfect gas*, where $\partial P/\partial V \to 0$, the isothermal compressibility goes to infinity. Thus

$$\frac{\sqrt{\langle(\Delta N)^2\rangle}}{N} \to \infty \qquad ; \text{ critical point} \tag{6.67}$$

One sees large density fluctuations near the critical point that show up, for example, in the phenomenon of critical opalescence.

[illegible]

$$[illegible]$$

[illegible] limited cases [illegible] (4.65)

[illegible] of the two [illegible] of the [illegible] number [illegible] the [illegible] values [illegible] and [illegible] with the exception the [illegible] number of [illegible] the [illegible] [illegible]

[illegible] The [illegible] The last [illegible] [illegible]

$$[illegible]$$

[illegible] (4.66)

[illegible] large density fluctuations [illegible] the critical point that show [illegible] plays [illegible] of critical [illegible]

Chapter 7

Applications of the Grand Canonical Ensemble

We present several applications of the grand canonical ensemble within the framework of the quantum statistics of identical, independent, non-localized systems.[1] First, recall the situation with classical statistics.

7.1 Boltzmann Statistics

The total number of complexions for identical, non-localized systems is

$$\Omega(E,V,N) = \frac{1}{N!}\sum_{\{n\}}\frac{N!}{\prod_i n_i!} \quad ; \text{ classical statistics}$$
$$E = \sum_i \varepsilon_i n_i \quad ; N = \sum_i n_i \tag{7.1}$$

Here the first sum goes over all distributions $\{n\} = (n_1, n_2, n_3, \cdots, n_\infty)$ consistent with the two constraints in the second line. As we have seen, this is a correct counting procedure if the mean occupation number of any level is much less than one

$$n_i^\star \ll 1 \tag{7.2}$$

From Eq. (2.144), this condition is

$$\frac{N}{(\text{p.f.})} = \frac{N}{V}\left(\frac{h^2}{2\pi m k_\text{B} T}\right)^{3/2} \ll 1 \tag{7.3}$$

[1]The inclusion of interactions between particles in assemblies with quantum statistics is the subject of the text [Fetter and Walecka (2003)].

The first of Eqs. (6.31) implies this is equivalent to the following condition on the absolute activity

$$\lambda \ll 1 \qquad ; \text{Boltzmann statistics} \tag{7.4}$$

If the condition in Eq. (7.4) is satisfied, we are in the domain of classical, or Boltzmann, statistics. The grand partition function with Boltzmann statistics was calculated in Eq. (6.30)

$$(\text{G.P.F.}) = \sum_N \lambda^N \frac{(\text{p.f.})^N}{N!} = \exp\left[\lambda\,(\text{p.f.})\right] \tag{7.5}$$

If we recall the definition of (p.f.) in Eqs. (2.39), this can be re-written as

$$(\text{G.P.F.})_{\text{Boltz}} = \prod_i \exp\left[\lambda e^{-\varepsilon_i/k_{\text{B}}T}\right] \tag{7.6}$$

Here the product goes over all the single-particle states. This is the result for Boltzmann statistics.

7.2 Quantum Statistics

If the condition in Eq. (7.4) is *not* satisfied, we are in the domain of *quantum statistics.* The correct quantum-mechanical counting procedure for an assembly of indistinguishable systems is to note that there is one state, or one many-body wave function, for each set of occupation numbers $\{n\} = (n_1, n_2, \cdots, n_\infty)$[2]

$$\Omega(E, V, N) = \sum_{\{n\}} 1 \qquad ; \text{quantum statistics}$$

$$E = \sum_i \varepsilon_i n_i \quad ; \; N = \sum_i n_i \tag{7.7}$$

There are two possibilities for the occupation numbers:

(1) With *bosons*, one can have any number of systems in any single-particle state, and therefore

$$n_i = 0, 1, 2, \cdots, \infty \qquad ; \text{bosons} \tag{7.8}$$

[2] See, for example, [Walecka (2008)].

(2) With *fermions*, the Pauli principle must be obeyed. There can be at most one system in any single-particle state. Therefore

$$n_i = 0, 1 \qquad ; \text{ fermions} \tag{7.9}$$

The grand partition function is readily calculated for either case.

7.2.1 *Grand Partition Function*

The grand partition function with quantum statistics is given by[3]

$$(\text{G.P.F.}) = \sum_{\{n\}} e^{-E/k_B T}\, e^{N\mu/k_B T} \tag{7.10}$$

Since the constraints of fixed $E = \sum_i \varepsilon_i n_i$ and fixed $N = \sum_i n_i$ have been eliminated, there are no longer any restrictions on the occupation numbers, except those imposed by the statistics. The sum over all distributions $\{n\} = (n_1, n_2, n_3, \cdots, n_\infty)$ reduces to *independent sums* over the individual occupation numbers. Thus for the grand partition function with quantum statistics

$$\begin{aligned}(\text{G.P.F.}) &= \sum_{n_1}\sum_{n_2}\sum_{n_3}\cdots e^{-(\sum_i n_i \varepsilon_i)/k_B T}\,\lambda^{(\sum_i n_i)} \\ &= \sum_{n_1}(\lambda e^{-\varepsilon_1/k_B T})^{n_1}\sum_{n_2}(\lambda e^{-\varepsilon_2/k_B T})^{n_2}\sum_{n_3}(\lambda e^{-\varepsilon_3/k_B T})^{n_3}\cdots\end{aligned}$$

$$; \text{ quantum statistics} \tag{7.11}$$

Hence we have the following results:

7.2.2 *Bose Statistics*

In the case of Bose statistics, the grand partition function is given by Eq. (7.11) as

$$(\text{G.P.F.}) = \sum_{n_1=0}^{\infty}\left(\lambda e^{-\varepsilon_1/k_B T}\right)^{n_1}\sum_{n_2=0}^{\infty}\left(\lambda e^{-\varepsilon_2/k_B T}\right)^{n_2}\cdots \tag{7.12}$$

---

[3] Note that, consistent with the previous discussion, each term in the summand of the grand partition function appears with unit strength in quantum statistics; see also Prob. 6.3.

Hence

$$(\text{G.P.F.})_{\text{Bose}} = \prod_i \left(\frac{1}{1 - \lambda e^{-\varepsilon_i/k_\text{B}T}} \right) \tag{7.13}$$

7.2.3 *Fermi Statistics*

In the case of Fermi statistics, the grand partition function follows from Eq. (7.11) as

$$(\text{G.P.F.}) = \sum_{n_1=0}^{1} \left(\lambda e^{-\varepsilon_1/k_\text{B}T} \right)^{n_1} \sum_{n_2=0}^{1} \left(\lambda e^{-\varepsilon_2/k_\text{B}T} \right)^{n_2} \cdots \tag{7.14}$$

Hence

$$(\text{G.P.F.})_{\text{Fermi}} = \prod_i \left(1 + \lambda e^{-\varepsilon_i/k_\text{B}T} \right) \tag{7.15}$$

7.2.4 *Distribution Numbers*

The distribution numbers can be obtained by computing the total particle number N from the second of Eqs. (6.28)

$$N = \lambda \frac{\partial}{\partial \lambda} \ln (\text{G.P.F.}) \tag{7.16}$$

and then defining

$$N \equiv \sum_i n_i^\star \tag{7.17}$$

It follows from Eqs. (7.6), (7.13), and (7.15) that

$$\begin{aligned} n_i^\star &= \lambda e^{-\varepsilon_i/k_\text{B}T} && ;\ \text{Boltzmann} \\ &= \frac{\lambda e^{-\varepsilon_i/k_\text{B}T}}{1 - \lambda e^{-\varepsilon_i/k_\text{B}T}} && ;\ \text{Bose-Einstein} \\ &= \frac{\lambda e^{-\varepsilon_i/k_\text{B}T}}{1 + \lambda e^{-\varepsilon_i/k_\text{B}T}} && ;\ \text{Fermi-Dirac} \end{aligned} \tag{7.18}$$

Several comments:

- Here we give the quantum statistics their full names (see [Bose (1924); Einstein (1924)] and [Fermi (1926); Dirac (1926)]).

- Within the framework of the grand canonical ensemble, the quantum statistics calculations have been carried out *exactly* for identical, independent, non-localized systems;
- If Eq. (7.4) is satisfied, then the three expressions in Eqs. (7.18) are identical;
- *Thus the distributions all reduce to the Boltzmann distribution in the limit* $\lambda \ll 1$, *where classical statistics holds;*
- *This, ultimately, is the underlying justification for the classical statistics of identical, independent non-localized systems;*
- The only difference between Bose and Fermi statistics is the sign in the denominator of the distribution numbers, but this makes a world of difference in the physical applications.

7.2.5 *Energy*

The total energy follows from the first of Eqs. (6.28)

$$E = k_{\rm B}T^2 \frac{\partial}{\partial T} \ln(\text{G.P.F.}) \tag{7.19}$$

In all three cases, it can be written in the following fashion (see Prob. 7.2)

$$E = \sum_i \varepsilon_i n_i^\star \tag{7.20}$$

7.3 Bosons

As a first application of Bose statistics, consider electromagnetic radiation in a cavity, which forms a gas of massless photons (Fig. 7.1).

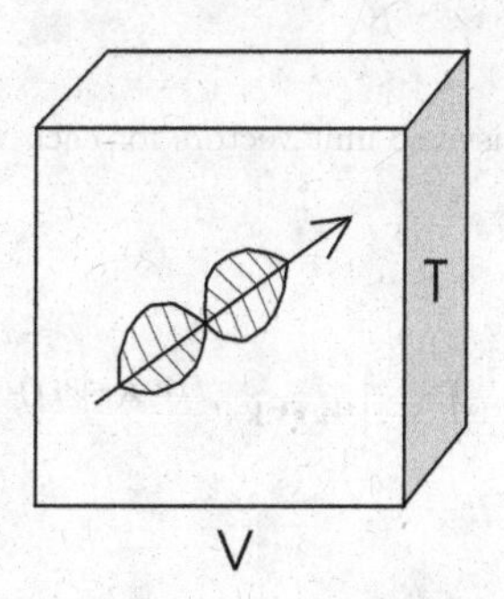

Fig. 7.1 Electromagnetic radiation in a large cavity of volume V with periodic boundary conditions. The walls of the cavity are held at a temperature T.

7.3.1 *Electromagnetic Radiation*

Maxwell's equations, written in terms of the vector potential $\mathbf{A}(\mathbf{x}, t)$, in the Coulomb gauge, read

$$\Box\mathbf{A} = 0 \qquad ; \text{ Maxwell's equations}$$
$$\boldsymbol{\nabla} \cdot \mathbf{A} = 0 \tag{7.21}$$

where $\Box = \nabla^2 - (1/c^2)\partial^2/\partial t^2$ is the wave operator, and c is the speed of light. The electric and magnetic fields are given in terms of the vector potential by

$$\mathbf{E} = -\frac{1}{c}\frac{\partial \mathbf{A}}{\partial t}$$
$$\mathbf{B} = \boldsymbol{\nabla} \times \mathbf{A} \tag{7.22}$$

7.3.1.1 *Normal Modes*

The normal-mode solutions to Maxwell's equations are transversely polarized plane waves. We impose the periodic boundary conditions of Eqs. (5.24)–(5.25).[4] Introduce two transverse unit vectors for each $\mathbf{k}$, as illustrated in Fig. 7.2. The vector potential can be expanded in terms of

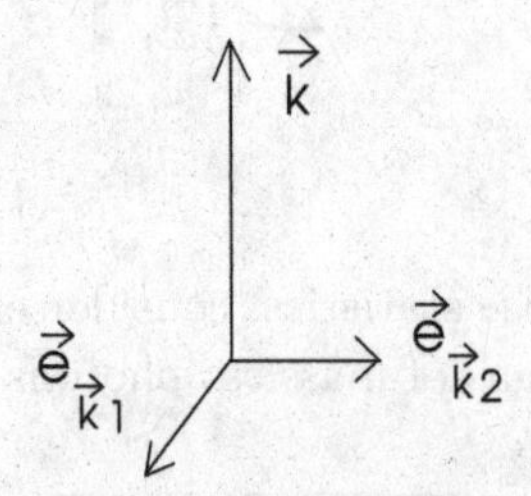

Fig. 7.2 Two transverse unit vectors for each wavenumber $\mathbf{k}$.

these normal modes as[5]

$$\mathbf{A}(\mathbf{x}, t) = \sum_{\mathbf{k}} \sum_{s=1}^{2} \left(\frac{2\pi\hbar c^2}{\omega_k V}\right)^{1/2} \left[a_{\mathbf{k}s}\mathbf{e}_{\mathbf{k}s}e^{i(\mathbf{k}\cdot\mathbf{x}-\omega_k t)} + a_{\mathbf{k}s}^{\dagger}\mathbf{e}_{\mathbf{k}s}e^{-i(\mathbf{k}\cdot\mathbf{x}-\omega_k t)}\right] \tag{7.23}$$

[4]We know that the *counting* of modes will not depend on the boundary conditions.
[5]This is in c.g.s. units (see appendix K of [Walecka (2008)]).

Here $a^\dagger_{\mathbf{k}s}$ denotes the hermitian conjugate of $a_{\mathbf{k}s}$, and

$$\omega_k = |\mathbf{k}|c \tag{7.24}$$

The total energy in the electromagnetic field is given by

$$E = \frac{1}{8\pi}\int_{\text{box}} (\mathbf{E}^2 + \mathbf{B}^2)\, d^3x \tag{7.25}$$

It is now a standard exercise to substitute Eqs. (7.22)–(7.23) into Eq. (7.25), do the integrals of the plane waves over the box, use the orthonormality of the unit vectors, and obtain the following normal-mode expansion[6]

$$E = \sum_{\mathbf{k}}\sum_{s=1}^{2} \hbar\omega_k \frac{1}{2}\left(a^\dagger_{\mathbf{k}s}a_{\mathbf{k}s} + a_{\mathbf{k}s}a^\dagger_{\mathbf{k}s}\right) \tag{7.26}$$

This represents an infinite set of uncoupled simple harmonic oscillators, and the oscillators are quantized in the canonical fashion. The quantized energy of the assembly is

$$E_{\text{tot}} = \sum_{\mathbf{k}}\sum_{s=1}^{2} \hbar\omega_k \left(n_{\mathbf{k}s} + \frac{1}{2}\right) \qquad ; \; n_{\mathbf{k}s} = 0, 1, 2, \cdots, \infty \tag{7.27}$$

The quanta are *photons*, each with energy $\varepsilon_{\mathbf{k}s} = \hbar\omega_k$ and transverse polarization s. Consistent with our usage, we refer to the photons as the systems. We always measure relative to the state with no photons present, the *vacuum*, and so the thermodynamic energy is actually the *difference*

$$E = E_{\text{tot}} - E_{\text{vac}} = \sum_{\mathbf{k}}\sum_{s=1}^{2} \hbar\omega_k\, n_{\mathbf{k}s} \tag{7.28}$$

7.3.1.2 *Chemical Potential*

Since photons can be created and destroyed, there is no longer any physical requirement that the total number of systems N_γ be *conserved*

$$N_\gamma = \sum_{\mathbf{k}}\sum_{s=1}^{2} n_{\mathbf{k}s} \qquad ; \; \underline{\text{not}} \text{ conserved} \tag{7.29}$$

[6]See Prob. 7.1.

This has the immediate consequence that in equilibrium the photon chemical potential must *vanish*

$$\mu_\gamma = 0 \qquad ; \text{ photon chemical potential} \qquad (7.30)$$

The absolute photon activity is therefore unity

$$\lambda_\gamma = e^{\mu_\gamma / k_B T} = 1 \qquad (7.31)$$

The condition $\lambda \ll 1$ for classical statistics is *never* satisfied for photons; one is *always* in the regime of quantum statistics for these systems.

To prove that $\mu_\gamma = 0$, compute the Helmholtz free energy $A(T, V, N_\gamma)$ for an assembly of N_γ photons. The general thermodynamic criterion for equilbrium at fixed (T, V) is that the Helmholtz free energy is *minimized*, and

$$(\delta A)_{T,V} = 0 \qquad ; \text{ equilbrium} \qquad (7.32)$$

However

$$(\delta A)_{T,V} = \left(\frac{\partial A}{\partial N_\gamma}\right)_{T,V} \delta N_\gamma \qquad (7.33)$$

Thus

$$\left(\frac{\partial A}{\partial N_\gamma}\right)_{T,V} \delta N_\gamma = 0 \qquad (7.34)$$

Since δN_γ is unconstrained, it follows that

$$\mu_\gamma = \left(\frac{\partial A}{\partial N_\gamma}\right)_{T,V} = 0 \qquad (7.35)$$

This is the stated result.[7]

7.3.1.3 *Spectral Weight*

The counting of normal modes proceeds exactly as in chapter 5. The only new feature is that there are now two transverse polarizations for each

[7] As in almost all our applications of statistical mechanics, we say nothing about the *mechanism* by which thermodynamic equilibrium is achieved; in this case it is by emission and absorption of photons in the walls of the container held at temperature T. All we ask is that statistical equilibrium be established in our assembly, in this case, the photon gas in the cavity.

wavenumber **k**, and thus there is a *degeneracy factor* of $g_s = 2$ in the counting of states[8]

$$\sum_i = g_s \sum_{n_x}\sum_{n_y}\sum_{n_z} \rightarrow \frac{g_s V}{(2\pi)^3}\int_{\text{all k}} d^3k$$

$$= \frac{4\pi g_s V}{c^3}\int_0^\infty \nu^2 d\nu \qquad ; \ g_s = 2 \qquad (7.36)$$

Here $\omega = |\mathbf{k}|c = 2\pi\nu$. Thus the *spectral weight* for electromagnetic radiation in a cavity is

$$\frac{1}{V}g(\nu) = \frac{8\pi\nu^2}{c^3} \qquad ; \text{ spectral weight} \qquad (7.37)$$

7.3.1.4 *Equation of State*

The spectral weight can be combined with the photon energy of $\varepsilon_i = \hbar\omega_i = h\nu_i$, and the Bose-Einstein distribution of occupation numbers $n_i^\star = (e^{\varepsilon_i/k_\mathrm{B}T} - 1)^{-1}$ obtained from the second of Eqs. (7.18) with $\lambda = 1$, to give the total internal thermodynamic energy of the electromagnetic radiation in a cavity

$$E = k_\mathrm{B}T^2\frac{\partial}{\partial T}\ln(\text{G.P.F.}) = \sum_i \varepsilon_i n_i^\star$$

$$\frac{E}{V} = \frac{8\pi}{c^3}\int_0^\infty \frac{h\nu}{e^{h\nu/k_\mathrm{B}T}-1}\nu^2 d\nu \qquad ; \text{ photon energy} \qquad (7.38)$$

The integrand in Eq. (7.38) is

$$\frac{dE_\gamma}{V} = \frac{8\pi\nu^2}{c^3}\frac{h\nu}{e^{h\nu/k_\mathrm{B}T}-1}d\nu$$

$$\equiv u(\nu,T)\,d\nu \qquad (7.39)$$

where the energy density $u(\nu, T)$, defined in the second line, is now the electromagnetic energy per unit volume per unit frequency interval

$$u(\nu,T) = \frac{8\pi\nu^2}{c^3}\frac{h\nu}{e^{h\nu/k_\mathrm{B}T}-1} \qquad ; \text{ energy density} \qquad (7.40)$$

This has two interesting limits:

---

[8]Compare Eq. (5.36), where there are two transverse modes and one longitudinal mode for the sound waves in a crystal.

(1) At high temperature, or long wavelength, where $h\nu/k_BT \ll 1$

$$u(\nu,T) = \frac{8\pi\nu^2}{c^3}k_BT \qquad ; \frac{h\nu}{k_BT} \ll 1 \tag{7.41}$$

(2) At low temperature, or short wavelength, where $h\nu/k_BT \gg 1$

$$u(\nu,T) = \frac{8\pi\nu^2}{c^3}h\nu\, e^{-h\nu/k_BT} \qquad ; \frac{h\nu}{k_BT} \gg 1 \tag{7.42}$$

The integrated energy density in the cavity follows from Eq. (7.38) as[9]

$$\begin{aligned} \frac{E}{V} &= \frac{(k_BT)^4}{\pi^2(\hbar c)^3}\int_0^\infty \frac{x^3\,dx}{e^x-1} \\ &= \left[\frac{\pi^2 k_B^4}{15(\hbar c)^3}\right]T^4 \qquad ; \text{ photon energy} \end{aligned} \tag{7.43}$$

The constant volume heat capacity per unit volume is the derivative of this expression with respect to temperature

$$\frac{C_V}{V} = \left[\frac{4\pi^2 k_B^4}{15(\hbar c)^3}\right]T^3 \qquad ; \text{ heat capacity} \tag{7.44}$$

The equation of state of the photon gas is obtained directly from the grand partition function

$$\begin{aligned} \frac{PV}{k_BT} &= \ln(\text{G.P.F.}) = -\sum_i \ln\left(1 - e^{-h\nu_i/k_BT}\right) \\ &= -\frac{8\pi V}{c^3}\int_0^\infty \nu^2\,d\nu\,\ln\left(1 - e^{-h\nu/k_BT}\right) \end{aligned}$$

$$\frac{PV}{k_BT} = -\frac{V(k_BT)^3}{\pi^2(\hbar c)^3}\int_0^\infty x^2\,dx\,\ln\left(1-e^{-x}\right) \tag{7.45}$$

The integral can be partially integrated by introducing

$$\begin{aligned} u &= \ln\left(1-e^{-x}\right) \qquad ; dv = x^2\,dx \\ du &= \frac{1}{e^x-1} \qquad ; v = \frac{x^3}{3} \end{aligned} \tag{7.46}$$

[9]Note $\int_0^\infty x^3dx/(e^x-1) = \pi^4/15$. Any good symbolic manipulation program, for example Mathcad11, now evaluates such definite integrals. Remember also here that $\hbar = h/2\pi$.

with the result that

$$PV = \frac{1}{3}\frac{V(k_{\rm B}T)^4}{\pi^2(\hbar c)^3}\int_0^\infty \frac{x^3\,dx}{e^x - 1}$$
$$= \frac{1}{3}E \qquad ; \text{ equation of state of photon gas} \qquad (7.47)$$

Here the last line follows from a comparison with the first of Eqs. (7.43).

7.3.1.5 *Discussion*

We proceed to discuss these results:

- Equation (7.41) is the classical *Rayleigh-Jeans* expression for the electromagnetic energy density in a cavity. It is obtained by multiplying the number of normal modes of the electromagnetic field per unit volume in a frequency interval between ν and $\nu + d\nu$ by the classical equipartition result of an energy $k_{\rm B}T$ per normal mode;
- The classical result for this energy density leads to the *ultraviolet catastrophe.* In contrast to the normal modes in a crystal, where there is a frequency cut-off, there is no maximum frequency, or minimum wavelength, for electromagnetic radiation in a cavity; the integrated energy density arising from the Raleigh-Jeans expression is *infinite!*
- The electromagnetic energy density in a black body can be measured by observing the energy streaming out of a small hole in the side of it. As illustrated in Fig. (7.3), experiment indicates the exponential decrease exhibited in Eq. (7.42);[10]

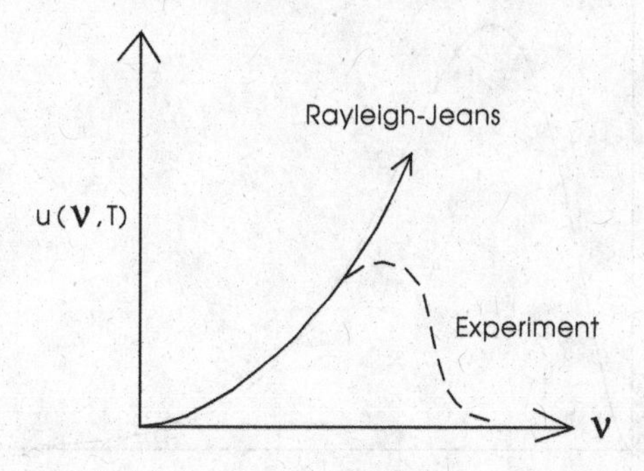

Fig. 7.3 Electromagnetic energy density in a cavity $u(\nu, T)$.

[10]The empirical exponential behavior is known as Wien's law; compare Prob. 7.5.

- It was Planck, at the beginning of the last century, who came up with the empirical *Planck distribution* for the quantity $\varepsilon_i n_i^\star$ appearing in Eq. (7.38)

$$\varepsilon_i n_i^\star = \frac{h\nu}{e^{h\nu/k_{\rm B}T} - 1} \qquad ; \text{ Planck distribution} \tag{7.48}$$

The quantity h is a phenomenological constant Planck introduced to fit the data, and this expression ultimately gave rise the the theory of quantum mechanics, which provides a marvelously successful description of the microscopic world. The Planck distribution here is just the Bose distribution of quantum statistics with $\lambda = 1$, weighted with the photon energy $h\nu$;

- Equation (7.40) can be re-cast in terms of the dimensionless variable $x \equiv h\nu/k_{\rm B}T$ according to

$$u(\nu, T) = h\left(\frac{k_{\rm B}T}{hc}\right)^3 u(x)$$
$$u(x) = \frac{8\pi x^3}{e^x - 1} \qquad ; \; x \equiv \frac{h\nu}{k_{\rm B}T} \tag{7.49}$$

The quantity $u(x)$ constitutes the *black-body spectrum.* It is drawn in Fig. 7.4. It yields, for example, an essentially perfect fit to the observed 2.73 °K cosmic microwave background (see [Ohanian (1995)]).

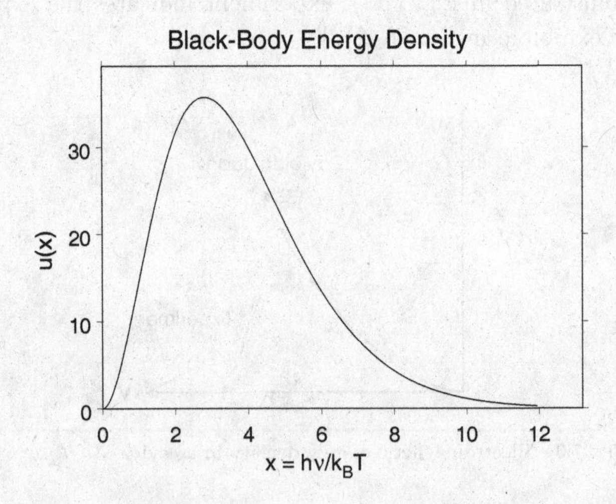

Fig. 7.4 Black-body energy density $u(\nu, T) = h(k_{\rm B}T/hc)^3 u(x)$ where $x \equiv h\nu/k_{\rm B}T$.

- Equation (7.43) for the integrated energy density, with its T^4 dependence, is known as the *Stefan-Boltzmann law*;
- Equation (7.44) indicates that one is *always in the T^3-regime* for the heat capacity of this photon gas;[11]
- Further discussion of the role of electromagnetic radiation in a cavity in the early days of modern physics can be found in [Walecka (2008)].

7.3.2 *Bose Condensation*

As a second application of bosons, consider an assembly of Bose atoms, the prime example being ^{4}He. The number of systems is now *conserved*, and in contrast to the case with photons, the number N cannot be varied in an unconstrained fashion in the Helmholtz free energy $A(T, V, N)$ of an isolated assembly. Consequently, the chemical potential does *not* vanish. Indeed, by varying the chemical potential, one varies the mean number of systems in an assembly in the grand canonical ensemble. We continue to focus on independent systems, and we shall find that Bose statistics, by itself, is sufficient to produce a phase transition.[12]

7.3.2.1 *Non-Relativistic Equation of State*

Consider N bosons of mass m in a volume V, and apply the periodic boundary conditions of Eqs. (5.24)–(5.25) [see Fig. 7.5]. Assume the assembly is in contact with a heat bath at temperature T.

The non-relativistic single-particle energy is given by

$$\varepsilon(k) = \frac{\mathbf{p}^2}{2m} = \frac{\hbar^2\mathbf{k}^2}{2m} \tag{7.50}$$

In a large volume V, with periodic boundary conditions, the sum over states can again be converted to an integral over wavenumbers using the relations

$$\sum_i \to g_s \int d^3n = \frac{g_s V}{(2\pi)^3} \int d^3k \qquad ; \text{ degeneracy } g_s \tag{7.51}$$

where g_s is the spin degeneracy of each level; in the case of a spin-zero boson such as ^{4}He, the degeneracy factor is $g_s = 1$.

---

[11]Compare Prob. 7.3; see also Prob. 7.28.

[12]The residual two-body interaction is sufficiently weak at the observed density of liquid ^{4}He that one can, to a first approximation, treat this as a collection of independent bosons. It is assumed in all these applications that any *internal* structure of the composite bosons (or fermions) is immaterial for the many-body physics of interest.

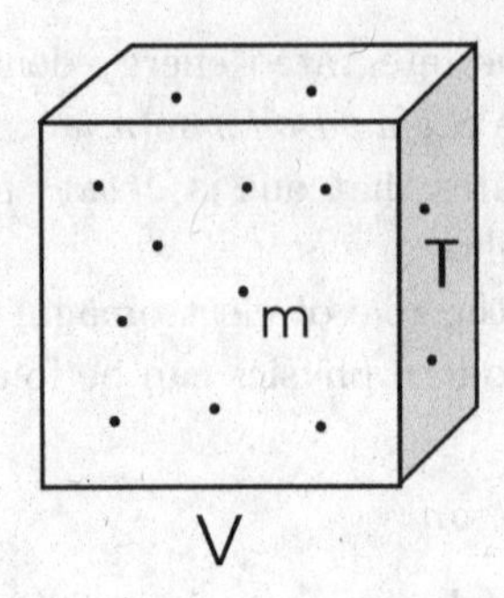

Fig. 7.5 Bose gas of N systems in a large volume V with periodic boundary conditions. The assembly is in contact with a heat bath at temperature T. The number density is N/V, and the systems have mass m.

When integrated over a function of $k \equiv |\mathbf{k}|$, the use of Eq. (7.50) allows the volume element in Eq. (7.51) to be expressed as

$$\frac{g_s V}{(2\pi)^3} 4\pi k^2 dk = \frac{g_s V}{2\pi^2}\left(\frac{2m}{\hbar^2}\right)^{3/2}\left(\frac{\varepsilon\, d\varepsilon}{2\sqrt{\varepsilon}}\right) = \frac{g_s V}{4\pi^2}\left(\frac{2m}{\hbar^2}\right)^{3/2}\sqrt{\varepsilon}\, d\varepsilon \tag{7.52}$$

The pressure of the Bose gas now follows from Eqs. (6.20) and (7.13)

$$\begin{aligned}\frac{PV}{k_{\mathrm{B}}T} &= \ln(\text{G.P.F.}) = -\sum_i \ln\left[1 - e^{(\mu-\varepsilon_i)/k_{\mathrm{B}}T}\right] \\ &= -\frac{g_s V}{4\pi^2}\left(\frac{2m}{\hbar^2}\right)^{3/2}\int_0^\infty \sqrt{\varepsilon}\, d\varepsilon\, \ln\left[1 - e^{(\mu-\varepsilon)/k_{\mathrm{B}}T}\right]\end{aligned} \tag{7.53}$$

This expression can be partially integrated using

$$\begin{aligned} du &= \sqrt{\varepsilon}\, d\varepsilon \quad &; \; v &= \ln\left[1 - e^{(\mu-\varepsilon)/k_{\mathrm{B}}T}\right] \\ u &= \frac{2}{3}\varepsilon^{3/2} \quad &; \; dv &= \frac{d\varepsilon}{k_{\mathrm{B}}T}\frac{1}{e^{(\varepsilon-\mu)/k_{\mathrm{B}}T} - 1}\end{aligned} \tag{7.54}$$

Thus

$$PV = \frac{2}{3}\frac{g_s V}{4\pi^2}\left(\frac{2m}{\hbar^2}\right)^{3/2}\int_0^\infty \varepsilon^{3/2} d\varepsilon\, \frac{1}{e^{(\varepsilon-\mu)/k_{\mathrm{B}}T} - 1} \tag{7.55}$$

The total energy for the Bose gas is obtained as

$$\begin{aligned} E &= \sum_i n_i^\star \varepsilon_i \\ n_i^\star &= \frac{1}{e^{(\varepsilon_i-\mu)/k_{\mathrm{B}}T} - 1}\end{aligned} \tag{7.56}$$

where the second line follows from the second of Eqs. (7.18). Thus

$$E = \frac{g_s V}{4\pi^2}\left(\frac{2m}{\hbar^2}\right)^{3/2} \int_0^\infty \varepsilon^{3/2} d\varepsilon \, \frac{1}{e^{(\varepsilon-\mu)/k_B T} - 1} \tag{7.57}$$

A comparison of Eqs. (7.55) and (7.57) produces the equation of state for the non-relativistic Bose gas

$$PV = \frac{2}{3}E \qquad ; \text{ N-R Bose gas} \tag{7.58}$$

This should be compared with the previous result for the relativistic photon gas of $PV = E/3$ in Eq. (7.47).

The number of particles is given by

$$\begin{aligned} N &= \sum_i n_i^\star \\ &= \frac{g_s V}{4\pi^2}\left(\frac{2m}{\hbar^2}\right)^{3/2} \int_0^\infty \varepsilon^{1/2} d\varepsilon \, \frac{1}{e^{(\varepsilon-\mu)/k_B T} - 1} \end{aligned} \tag{7.59}$$

Thus, in *summary*, for a collection of independent, non-relativistic bosons

$$\begin{aligned} P &= \frac{2}{3}\frac{g_s}{4\pi^2}\left(\frac{2m}{\hbar^2}\right)^{3/2} \int_0^\infty \frac{\varepsilon^{3/2}\, d\varepsilon}{e^{(\varepsilon-\mu)/k_B T} - 1} \qquad ; \text{ N-R bosons} \\ \frac{E}{V} &= \frac{g_s}{4\pi^2}\left(\frac{2m}{\hbar^2}\right)^{3/2} \int_0^\infty \frac{\varepsilon^{3/2}\, d\varepsilon}{e^{(\varepsilon-\mu)/k_B T} - 1} \\ \frac{N}{V} &= \frac{g_s}{4\pi^2}\left(\frac{2m}{\hbar^2}\right)^{3/2} \int_0^\infty \frac{\varepsilon^{1/2}\, d\varepsilon}{e^{(\varepsilon-\mu)/k_B T} - 1} \end{aligned} \tag{7.60}$$

7.3.2.2 *Transition Temperature*

We make several points concerning the above results:

(1) The last of Eqs. (7.60) can, in principle, be inverted to determine the chemical potential in term of a specified particle density N/V and temperature, $\mu(N/V, T)$;[13]

(2) Equations (7.59) only make sense if $\varepsilon - \mu \geq 0$, otherwise $n_i^\star < 0$ for some ε_i. The value $\varepsilon_0 = 0$ is an allowed single-particle energy in a big box with p.b.c. Thus the chemical potential must be non-positive for a non-relativistic Bose gas

$$\mu \leq 0 \qquad ; \text{ N-R Bose gas} \tag{7.61}$$

[13] See Prob. 7.8.

(3) Recall from Eqs. (6.31) and (7.3) that with Boltzmann statistics the chemical potential of a perfect gas takes the form[14]

$$\lambda_c = e^{\mu_c/k_B T} = \frac{N}{V}\left(\frac{h^2}{2\pi m k_B T}\right)^{3/2}$$
$$\frac{\mu_c}{k_B T} = \ln\left[\frac{N}{V}\left(\frac{h^2}{2\pi m k_B T}\right)^{3/2}\right] \tag{7.62}$$

This result is sketched as a function of T for fixed N/V in Fig. 7.6.

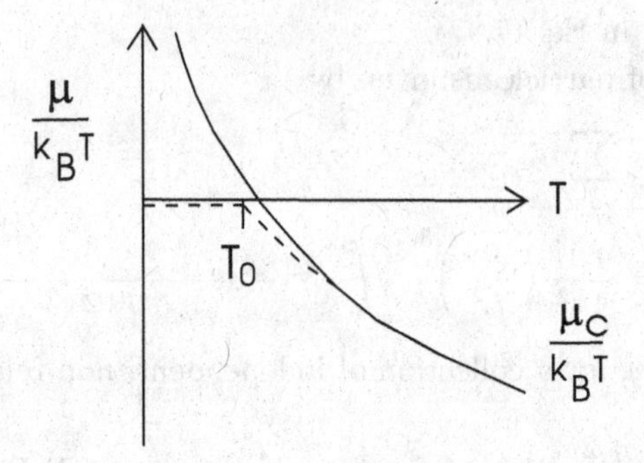

Fig. 7.6 Sketch of the chemical potential μ/k_BT of a Bose gas as a function of temperature for given N/V (dashed line); for $T < T_0$, one has $\mu = 0^-$. Here μ_c/k_BT is the result for a perfect gas with Boltzmann statistics in Eq. (7.62). (Compare Prob. 7.8.)

The actual result for the chemical potential with Bose statistics, which must continue to satisfy the condition in Eq. (7.61) as T is lowered, is indicated with the dashed line;

(4) Lower the temperature, and define T_0 to be that temperature where the chemical potential first reaches zero. From the last of Eqs. (7.60), this temperature is given by

$$\frac{N}{V} = \frac{g_s}{4\pi^2}\left(\frac{2m}{\hbar^2}\right)^{3/2}\int_0^\infty \frac{\varepsilon^{1/2}\,d\varepsilon}{e^{\varepsilon/k_B T_0}-1} \quad ;\text{ determines } T_0 \tag{7.63}$$

Introduce the dimensionless variable $x = \varepsilon/k_BT_0$, then

$$\frac{N}{V} = \frac{g_s}{4\pi^2}\left(\frac{2mk_BT_0}{\hbar^2}\right)^{3/2}\int_0^\infty \frac{x^{1/2}\,dx}{e^x-1} \tag{7.64}$$

[14]Remember that quantum statistics and Boltzmann statistics coincide at high T.

Since integrals of this type occur frequently in the analysis, we spend a little time on them. Consider

$$f(z) \equiv \int_0^\infty \frac{x^{z-1}\,dx}{e^x - 1} \qquad ; \ \mathrm{Re}\, z > 1 \tag{7.65}$$

Rewrite this as

$$\begin{aligned} f(z) &= \int_0^\infty x^{z-1}e^{-x}\left(\frac{1}{1-e^{-x}}\right)dx \\ &= \int_0^\infty x^{z-1}e^{-x}\left(\sum_{n=0}^{\infty} e^{-nx}\right)dx \\ &= \sum_{n=0}^{\infty}\int_0^\infty x^{z-1}e^{-(n+1)x}\,dx \\ &= \sum_{n=0}^{\infty}\frac{1}{(n+1)^z}\int_0^\infty t^{z-1}e^{-t}\,dt \end{aligned} \tag{7.66}$$

Now identify the Γ-function from Eq. (2.176), and relabel the summation variable $n+1 \to n$, to give

$$f(z) = \Gamma(z)\sum_{n=1}^{\infty}\frac{1}{n^z} \equiv \Gamma(z)\zeta(z) \qquad ; \ \text{Riemann zeta-function} \tag{7.67}$$

where $\zeta(z)$ is the Riemann zeta-function. In *summary*

$$\int_0^\infty \frac{x^{z-1}\,dx}{e^x-1} = \Gamma(z)\sum_{n=1}^{\infty}\frac{1}{n^z} \equiv \Gamma(z)\zeta(z) \qquad ; \ \mathrm{Re}\, z > 1 \tag{7.68}$$

A few useful numerical values are

$$\begin{aligned} \zeta(3/2) &= 2.612 \qquad &; \ \Gamma(3/2) = \sqrt{\pi}/2 \\ \zeta(5/2) &= 1.341 \qquad &; \ \Gamma(5/2) = 3\sqrt{\pi}/4 \end{aligned} \tag{7.69}$$

(5) Insertion of these results in Eq. (7.64) gives

$$\begin{aligned} \frac{N}{V} &= \frac{g_s}{4\pi^2}\left(\frac{2mk_{\mathrm{B}}T_0}{\hbar^2}\right)^{3/2}\int_0^\infty \frac{x^{1/2}\,dx}{e^x-1} \\ &= \frac{g_s}{4\pi^2}\left(\frac{2mk_{\mathrm{B}}T_0}{\hbar^2}\right)^{3/2}\Gamma(3/2)\zeta(3/2) \end{aligned} \tag{7.70}$$

The solution for T_0 is then

$$T_0 = \frac{\hbar^2}{2mk_\mathrm{B}}\left[\frac{4\pi^2}{g_s\,\Gamma(3/2)\zeta(3/2)}\right]^{2/3}\left(\frac{N}{V}\right)^{2/3}$$

$$T_0 = \frac{3.31}{g_s^{2/3}}\frac{\hbar^2}{mk_\mathrm{B}}\left(\frac{N}{V}\right)^{2/3} \tag{7.71}$$

(6) What happens for $T < T_0$? Clearly, all the bosons start going into the lowest-energy single-particle state.[15] For $\mu = 0$ and $T < T_0$, one gets something *less* than the specified N/V from the last of Eqs. (7.60). *Where have all the particles gone in this approach?* The difficulty arises in replacing a sum by an integral. Go back to

$$N = \sum_i n_i^\star$$

$$= \sum_i \frac{1}{e^{(\varepsilon_i-\mu)/k_\mathrm{B}T}-1} \tag{7.72}$$

Now observe:

- As $\mu \to 0$, all terms with $\varepsilon_i > 0$ go to a finite limit given by the integral in Eq. (7.59);
- The first term with $\varepsilon_0 = 0$ has been lost due to the $\sqrt{\varepsilon}$ in the continuum density of states in that integral;
- In fact, if $\mu = 0$ and $\varepsilon_0 = 0$, *the first term in the sum is infinite!*
- If μ is kept very small and negative, then the first term in the sum will be finite, and can make up the rest of the particles.[16] Therefore, take (see Fig. 7.6)

$$\mu = 0^- \qquad ;\ \text{for } T < T_0 \tag{7.73}$$

Thus for $T < T_0$:

---

[15] In the Hartree approximation, with spin-zero bosons, all the systems occupy the same lowest-energy single-particle spatial state at $T = 0$. This then provides a macroscopic quantum wave function which allows one to understand such diverse phenomena as quantized circulation, the spatial structure of vortices, and quantized flux in type-II superconductors arising from the flow of Cooper pairs near the Fermi surface (see, for example, [Walecka (2008)]).

[16] We forgo the mathematical niceties of this argument.

(a) If $\varepsilon > 0$, then the particle number is given by

$$\frac{dN_{\varepsilon>0}}{V} = \frac{g_s}{4\pi^2}\left(\frac{2m}{\hbar^2}\right)^{3/2} \frac{\sqrt{\varepsilon}\,d\varepsilon}{e^{\varepsilon/k_{\rm B}T}-1}$$

$$\frac{N_{\varepsilon>0}}{V} = \frac{g_s}{4\pi^2}\left(\frac{2mk_{\rm B}T}{\hbar^2}\right)^{3/2} \int_0^\infty \frac{x^{1/2}\,dx}{e^x-1}$$

$$= \frac{N}{V}\left(\frac{T}{T_0}\right)^{3/2} \tag{7.74}$$

where Eq. (7.70) has been used to obtain the last equality;

(b) If $\varepsilon = 0$, the contribution must make up the rest of the particles

$$\frac{N_{\varepsilon=0}}{V} = \frac{N}{V}\left[1-\left(\frac{T}{T_0}\right)^{3/2}\right] \qquad ;\ \text{Bose gas } T < T_0 \tag{7.75}$$

(7) The energy only receives a contribution from those particles with $\varepsilon_i > 0$, and therefore

$$\frac{E}{V} = \frac{g_s}{4\pi^2}\left(\frac{2mk_{\rm B}T}{\hbar^2}\right)^{3/2} k_{\rm B}T \int_0^\infty \frac{x^{3/2}\,dx}{e^x-1}$$

$$= \frac{g_s}{4\pi^2}\left(\frac{2mk_{\rm B}T}{\hbar^2}\right)^{3/2} k_{\rm B}T\,\Gamma(5/2)\zeta(5/2) \tag{7.76}$$

Thus, from the ratio to Eq. (7.70),

$$\frac{E}{N} = \frac{\Gamma(5/2)\zeta(5/2)}{\Gamma(3/2)\zeta(3/2)} k_{\rm B}T \left(\frac{T}{T_0}\right)^{3/2}$$

$$\frac{E}{N} = 0.770\, k_{\rm B}T\left(\frac{T}{T_0}\right)^{3/2} \qquad ;\ \text{Bose gas } T < T_0 \tag{7.77}$$

(8) The constant-volume heat capacity is therefore

$$\frac{C_V}{N} = \frac{5}{2}\left[0.770\, k_{\rm B}\left(\frac{T}{T_0}\right)^{3/2}\right] \qquad ;\ \text{Bose gas } T < T_0 \tag{7.78}$$

(9) The pressure for $T < T_0$ follows from the equation of state and Eq. (7.76)

$$P = \frac{2}{3}\frac{E}{V}$$

$$= \frac{2}{3}\frac{g_s}{4\pi^2}\left(\frac{2mk_{\rm B}T}{\hbar^2}\right)^{3/2} k_{\rm B}T\,\Gamma(5/2)\zeta(5/2) \tag{7.79}$$

Hence

$$P = 0.0851\, g_s \frac{m^{3/2}(k_B T)^{5/2}}{\hbar^3} \qquad ; \text{ Bose gas } T < T_0 \tag{7.80}$$

Three comments on these results:

- The heat capacity is proportional to $T^{3/2}$ below T_0;
- The pressure is proportional to $T^{5/2}$ below T_0. At $T = 0$, all the bosons are in the state with $\varepsilon_0 = 0$, which has $\mathbf{k} = 0$, and there is no pressure on the walls;
- Note the pressure below T_0 is independent of the particle density N/V.

7.3.2.3 *Discontinuity in Slope of* C_V

Let us investigate the thermodynamic properties of an assembly of bosons, in particular the constant-volume heat capacity, in the vicinity of T_0. Suppose there is sufficient energy for a small, positive $T - T_0 > 0$. In this region, the chemical potential is very small and negative, $\mu < 0$ (Fig. 7.6). Define the particle number calculated at this temperature with $\mu = 0$ by

$$N_0(T) \equiv \frac{g_s V}{4\pi^2}\left(\frac{2m}{\hbar^2}\right)^{3/2} \int_0^\infty \frac{\sqrt{\varepsilon}\, d\varepsilon}{e^{\varepsilon/k_B T} - 1} \qquad ; T > T_0 \tag{7.81}$$

This number will be larger than the actual specified number of particles N, but reduces to N at T_0. Thus

$$\frac{N_0(T)}{N_0(T_0)} = \frac{N_0(T)}{N} = \left(\frac{T}{T_0}\right)^{3/2} \qquad ; T > T_0 \tag{7.82}$$

Now take the difference with N calculated with the correct μ

$$N - N_0(T) = \frac{g_s V}{4\pi^2}\left(\frac{2m}{\hbar^2}\right)^{3/2} \int_0^\infty \sqrt{\varepsilon}\, d\varepsilon \left[\frac{1}{e^{(\varepsilon-\mu)/k_B T} - 1} - \frac{1}{e^{\varepsilon/k_B T} - 1}\right] \qquad ; \text{ determines } \mu \tag{7.83}$$

This equation determines $\mu(N/V, T)$. Since μ is small and negative, only small ε contributes significantly in the integrand, otherwise the two terms in square brackets cancel. The term in square brackets can then be expanded as

$$\left[\frac{1}{e^{(\varepsilon-\mu)/k_{\rm B}T}-1}-\frac{1}{e^{\varepsilon/k_{\rm B}T}-1}\right]=k_{\rm B}T\left[\frac{1}{\varepsilon-\mu}-\frac{1}{\varepsilon}\right]+\cdots$$
$$=\frac{k_{\rm B}T\mu}{\varepsilon(\varepsilon-\mu)}+\cdots \tag{7.84}$$

Now use

$$\int_0^\infty \frac{d\varepsilon}{\sqrt{\varepsilon}(\varepsilon+|\mu|)}=2\int_0^\infty \frac{dx}{x^2+|\mu|}=\frac{\pi}{\sqrt{|\mu|}} \qquad ; \varepsilon \equiv x^2 \tag{7.85}$$

Hence Eq. (7.83) becomes

$$N-N_0(T)=\frac{g_sV}{4\pi^2}\left(\frac{2m}{\hbar^2}\right)^{3/2}\pi k_{\rm B}T\frac{\mu}{\sqrt{|\mu|}} \qquad ; T>T_0 \tag{7.86}$$

Then from Eq. (7.70)

$$\frac{1}{N}[N-N_0(T)]=\frac{\pi}{\Gamma(3/2)\zeta(3/2)}\frac{k_{\rm B}T}{(k_{\rm B}T_0)^{3/2}}\frac{\mu}{\sqrt{|\mu|}} \tag{7.87}$$

The square of this result allows one to solve for $|\mu|$, and affixing the correct sign gives

$$-\mu=\left[\frac{\Gamma(3/2)\zeta(3/2)}{\pi}\right]^2\frac{(k_{\rm B}T_0)^3}{(k_{\rm B}T)^2}\left[\frac{N_0(T)-N}{N}\right]^2 \tag{7.88}$$

Now employ the last of Eqs. (7.82), and evaluate the factor in front at T_0,

$$-\mu=k_{\rm B}T_0\left[\frac{\Gamma(3/2)\zeta(3/2)}{\pi}\right]^2\left[\left(\frac{T}{T_0}\right)^{3/2}-1\right]^2 ; T>T_0 \tag{7.89}$$

This relation for the small, negative chemical potential is valid as $T\to T_0$ from above (Fig. 7.6).

Recall the following thermodynamic relations for the non-relativistic Bose gas

$$E=\frac{3}{2}PV$$
$$\left(\frac{\partial E}{\partial\mu}\right)_{T,V}=\frac{3}{2}\left[\frac{\partial(PV)}{\partial\mu}\right]_{T,V}=\frac{3}{2}N \tag{7.90}$$

Here the last equality follows from Eqs. (6.18). Thus

$$
\begin{aligned}
E &= E_0 + \frac{3}{2}N\mu \qquad &; T > T_0 \\
E &= E_0 \qquad &; T < T_0
\end{aligned}
\tag{7.91}
$$

where E_0 is the energy computed with $\mu = 0$ at a given (T, V); it is just the result in Eqs. (7.76)

$$
E_0 = \frac{g_s V}{4\pi^2}\left(\frac{2mk_BT}{\hbar^2}\right)^{3/2} k_BT\,\Gamma(5/2)\zeta(5/2) \tag{7.92}
$$

With the use of Eq. (7.89), Eqs. (7.91) become

$$
\begin{aligned}
E(T) &= E_0(T) - \frac{3}{2}Nk_BT_0\left[\frac{\Gamma(3/2)\zeta(3/2)}{\pi}\right]^2\left[\left(\frac{T}{T_0}\right)^{3/2} - 1\right]^2 \qquad &; T > T_0 \\
E &= E_0(T) \qquad &; T < T_0
\end{aligned}
\tag{7.93}
$$

Now fix (N, V), and note that $E_0(T)$ is smooth and continuous through T_0. Then

(1) $E(T)$ is continuous at T_0;
(2) The constant-volume heat capacity $C_V = (\partial E/\partial T)_{N,V}$ is continuous at T_0;
(3) The *derivative* of the heat capacity is *discontinuous* at T_0.

The situation is illustrated in Fig. 7.7.

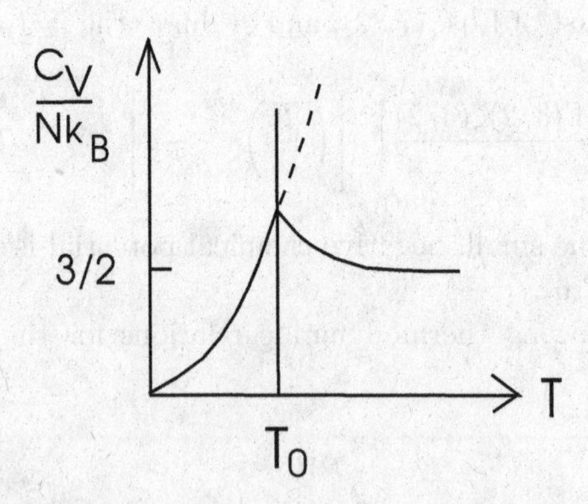

Fig. 7.7 Constant-volume heat capacity $C_V = (\partial E/\partial T)_{N,V}$ of the Bose gas. (Compare Prob. 7.7.)

To compute the change in slope of the heat capacity at T_0 in Fig. 7.7 we need to calculate

$$\Delta\left[\frac{\partial C_V}{\partial T}\right] = \Delta\left[\frac{\partial^2 E(T)}{\partial T^2}\right]$$
$$= -\frac{3}{2}\frac{Nk_{\rm B}T_0}{T_0^2}\left[\frac{\Gamma(3/2)\zeta(3/2)}{\pi}\right]^2\left[\frac{d^2}{dx^2}(x^{3/2}-1)^2\right]_{x=1} \qquad (7.94)$$

where $x \equiv T/T_0$. The required derivative is readily evaluated as

$$\left[\frac{d^2}{dx^2}(x^{3/2}-1)^2\right]_{x=1} = \left[2\frac{3}{2}\frac{d}{dx}x^{1/2}(x^{3/2}-1)\right]_{x=1} = 2\frac{3}{2}\frac{3}{2} \qquad (7.95)$$

Hence

$$\Delta\left[\frac{\partial C_V}{\partial T}\right] = -2\left(\frac{3}{2}\right)^3\left[\frac{\Gamma(3/2)\zeta(3/2)}{\pi}\right]^2\frac{Nk_{\rm B}}{T_0}$$
$$= -3.66\frac{Nk_{\rm B}}{T_0} \qquad (7.96)$$

Several comments:

- The assembly of independent bosons clearly exhibits a *phase transition* at T_0, where the second derivative of the energy with respect to temperature is discontinuous;
- Below T_0, the particles macroscopically occupy the lowest-energy single-particle state, with $\mathbf{k} = 0$;
- The assembly becomes ordered in *momentum space*, and not in coordinate space;
- This phenomenon is known as *Bose-Einstein condensation*

7.3.2.4 *Liquid* ^{4}He

Liquid ^{4}He forms a relatively low-density assembly of spin-zero bosons.[17] At atmospheric pressure, it remains a liquid down to $T = 0$. The observed mass density of liquid ^{4}He can be used in Eq. (7.71) to obtain a prediction for the transition temperature for this assembly

$$\rho_{\rm He} = 0.145\,\frac{\rm gm}{\rm cm^3} \qquad ;\ \text{observed mass density of liquid } ^4\text{He}$$
$$T_0 = 3.1\,^{\rm o}{\rm K} \qquad ;\ \text{predicted value} \qquad (7.97)$$

[17] See [Fetter and Walecka (2003)] for a discussion of liquid ^{4}He.

The above picture indicates that below T_0, the Bose gas should consist of two components, one a normal gas, and one a condensate that carries no energy.

Now it is a series of experimental facts that:

- Liquid ^{4}He exhibits a phase transition from He-I to He-II at a lambda-point temperature at zero pressure of

$$T_\lambda = 2.17\,^\circ\text{K} \qquad ; \text{ observed value} \tag{7.98}$$

- Below this temperature, He-II acts as a mixture of two fluids, a normal fluid and a *superfluid*;
- The superfluid has no heat capacity or viscosity;
- The fraction of normal fluid vanishes as $T \to 0$.

The ideal Bose gas provides a qualitative picture of what is going on in liquid ^{4}He; however, in contrast to the behavior illustrated in Fig. 7.7, the heat capacity of liquid ^{4}He becomes *logarithmically infinite* at the λ-point.

Spatially confined, laser-cooled Bose condensates from a variety of materials currently form the basis for an important branch of physics research [Colorado (2011)].

7.4 Fermions

We turn next to the quantum statistics of independent fermions. There are many important applications here, as this provides a good first approximation to assemblies as diverse as

- Electrons in metals
- Electrons in atoms
- Nucleons in nuclei
- Liquid ^{3}He
- Electrons in white dwarf stars
- Quarks in a quark-gluon plasma
- *etc.*

The analysis follows as in the Bose case.

7.4.1 *General Considerations*

The grand partition function and distribution numbers are given by Eq. (7.15) and the last of Eqs. (7.18), where $\lambda = e^{\mu/k_{\rm B}T}$

$$(\text{G.P.F.}) = \prod_i \left[1 + e^{(\mu-\varepsilon_i)/k_{\rm B}T}\right]$$

$$n_i^\star = \frac{1}{e^{(\varepsilon_i-\mu)/k_{\rm B}T} + 1} \tag{7.99}$$

As before, we assume non-relativistic single-particle energies and a big box with the periodic boundary conditions of Eqs. (5.24)–(5.25)[18]

$$\varepsilon(k) = \frac{\mathbf{p}^2}{2m} = \frac{\hbar^2\mathbf{k}^2}{2m} \tag{7.100}$$

The sum over states then follows as in Eqs. (7.51)–(7.52)

$$\sum_i \to \frac{g_s V}{4\pi^2}\left(\frac{2m}{\hbar^2}\right)^{3/2} \int \sqrt{\varepsilon}\, d\varepsilon \tag{7.101}$$

For spin-1/2 fermions, the spin degeneracy is $g_s = 2$

$$g_s = 2 \qquad ; \text{ spin-1/2} \tag{7.102}$$

7.4.1.1 *Non-Relativistic Equation of State*

The energy follows from Eqs. (7.19)–(7.20)[19]

$$\begin{aligned} E &= k_{\rm B}T^2 \frac{\partial}{\partial T} \ln(\text{G.P.F.}) = \sum_i \varepsilon_i n_i^\star \\ &= \frac{g_s V}{4\pi^2}\left(\frac{2m}{\hbar^2}\right)^{3/2} \int_0^\infty \frac{\varepsilon^{3/2}\, d\varepsilon}{e^{(\varepsilon-\mu)/k_{\rm B}T} + 1} \end{aligned} \tag{7.103}$$

The pressure follows as in Eq. (7.53)

$$\begin{aligned} PV &= k_{\rm B}T \ln(\text{G.P.F.}) \\ &= k_{\rm B}T\, \frac{g_s V}{4\pi^2}\left(\frac{2m}{\hbar^2}\right)^{3/2} \int_0^\infty \sqrt{\varepsilon}\, d\varepsilon \ln\left[1 + e^{(\mu-\varepsilon)/k_{\rm B}T}\right] \end{aligned} \tag{7.104}$$

[18]Compare Fig. 7.5.
[19]Remember (V, λ) are kept fixed here.

This last expression can be partially integrated with

$$
\begin{aligned}
du &= \sqrt{\varepsilon}\, d\varepsilon & ;\ v &= \ln\left[1 + e^{(\mu-\varepsilon)/k_{\rm B}T}\right] \\
u &= \frac{2}{3}\varepsilon^{3/2} & ;\ dv &= -\frac{d\varepsilon}{k_{\rm B}T}\frac{1}{e^{(\varepsilon-\mu)/k_{\rm B}T}+1}
\end{aligned}
\tag{7.105}
$$

The result is

$$
\begin{aligned}
PV &= \frac{2}{3}\frac{g_s V}{4\pi^2}\left(\frac{2m}{\hbar^2}\right)^{3/2}\int_0^\infty \frac{\varepsilon^{3/2}\, d\varepsilon}{e^{(\varepsilon-\mu)/k_{\rm B}T}+1} \\
&= \frac{2}{3}E \qquad ;\ \text{N-R Fermi gas}
\end{aligned}
\tag{7.106}
$$

This is the equation of state of a non-relativistic Fermi gas. The particle number is

$$
\begin{aligned}
N &= \sum_i n_i^\star \\
&= \frac{g_s V}{4\pi^2}\left(\frac{2m}{\hbar^2}\right)^{3/2}\int_0^\infty \frac{\varepsilon^{1/2}\, d\varepsilon}{e^{(\varepsilon-\mu)/k_{\rm B}T}+1}
\end{aligned}
\tag{7.107}
$$

Thus, in *summary*, for a collection of independent, non-relativistic fermions

$$
\begin{aligned}
P &= \frac{2}{3}\frac{g_s}{4\pi^2}\left(\frac{2m}{\hbar^2}\right)^{3/2}\int_0^\infty \frac{\varepsilon^{3/2}\, d\varepsilon}{e^{(\varepsilon-\mu)/k_{\rm B}T}+1} \qquad ;\ \text{N-R fermions} \\
\frac{E}{V} &= \frac{g_s}{4\pi^2}\left(\frac{2m}{\hbar^2}\right)^{3/2}\int_0^\infty \frac{\varepsilon^{3/2}\, d\varepsilon}{e^{(\varepsilon-\mu)/k_{\rm B}T}+1} \\
\frac{N}{V} &= \frac{g_s}{4\pi^2}\left(\frac{2m}{\hbar^2}\right)^{3/2}\int_0^\infty \frac{\varepsilon^{1/2}\, d\varepsilon}{e^{(\varepsilon-\mu)/k_{\rm B}T}+1}
\end{aligned}
\tag{7.108}
$$

The only difference from the analogous relations for bosons in Eqs. (7.60) is the sign in the denominators!

As before, the last relation can, in principle, be inverted to obtain the chemical potential as a function of a specified density and temperature $\mu(N/V, T)$.

7.4.1.2 *Distribution Numbers*

Fix the particle density N/V. Now consider the distribution numbers $n^\star(\varepsilon)$ for the Fermi gas at different temperatures

$$n^\star(\varepsilon) = \frac{1}{e^{(\varepsilon-\mu)/k_{\rm B}T} + 1} \tag{7.109}$$

In contrast to the situation with bosons, the condition $n^\star \leq 1$ is *built in for all values of μ.* What do we expect μ to look like as a function of temperature for a given density N/V? The extention of Fig. 7.6 to the fermion case is indicated by the second dashed line in Fig. 7.8.

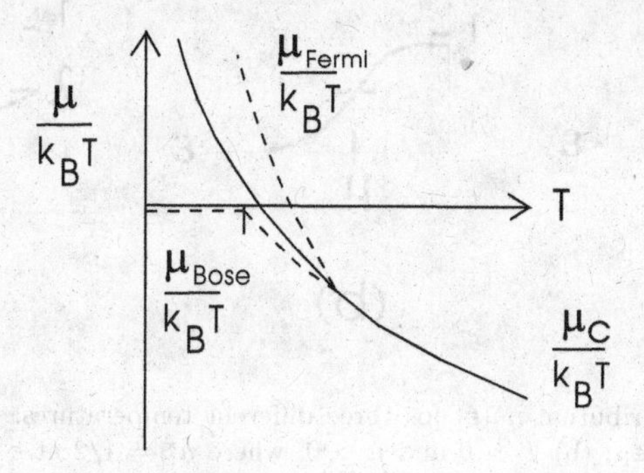

Fig. 7.8 Sketch of the chemical potentials $\mu/k_{\rm B}T$ of Bose and Fermi gases (dashed lines) as functions of temperature for given N/V. Here $\mu_c/k_{\rm B}T$ is again the result for a perfect gas with Boltzmann statistics. As T is lowered, the Fermi chemical potential $\mu_{\rm Fermi}$ becomes positive, and as $T \to 0$, it goes to the Fermi energy $\varepsilon_{\rm F}$.

Consider the three cases illustrated in Fig. 7.9 :

(1) For large T, one has $\mu/k_{\rm B}T \to -\infty$, and thus the particle distribution reduces to that of Boltzmann statistics [Fig. 7.9(c)]

$$n^\star(\varepsilon) = \lambda e^{-\varepsilon/k_{\rm B}T} = e^{\mu/k_{\rm B}T}e^{-\varepsilon/k_{\rm B}T} \qquad ; T \to \infty \tag{7.110}$$

(2) At lower temperature where μ becomes positive, the distribution $n^\star(\varepsilon)$ falls to the value $n^\star = 1/2$ when $\varepsilon = \mu$ [Fig. 7.9(b)];

(3) As $T \to 0$, the chemical potential approaches some positive value $\mu \equiv \varepsilon_{\rm F} > 0$

$$\mu \equiv \varepsilon_F \qquad ; T \to 0 \tag{7.111}$$

The particle distribution $n^\star(\varepsilon)$ then becomes a *step-function* [Fig. 7.9(a)]

$$\begin{aligned} n^\star(\varepsilon) &= 0 \qquad ; \varepsilon > \varepsilon_F \\ &= 1 \qquad ; \varepsilon < \varepsilon_F \end{aligned} \tag{7.112}$$

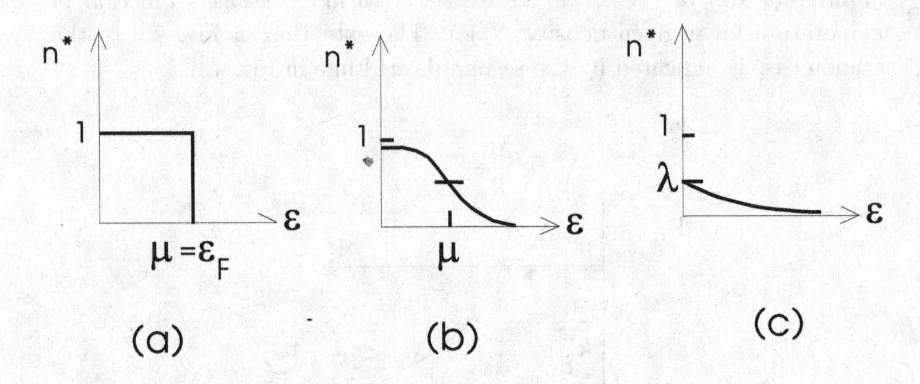

Fig. 7.9 Fermi distribution $n^\star(\varepsilon)$ for three different temperatures: (a) $T = 0$, where $n^\star = 1$ out to $\mu = \varepsilon_{\rm F}$; (b) $T > 0$ and $\mu > 0$, where $n^\star = 1/2$ at $\varepsilon = \mu$; (c) $T \to \infty$, where $\mu/k_{\rm B}T \to -\infty$ and $n^\star = \lambda = e^{\mu/k_{\rm B}T}$ at $\varepsilon = 0$.

7.4.1.3 *Zero Temperature*

Consider the zero-temperature case, where the assembly of independent fermions is in its ground state. Here the single-particle levels are occupied up to the Fermi energy as indicated in Fig. 7.10, with the Fermi wavenumber $k_{\rm F}$ defined by

$$\mu = \varepsilon_{\rm F} \equiv \frac{\hbar^2 k_{\rm F}^2}{2m} \qquad ; T = 0 \tag{7.113}$$

The counting of states is done most directly from Eq. (7.51)

$$\sum_i \to \frac{g_s V}{(2\pi)^3} \int d^3k \tag{7.114}$$

The total number of particles occupying the filled states in Fig. 7.10 is therefore

$$N = \frac{g_s V}{(2\pi)^3} \int_0^{k_F} d^3k \tag{7.115}$$

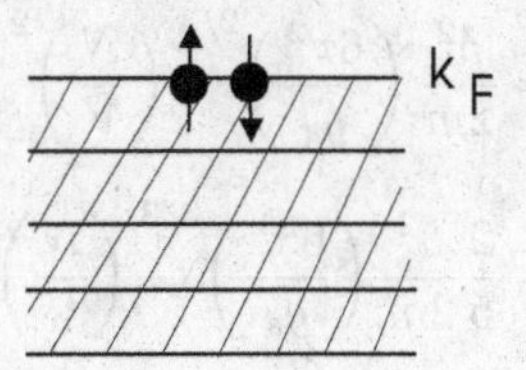

Fig. 7.10 Ground state of a spin-1/2 Fermi gas, where $g_s = 2$. The Fermi energy and Fermi wavenumber are related through Eq. (7.113).

Hence the particle density is

$$\frac{N}{V} = \frac{g_s}{(2\pi)^3} \frac{4\pi k_F^3}{3} = \frac{g_s k_F^3}{6\pi^2} \qquad ; \text{ particle density} \tag{7.116}$$

The total energy is obtained by summing the energies of all the occupied states

$$E = \frac{g_s V}{(2\pi)^3} \int_0^{k_F} \frac{\hbar^2 k^2}{2m} d^3k \tag{7.117}$$

Hence the energy density is

$$\begin{aligned} \frac{E}{V} &= \frac{3}{5} \frac{\hbar^2 k_F^2}{2m} \frac{g_s k_F^3}{6\pi^2} \\ &= \frac{3}{5} \varepsilon_F \left(\frac{N}{V} \right) \qquad ; \text{ energy density} \end{aligned} \tag{7.118}$$

The energy per particle in the Fermi gas is therefore 3/5 of the Fermi energy

$$\frac{E}{N} = \frac{3}{5} \varepsilon_F \qquad ; \text{ energy/particle} \tag{7.119}$$

The pressure follows from the equation of state

$$PV = \frac{2}{3} E = \frac{2}{5} N \varepsilon_F \tag{7.120}$$

The pressure is therefore related to the particle density by

$$P = \frac{2}{5} \varepsilon_F \left(\frac{N}{V} \right) \qquad ; \text{ pressure} \tag{7.121}$$

It is very useful to write everything in terms of the particle density, and thus

$$k_{\rm F} = \left(\frac{6\pi^2}{g_s}\right)^{1/3}\left(\frac{N}{V}\right)^{1/3}$$

$$\varepsilon_{\rm F} = \frac{\hbar^2}{2m}\left(\frac{6\pi^2}{g_s}\right)^{2/3}\left(\frac{N}{V}\right)^{2/3}$$

$$P = \frac{2}{5}\frac{\hbar^2}{2m}\left(\frac{6\pi^2}{g_s}\right)^{2/3}\left(\frac{N}{V}\right)^{5/3} \tag{7.122}$$

It is important to realize why a Fermi gas exerts a pressure at zero temperature.[20] The wavenumbers of all the occupied states must fit into the big box (in contrast to the Bose gas, the fermions cannot all occupy the $\mathbf{k} = 0$ state). If the size of the box is decreased slightly, each wavenumber increases slightly, and so does the energy of each filled level. Thus the total energy of the assembly increases as the volume of the assembly decreases, and by the first law of thermodynamics, this corresponds to a pressure.

7.4.2 *Low-Temperature C_V*

We have analyzed the Fermi gas at $T = 0$. The next task is to look at the thermodynamic properties of the Fermi gas at finite, but small T. To accomplish this, we carry out a consistent expansion in powers of $k_{\rm B}T/\mu$, where the chemical potential μ is only slightly smaller than the Fermi energy $\varepsilon_{\rm F}$.[21] Call

$$x \equiv \frac{(\varepsilon - \mu)}{k_{\rm B}T} \qquad ; \alpha \equiv \frac{\mu}{k_{\rm B}T} \tag{7.123}$$

Then from Eq. (7.106)

$$PV = \frac{2}{3}\frac{g_s V}{4\pi^2}\left(\frac{2m}{\hbar^2}\right)^{3/2}(k_{\rm B}T)^{5/2}\int_{-\mu/k_{\rm B}T}^{\infty}\frac{(x+\mu/k_{\rm B}T)^{3/2}\,dx}{e^x+1} \tag{7.124}$$

[20] One can make great use of the fact that this "quantum pressure" is directly proportional to a power of the local particle density in the ground state of a Fermi gas; see, for example, Probs. 7.14, 7.24–7.27. This is generically known as *Thomas-Fermi theory* [Thomas (1927); Fermi (1927)].

[21] See the result in Eq. (7.139).

Consider the integral

$$I(\alpha) \equiv \int_{-\alpha}^{\infty} \frac{(x+\alpha)^{3/2}\,dx}{e^x+1}$$

$$= \int_{-\alpha}^{0} \frac{(\alpha+x)^{3/2}\,dx}{e^x+1} + \int_{0}^{\infty} \frac{(\alpha+x)^{3/2}\,dx}{e^x+1} \tag{7.125}$$

We are interested in the behavior of $I(\alpha)$ as $\alpha \to \infty$.

Let $x \to -x$ in the first integral in the second line, then

$$I(\alpha) = \int_{0}^{\alpha} \frac{(\alpha-x)^{3/2}\,dx}{e^{-x}+1} + \int_{0}^{\infty} \frac{(\alpha+x)^{3/2}\,dx}{e^x+1} \tag{7.126}$$

In the first integral, write

$$\frac{1}{e^{-x}+1} = \frac{1+e^{-x}-e^{-x}}{e^{-x}+1} = 1 - \frac{1}{e^x+1} \tag{7.127}$$

Then

$$I(\alpha) = \int_{0}^{\alpha} (\alpha-x)^{3/2}dx + \int_{0}^{\infty} \frac{dx}{e^x+1}\left[(\alpha+x)^{3/2} - (\alpha-x)^{3/2}\right]$$

$$+ \int_{\alpha}^{\infty} \frac{(\alpha-x)^{3/2}dx}{e^x+1} \tag{7.128}$$

The first integral is immediately evaluated

$$\int_{0}^{\alpha} (\alpha-x)^{3/2}dx = \frac{2}{5}\alpha^{5/2} \tag{7.129}$$

Expand the term in brackets in the second integral

$$\left[(\alpha+x)^{3/2} - (\alpha-x)^{3/2}\right] = \alpha^{3/2}\left[1 + \frac{3}{2}\frac{x}{\alpha} + \cdots - \left(1 - \frac{3}{2}\frac{x}{\alpha} + \cdots\right)\right] \tag{7.130}$$

Then[22]

$$\int_{0}^{\infty} \frac{dx}{e^x+1}\left[(\alpha+x)^{3/2} - (\alpha-x)^{3/2}\right] = 3\alpha^{1/2}\int_{0}^{\infty} \frac{x\,dx}{e^x+1} + O(\alpha^{-3/2})$$

$$= \frac{\pi^2}{4}\alpha^{1/2} + O(\alpha^{-3/2}) \tag{7.131}$$

---

[22] Use $\int_0^\infty x\,dx/(e^x+1) = \pi^2/12$.

The last integral in Eq. (7.128) is exponentially small for large α [23]

$$\int_{\alpha}^{\infty} \frac{(\alpha - x)^{3/2} dx}{e^x + 1} = O(e^{-\alpha}) \tag{7.132}$$

A combination of these results gives

$$\int_{-\alpha}^{\infty} \frac{(x+\alpha)^{3/2}\, dx}{e^x + 1} = \frac{2}{5}\alpha^{5/2} + \frac{\pi^2}{4}\alpha^{1/2} + O(\alpha^{-3/2}) \qquad ; \alpha \to \infty \tag{7.133}$$

Re-expressed in terms of the chemical potential, the result for large $\mu/k_{\rm B}T$ is

$$\int_{-\mu/k_{\rm B}T}^{\infty} \frac{(x + \mu/k_{\rm B}T)^{3/2} dx}{e^x + 1} = \frac{1}{(k_{\rm B}T)^{5/2}} \left[\frac{2}{5}\mu^{5/2} + \frac{\pi^2}{4}(k_{\rm B}T)^2\mu^{1/2} + \cdots\right] \tag{7.134}$$

Substitution of this result into Eq. (7.124) provides the low-temperature expansion of the thermodynamic potential

$$PV = \frac{g_s V}{4\pi^2}\left(\frac{2m}{\hbar^2}\right)^{3/2} \frac{2}{3}\left[\frac{2}{5}\mu^{5/2} + \frac{\pi^2}{4}(k_{\rm B}T)^2\mu^{1/2} + \cdots\right] \qquad ; T \to 0 \tag{7.135}$$

This is a very useful result:

- It provides an explicit low-temperature expression for $PV(\mu, V, T)$;
- The correction terms are higher order in $(k_{\rm B}T/\mu)^2$
- The particle number is obtained immediately from Eq. (6.18)[24]

$$N = \left[\frac{\partial(PV)}{\partial\mu}\right]_{V,T} = \frac{g_s V}{4\pi^2}\left(\frac{2m}{\hbar^2}\right)^{3/2} \frac{2}{3}\mu^{3/2}\left[1 + \frac{\pi^2}{8}\left(\frac{k_{\rm B}T}{\mu}\right)^2 + \cdots\right] \tag{7.136}$$

  The inversion of this relation gives $\mu(N/V, T)$ at low temperature;
- The particle density is expressed in terms of $\varepsilon_{\rm F}$ through the second of Eqs. (7.122)

$$\frac{N}{V} = \frac{g_s}{4\pi^2}\left(\frac{2m}{\hbar^2}\right)^{3/2} \frac{2}{3}\varepsilon_{\rm F}^{3/2} \tag{7.137}$$

---

[23] That it is imaginary is irrelevant.

[24] Note carefully how the variables enter here, and in the following.

Substitution into Eq. (7.136) gives

$$\varepsilon_{\rm F} = \mu \left[1 + \frac{\pi^2}{8}\left(\frac{k_{\rm B}T}{\mu}\right)^2 + \cdots\right]^{2/3}$$

$$\Longrightarrow \qquad \mu = \varepsilon_{\rm F} \left[1 + \frac{\pi^2}{8}\left(\frac{k_{\rm B}T}{\mu}\right)^2 + \cdots\right]^{-2/3} \tag{7.138}$$

One iteration in $(k_{\rm B}T/\mu)^2$ then expresses the finite-temperature chemical potential in terms of the Fermi energy

$$\mu = \varepsilon_{\rm F} \left[1 - \frac{\pi^2}{12}\left(\frac{k_{\rm B}T}{\varepsilon_{\rm F}}\right)^2 + \cdots\right] \tag{7.139}$$

- The entropy is also obtained from the thermodynamic potential through Eqs. (6.18)

$$S = \left[\frac{\partial(PV)}{\partial T}\right]_{\mu,V}$$

$$= k_{\rm B}\frac{g_s V}{4\pi^2}\left(\frac{2m}{\hbar^2}\right)^{3/2}\frac{2}{3}\mu^{3/2}\left[\frac{\pi^2}{2}\left(\frac{k_{\rm B}T}{\mu}\right) + \cdots\right] \tag{7.140}$$

To leading order we have $\mu = \varepsilon_{\rm F}$, and thus with the aid of Eq. (7.137)

$$S = Nk_{\rm B}\frac{\pi^2}{2}\left(\frac{k_{\rm B}T}{\varepsilon_{\rm F}}\right) \qquad ; T \to 0 \tag{7.141}$$

This is a nice result; we have obtained the low-temperature entropy $S(N, V, T)$ of a Fermi gas, which is linear in T;

- The constant volume heat capacity follows from $S(N, V, T)$ through the first and second laws of thermodynamics

$$C_V = T\left(\frac{\partial S}{\partial T}\right)_{N,V}$$

$$= Nk_{\rm B}\frac{\pi^2}{2}\left(\frac{k_{\rm B}T}{\varepsilon_{\rm F}}\right) \qquad ; T \to 0 \tag{7.142}$$

C_V is also linear in the temperature. It is sketched as a function of temperature in Fig. 7.11. In contrast to the Bose case, there are no discontinuities here.

- The electronic contribution to the low-temperature specific heat of metals is small, since $\varepsilon_{\rm F}$ is a relatively large energy in that case (see, for example, [Davidson (2003)]).

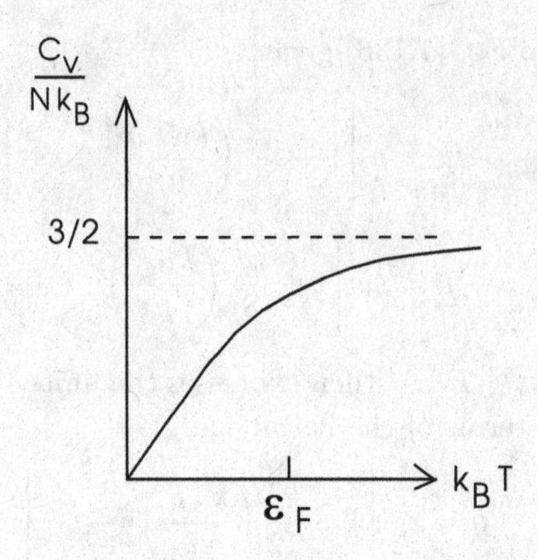

Fig. 7.11 Sketch of the constant volume heat capacity of a Fermi gas. (Compare Prob. 7.7.)

7.4.3 *Pauli Spin Paramagnetism*

Consider a Fermi gas with spin-1/2 in an external magnetic field (Fig. 7.12). There will be an additional interaction of the spin magnetic moment with the field of the form[25]

$$\begin{aligned} \varepsilon_{\text{mag}} &= -\boldsymbol{\mu} \cdot \mathbf{B} = -\mu_0 B \qquad ; \text{ spin } \uparrow \\ &= +\mu_0 B \qquad ; \text{ spin } \downarrow \end{aligned} \tag{7.143}$$

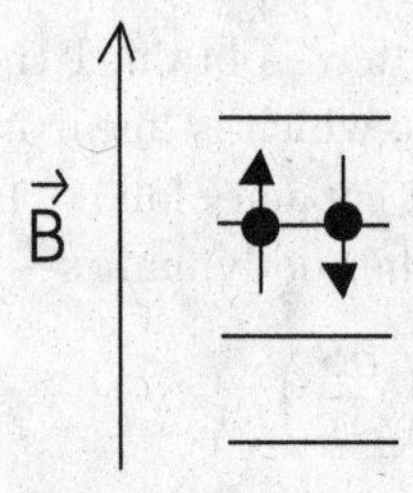

Fig. 7.12 Spin-1/2 Fermi gas in an external magnetic field.

For electrons

$$\mu_0 = \frac{e\hbar}{2m_e c} \qquad ; \text{ electrons with } e < 0 \tag{7.144}$$

[25]See [Pauli (2000)].

Here e is algebraic. For the assembly, the interaction hamiltonian takes the form

$$\begin{aligned} H' &= -\mu_0 B(N_\uparrow - N_\downarrow) \\ &= -\mu_0 B\left(\sum_i n_{i\uparrow} - \sum_i n_{i\downarrow}\right) \end{aligned} \tag{7.145}$$

We shall employ the grand canonical ensemble. How do we deal with the chemical potential? There are some options:

(1) One way is to say that there is one species of fermion (say, electron) that can go into any one of the different energy states. There is then just one chemical potential μ for the ensemble;
(2) Another way is to say that there are now *two* different species, spin-$\uparrow$ and spin-$\downarrow$, with *two* distinct chemical potentials $\mu_\uparrow$ and $\mu_\downarrow$.[26] From our previous discussion of chemical equilibrium in thermodynamics, we then know that

$$\mu_\uparrow = \mu_\downarrow \qquad ; \text{ chemical equilbrium} \tag{7.146}$$

7.4.3.1 *Grand Partition Function*

We shall compute the grand partition function for this problem using the first approach, with one species and a single chemical potential. We use the notation

$$\varepsilon_i \equiv \varepsilon(k) = \frac{\hbar^2 \mathbf{k}^2}{2m} \tag{7.147}$$

Because of the additional magnetic energy, the following replacement must be made in the assembly energy E in computing (G.P.F.)

$$\begin{aligned} E &= \sum_i n_{i\uparrow}\varepsilon_{i\uparrow} + \sum_i n_{i\downarrow}\varepsilon_{i\downarrow} \\ &\to \sum_i n_{i\uparrow}(\varepsilon_i - \mu_0 B) + \sum_i n_{i\downarrow}(\varepsilon_i + \mu_0 B) \end{aligned} \tag{7.148}$$

Here we explicitly exhibit the spin-up and spin-down components of the spatial state i. The particle number is similarly

$$N = \sum_i n_{i\uparrow} + \sum_i n_{i\downarrow} \tag{7.149}$$

[26]Compare Prob. 7.10.

The grand partition function is then given by

$$(\text{G.P.F.}) = \sum_{n_1}\sum_{n_2}\sum_{n_3}\cdots e^{-E/k_BT}e^{N\mu/k_BT}$$
$$= \sum_{n_{1\uparrow}}\sum_{n_{1\downarrow}}\sum_{n_{2\uparrow}}\sum_{n_{2\downarrow}}\cdots e^{-E/k_BT}e^{N\mu/k_BT} \tag{7.150}$$

Substitution of Eqs. (7.148)–(7.149), and evaluation of the sums for fermions, gives

$$(\text{G.P.F.})_B = \prod_i \left[1+e^{(\mu-\varepsilon_i+\mu_0 B)/k_BT}\right]\left[1+e^{(\mu-\varepsilon_i-\mu_0 B)/k_BT}\right] \tag{7.151}$$

The particle number again follows from the second of Eq. (6.28), and conversion of a sum to an integral with the aid of Eq. (7.101), gives[27]

$$\frac{N}{V} = \frac{1}{4\pi^2}\left(\frac{2m}{\hbar^2}\right)^{3/2}\int_0^\infty \sqrt{\varepsilon}\,d\varepsilon$$
$$\times\left\{\frac{1}{e^{[\varepsilon-(\mu+\mu_0 B)]/k_BT}+1}+\frac{1}{e^{[\varepsilon-(\mu-\mu_0 B)]/k_BT}+1}\right\} \tag{7.152}$$

7.4.3.2 *Magnetization*

The induced magnetization in the direction of $\mathbf{B}$ is calculated directly as

$$\mathcal{M} = \mu_0\left(\sum_i n_{i\uparrow} - \sum_i n_{i\downarrow}\right) \tag{7.153}$$

Thus

$$\frac{\mathcal{M}}{V} = \frac{\mu_0}{4\pi^2}\left(\frac{2m}{\hbar^2}\right)^{3/2}\int_0^\infty \sqrt{\varepsilon}\,d\varepsilon$$
$$\times\left\{\frac{1}{e^{[\varepsilon-(\mu+\mu_0 B)]/k_BT}+1}-\frac{1}{e^{[\varepsilon-(\mu-\mu_0 B)]/k_BT}+1}\right\} \tag{7.154}$$

The low-temperature behavior of the required integrals follows from the result in Eq. (7.136), where each spin component is now treated separately

$$\frac{N}{V} = \frac{1}{4\pi^2}\left(\frac{2m}{\hbar^2}\right)^{3/2}\frac{2}{3}\left\{(\mu+\mu_0 B)^{3/2}+(\mu-\mu_0 B)^{3/2}\right.$$
$$\left.+\frac{\pi^2}{8}(k_BT)^2\left[\frac{1}{(\mu+\mu_0 B)^{1/2}}+\frac{1}{(\mu-\mu_0 B)^{1/2}}\right]\right\} \tag{7.155}$$

[27]Note that here $g_s = 1$, since we are now explicitly including the spin components; they are no longer degenerate.

and

$$\frac{\mathcal{M}}{V} = \frac{\mu_0}{4\pi^2}\left(\frac{2m}{\hbar^2}\right)^{3/2}\frac{2}{3}\left\{(\mu+\mu_0 B)^{3/2} - (\mu-\mu_0 B)^{3/2} + \frac{\pi^2}{8}(k_{\rm B}T)^2\left[\frac{1}{(\mu+\mu_0 B)^{1/2}} - \frac{1}{(\mu-\mu_0 B)^{1/2}}\right]\right\} \tag{7.156}$$

These expressions can now be expanded in B to obtain the magnetic susceptibility.

(1) It is evident from Eq. (7.155) that N/V is even in B. Thus

$$\frac{N}{V} = \left(\frac{N}{V}\right)_0 + O(B^2) \tag{7.157}$$

To first order, one can set $B = 0$ in Eq. (7.155), reproducing the previous result for $\mu(N/V, T)$ in Eq. (7.139)

$$\mu = \varepsilon_{\rm F}\left[1 - \frac{\pi^2}{12}\left(\frac{k_{\rm B}T}{\varepsilon_{\rm F}}\right)^2\right] \qquad ; \; T \to 0 \tag{7.158}$$

(2) The expansion of Eq. (7.156) to order B gives

$$\frac{\mathcal{M}}{V} = \frac{\mu_0}{4\pi^2}\left(\frac{2m}{\hbar^2}\right)^{3/2}\frac{2}{3}\left[3\mu^{3/2} - \frac{\pi^2}{8}(k_{\rm B}T)^2\mu^{-1/2}\right]\left(\frac{\mu_0 B}{\mu}\right) \tag{7.159}$$

Hence

$$\frac{\mathcal{M}}{V} = \frac{1}{6\pi^2}\left(\frac{2m}{\hbar^2}\right)^{3/2}(3\mu_0^2 B)\,\mu^{1/2}\left[1 - \frac{\pi^2}{24}\left(\frac{k_{\rm B}T}{\mu}\right)^2\right] \tag{7.160}$$

Now insert Eq. (7.158), and expand the last two factors consistently through $O[(k_{\rm B}T/\varepsilon_{\rm F})^2]$[28]

$$\begin{aligned}\mu^{1/2}\left[1 - \frac{\pi^2}{24}\left(\frac{k_{\rm B}T}{\mu}\right)^2\right] &= \varepsilon_{\rm F}^{1/2}\left[1 - \frac{\pi^2}{24}\left(\frac{k_{\rm B}T}{\varepsilon_{\rm F}}\right)^2 + \cdots\right] \\ &\quad\times\left[1 - \frac{\pi^2}{24}\left(\frac{k_{\rm B}T}{\varepsilon_{\rm F}}\right)^2 + \cdots\right] \\ &= \varepsilon_{\rm F}^{1/2}\left[1 - \frac{\pi^2}{12}\left(\frac{k_{\rm B}T}{\varepsilon_{\rm F}}\right)^2 + \cdots\right]\end{aligned} \tag{7.161}$$

[28]Note carefully the factors of 2 involved here, and in Eq. (7.162).

Equation (7.137), with $g_s = 2$, expresses the particle density in terms of $\varepsilon_{\rm F}$

$$\frac{N}{V} = \frac{1}{3\pi^2}\left(\frac{2m}{\hbar^2}\right)^{3/2} \varepsilon_{\rm F}^{3/2} \tag{7.162}$$

A combination of these results reduces Eq. (7.160) to

$$\frac{\mathcal{M}}{V} = \frac{1}{2}\left(\frac{N}{V}\right)\left(\frac{3\mu_0^2 B}{\varepsilon_{\rm F}}\right)\left[1 - \frac{\pi^2}{12}\left(\frac{k_{\rm B}T}{\varepsilon_{\rm F}}\right)^2 + \cdots\right] \quad ; T \to 0 \quad ; B \to 0 \tag{7.163}$$

This result holds through order $(k_{\rm B}T/\varepsilon_{\rm F})^2$ as $B \to 0$.

The magnetic susceptiblity is defined in terms of the magnetic field by[29]

$$\frac{\mathcal{M}}{V} \equiv \kappa_m B \qquad ; \text{ magnetic susceptibility} \tag{7.164}$$

The low-temperature magnetic susceptibility is thus identified as

$$\kappa_{\rm Pauli} = \left(\frac{3\mu_0^2}{2\varepsilon_{\rm F}}\right)\left(\frac{N}{V}\right)\left[1 - \frac{\pi^2}{12}\left(\frac{k_{\rm B}T}{\varepsilon_{\rm F}}\right)^2 + \cdots\right] \quad ; T \to 0 \tag{7.165}$$

We comment on this result:

- This is the low-temperature *Pauli paramagnetic spin susceptibility* of a spin-1/2 Fermi gas;
- Equation (7.165) can be compared with the previous classical Boltzmann expression for a gas of permanent dipoles with $J = 1/2$ [30]

$$\kappa_{\rm Boltz} = \left(\frac{\mu_0^2}{k_{\rm B}T}\right)\left(\frac{N}{V}\right) \tag{7.166}$$

- The two results are sketched in Fig. 7.13;
- The Pauli principle cuts down on the magnetization at low temperature since only those particles close to the Fermi surface are free to re-orient their spins;
- Note that the magnetization in Eq. (7.153) is obtained from the grand partition function through a derivative with respect to the field (see Prob. 7.20)

$$\mathcal{M} = k_{\rm B}T\left[\frac{\partial}{\partial B}\ln{(\text{G.P.F.})_B}\right]_{T,V;\mu} \tag{7.167}$$

---

[29] Here $\mathbf{B} \equiv \mathbf{H}$.

[30] This expression is obtained by setting $J = 1/2$ and using $\boldsymbol{\mu}^2 = \mu_0^2$ in Eq. (3.173).

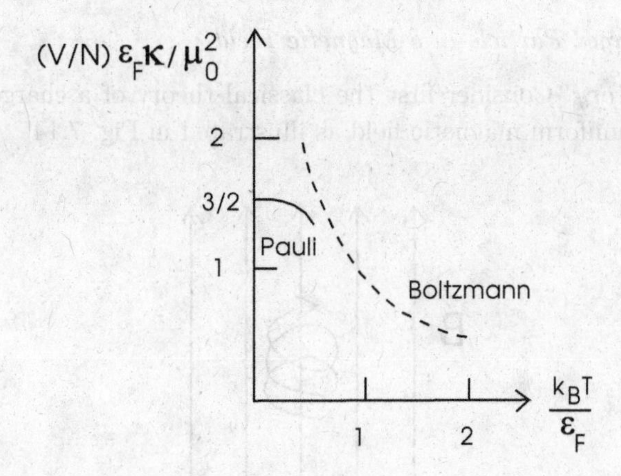

Fig. 7.13 Comparison of κ_{Pauli} and κ_{Boltz} as a function of temperature.

The total differential of the thermodynamic potential $\Phi(T, V, \mu, B)$ in the presence of the field is thus

$$d\Phi = -SdT - PdV - Nd\mu - \mathcal{M}dB \tag{7.168}$$

where, again, $\mathbf{B} = \mathbf{H}$ for the problem at hand. The field contribution then reproduces our previous result in Eq. (3.180).

7.4.4 *Landau Diamagnetism*

The basic idea here is that if a particle goes around in a circle in a uniform magnetic field, the resulting current loop creates a magnetic dipole moment that opposes the field—this is *diamagnetism.* The *Bohr-Van Leeuwen theorem,* proven in Prob. 7.17, states that an assembly of charged particles obeying classical mechanics and classical statistics has *vanishing diamagnetic susceptibility.* The result holds even in the presence of two-body interactions $V(ij)$. The reason for this is not hard to find. In a big box with a uniform magnetic field and periodic boundary conditions, there is no net current flowing anywhere in the assembly. We shall show that quantum mechanics implies a non-zero diamagnetic susceptibility; this is *Landau diamagnetism.*[31] Fortunately, the quantum mechanics problem of a charged particle moving in a uniform magnetic field can be solved analytically, and we first review that solution.

---

[31] See [Landau and Lifshitz (1980); Landau and Lifshitz (1980a)].

7.4.4.1 *Charged Particle in a Magnetic Field*

Classical Theory: Consider first the classical theory of a charged particle moving in a uniform magnetic field as illustrated in Fig. 7.14.

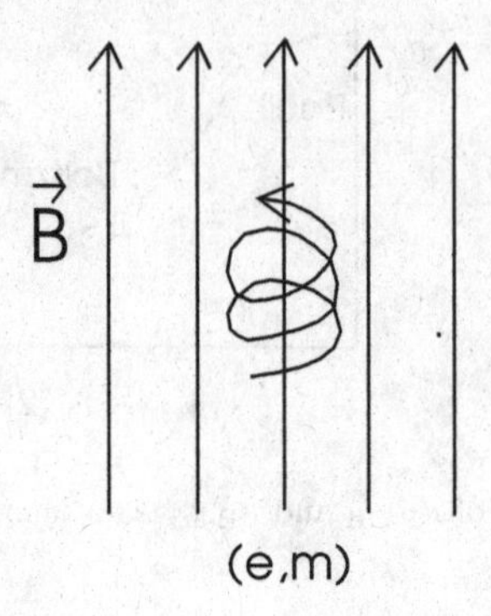

Fig. 7.14 Charged particle with (e, m) moving in a uniform magnetic field $\mathbf{B}$; here $e < 0$.

Define the kinetic momentum by

$$\boldsymbol{\pi} \equiv m\mathbf{v} \qquad ; \text{kinetic momentum} \tag{7.169}$$

Newton's second law and the Lorentz force equation then give

$$\frac{d\boldsymbol{\pi}}{dt} = \frac{e}{c}\mathbf{v} \times \mathbf{B} = \frac{e}{mc}\boldsymbol{\pi} \times \mathbf{B} \qquad ; \text{Newton's 2nd and Lorentz force} \tag{7.170}$$

Decompose the vectors as

$$\begin{aligned} \mathbf{B} &= B\mathbf{e}_z \\ \boldsymbol{\pi} &= \pi_z\mathbf{e}_z + \boldsymbol{\pi}_\perp = \boldsymbol{\pi}_z + \boldsymbol{\pi}_\perp \end{aligned} \tag{7.171}$$

where $\mathbf{e}_z$ is a unit vector in the z-direction. Then from Eqs. (7.170)

$$\frac{d\boldsymbol{\pi}_z}{dt} = 0 \qquad ; \text{constant of motion} \tag{7.172}$$

The kinetic momentum of the particle in the z-direction $\boldsymbol{\pi}_z = \pi_z\mathbf{e}_z$ is a constant of the motion. The equation of motion for the transverse momentum $\boldsymbol{\pi}_\perp$ decouples, and it is given by

$$\frac{d\boldsymbol{\pi}_\perp}{dt} = \frac{e}{mc}\left(\boldsymbol{\pi}_\perp \times \mathbf{B}\right) \tag{7.173}$$

If this equation is dotted into $\boldsymbol{\pi}_\perp$, one obtains

$$\frac{d}{dt}\left(\boldsymbol{\pi}_\perp^2\right) = 0 \qquad ; \text{ constant of motion} \tag{7.174}$$

The square of the magnitude of the transverse kinetic momentum $\boldsymbol{\pi}_\perp^2 = \pi_x^2 + \pi_y^2$ is also a constant of the motion.

Let us write out Eqs. (7.170) in terms of components

$$\begin{aligned}\left[\frac{d\pi_x}{dt}, \frac{d\pi_y}{dt}, \frac{d\pi_z}{dt}\right] &= \frac{e}{mc}\left[B\pi_y, -B\pi_x, 0\right] \\ &= \frac{eB}{c}\left[v_y, -v_x, 0\right] \\ &= \frac{eB}{c}\left[\frac{dy}{dt}, -\frac{dx}{dt}, 0\right]\end{aligned} \tag{7.175}$$

Thus

$$\frac{d}{dt}\left(\pi_x - \frac{eB}{c}y\right) = \frac{d}{dt}\left(\pi_y + \frac{eB}{c}x\right) = 0 \tag{7.176}$$

We can therefore identify two additional constants of the motion

$$\begin{aligned}x_0 &\equiv x + \frac{c}{eB}\pi_y \\ y_0 &\equiv y - \frac{c}{eB}\pi_x \qquad ; \text{ constants of motion}\end{aligned} \tag{7.177}$$

It follows from these equations that

$$(x - x_0)^2 + (y - y_0)^2 = \left(\frac{c}{eB}\right)^2 \boldsymbol{\pi}_\perp^2 = \text{constant} \tag{7.178}$$

This is the equation of a circle in the (x, y)-plane, centered on the point (x_0, y_0). Furthermore, since the particles move with a uniform momentum in the z-direction along the field, we conclude that *the particles follow spiral orbits along the field, centered on the point* (x_0, y_0) *in the* (x, y) *plane* (see Fig. 7.14).[32]

Quantum Mechanics: First, let us choose a convenient gauge and take

$$\begin{aligned}\mathbf{A}(\mathbf{x}) &= [-yB, 0, 0] \qquad ; \text{ choice of gauge} \\ \boldsymbol{\nabla} \times \mathbf{A} &= [0, 0, B]\end{aligned} \tag{7.179}$$

[32]Note that nowhere in this classical analysis did we assume that the motion of the particle is non-relativistic.

The kinetic momentum is then given by

$$\boldsymbol{\pi} = \mathbf{p} - \frac{e}{c}\mathbf{A}(\mathbf{x}) = \left[p_x + \frac{eB}{c}y,\, p_y,\, p_z\right] \tag{7.180}$$

The hamiltonian for a non-relativistic particle moving in the constant, uniform magnetic field $\mathbf{B} = B\mathbf{e}_z$ is now[33]

$$\begin{aligned} H &= \frac{1}{2m}\boldsymbol{\pi}^2 = \frac{1}{2m}\left[\mathbf{p} - \frac{e}{c}\mathbf{A}(\mathbf{x})\right]^2 \\ &= \frac{1}{2m}p_z^2 + \frac{1}{2m}\left[\left(p_x + \frac{eB}{c}y\right)^2 + p_y^2\right] \end{aligned} \tag{7.181}$$

We observe the following:

- The first of Hamilton's equations of motion says that

$$\frac{dx_i}{dt} = \frac{\partial H}{\partial p_i} = \frac{\pi_i}{m} \qquad ; \; i = x, y, z$$
$$; \text{ kinetic momentum} \tag{7.182}$$

  Hence, $\boldsymbol{\pi}$ is indeed the kinetic momentum;
- In quantum mechanics, the quantities (x_0, y_0) become *operators*

$$\begin{aligned} x_0 &= x + \frac{c}{eB}\pi_y = x + \frac{c}{eB}p_y \\ y_0 &= y - \frac{c}{eB}\pi_x = -\frac{c}{eB}p_x \end{aligned} \tag{7.183}$$

- The canonical commutation relations in quantum mechanics are

$$[p_i, x_j] = \frac{\hbar}{i}\delta_{i,j} \qquad ; \; (i,j) = x, y, z \tag{7.184}$$

- These relations imply that (x_0, y_0) both commute with the hamiltonian

$$[H, y_0] = [H, x_0] = 0 \tag{7.185}$$

  Thus, both are constants of the motion;
- However, (x_0, y_0) *do not* commute with each other!

$$[x_0, y_0] \neq 0 \tag{7.186}$$

  Hence, they cannot be simultaneously specified;

[33]It is interesting to note that the relativistic Dirac equation can also be solved analytically in the presence of a uniform magnetic field. Note that the choice of gauge in Eq. (7.179) implies that the (x, y) coordinates will now be treated differently in these problems.

- The challenge is to pick a complete set of mutually commuting hermitian operators for this problem; as for a free particle, there will be three of them. Take

$$H,\ p_x\,(\text{or}\ y_0),\ p_z \qquad ;\ \text{complete set of commuting operators} \tag{7.187}$$

As usual, we will work in the coordinate representation where

$$\mathbf{p} = \frac{\hbar}{i}\boldsymbol{\nabla} \tag{7.188}$$

The time-independent one-body Schrödinger equation is

$$H\psi = \varepsilon\psi \qquad ;\ \text{S-eqn} \tag{7.189}$$

Let us look for solutions to this equation of the form

$$\psi(x,y,z) = e^{(i/\hbar)p_1 x}e^{(i/\hbar)p_3 z}\psi(y) \tag{7.190}$$

Here $(p_1, p_3)/\hbar = (k_1, k_3)$ are eigenvalues. If we impose periodic boundary conditions along the direction of the field, then

$$k_3 = \frac{2\pi l}{L} \qquad ; l = 0, \pm 1, \pm 2, \cdots \qquad ;\ \text{p.b.c.} \tag{7.191}$$

The use of Eq. (7.181) then reduces the Schrödinger equation to

$$\frac{1}{2m}\left[p_3^2 + \left(p_1 + \frac{eB}{c}y\right)^2 + p_y^2\right]\psi(y) = \varepsilon\psi(y) \tag{7.192}$$

In line with the second of Eqs. (7.183), call

$$y_0 = -\frac{c}{eB}p_1 \tag{7.193}$$

Then Eq. (7.192) becomes

$$\left[-\hbar^2\frac{\partial^2}{\partial y^2} + \left(\frac{eB}{c}\right)^2(y-y_0)^2\right]\psi(y) = 2m(\varepsilon - \varepsilon_3)\psi(y) \tag{7.194}$$

where ε_3 is the kinetic energy in the z-direction

$$\varepsilon_3 \equiv \frac{p_3^2}{2m} \tag{7.195}$$

Now introduce a dimensionless variable

$$\xi \equiv \left(\frac{|e|B}{\hbar c}\right)^{1/2}(y - y_0) \tag{7.196}$$

Equation (7.194) then takes the form

$$\left(-\frac{\partial^2}{\partial\xi^2}+\xi^2\right)\psi(\xi)=2\epsilon\psi(\xi)$$

$$\epsilon\equiv\frac{mc(\varepsilon-\varepsilon_3)}{|e|\hbar B} \tag{7.197}$$

This is just the dimensionless Schrödinger equation for the one-dimensional simple harmonic oscillator in the coordinate representation [Schiff (1968)]! The eigenvalues are well known

$$\epsilon=\left(n+\frac{1}{2}\right) \qquad ;\ n=0,1,2,\cdots \tag{7.198}$$

A combination of Eqs. (7.191), (7.195), (7.197), and (7.198) then gives the eigenvalue spectrum for a non-relativistic particle in a uniform magnetic field

$$\varepsilon_{n,l}=\frac{4\pi^2\hbar^2l^2}{2mL^2}+\frac{|e|\hbar B}{mc}\left(n+\frac{1}{2}\right) \qquad ;\ l=0,\pm1,\pm2,\cdots \quad n=0,1,2,\cdots \tag{7.199}$$

The eigenvalue is independent of $p_1=-(eB/c)y_0$, that is, of just where the center of the orbit is located along the y-axis.[34]

7.4.4.2 *Counting of States*

We need to be able to perform sums over single-particle states $\sum_i$ of expressions involving the energies $\varepsilon_{n,l}$ in Eq. (7.199). The state is specified with the quantum numbers (p_1,l,n), and the quantum number p_1 does not appear in $\varepsilon_{n,l}$. Since $(p_x,x)\rightarrow(p_1,x)$ form a continuous, canonical pair of variables, we can use our basic result from statistical mechanics that the number of states in the differential region $dx\,dp_1$ in phase space is

$$dN=\frac{dx\,dp_1}{h} \qquad ;\ \text{number of states}$$

$$\rightarrow\frac{L\,dp_1}{2\pi\hbar} \tag{7.200}$$

The last relation comes from integrating over the position x, which can be anywhere in the box, and writing $h=2\pi\hbar$. Now use Eq. (7.193) to relate

[34]The wave functions $\psi_n(\xi)$ are examined in Prob. 7.21.

p_1 to the y_0-position of the orbit[35]

$$dp_1 = \frac{|e|B}{c} dy_0 \tag{7.201}$$

Then

$$\begin{aligned} dN &= \frac{|e|B}{c} \frac{L}{2\pi\hbar} dy_0 \\ &\to \frac{|e|B}{c} \frac{L^2}{2\pi\hbar} \end{aligned} \tag{7.202}$$

The last relation follows since y_0 can also be anywhere inside the box. Thus we arrive at the following expression for the sum over states

$$\sum_i \to g_s \frac{|e|B}{c} \frac{L^2}{2\pi\hbar} \sum_l \sum_n \tag{7.203}$$

Now the quantum number l enters through Eq. (7.191), where $p_3/\hbar = k_3 = 2\pi l/L$. In the limit $L \to \infty$, the sum over l in Eq. (7.203) also becomes an integral, and as previously

$$\sum_l \to \frac{L}{2\pi\hbar} \int dp_3 \qquad ; \; p_3 = \hbar k_3 \qquad ; \; L \to \infty \tag{7.204}$$

Thus, in *summary*, for a non-relativistic charged particle moving in a uniform magnetic field, as $L \to \infty$,

$$\begin{aligned} \sum_i &\to g_s \frac{|e|B}{c} \frac{L^3}{(2\pi\hbar)^2} \int dp_3 \sum_n \qquad ; \; L \to \infty \\ \varepsilon(p_3, n) &= \frac{p_3^2}{2m} + \frac{|e|\hbar B}{mc}\left(n + \frac{1}{2}\right) \end{aligned} \tag{7.205}$$

where we have relabeled the eigenvalue appropriately. Note that $L^3 = V$. Note also that the density of states is proportional to the magnetic field B!

[35] To simplify matters, one can assume the particles have $e < 0$ (electrons, for example); however, what one needs here is really just the magnitude of dp_1.

7.4.4.3 *Grand Partition Function and Magnetization*

Assume the particles are fermions. The grand partition function then follows exactly as in Eqs. (7.99)

$$\ln(\text{G.P.F.})_{\text{B}} = \sum_i \ln\left[1 + e^{(\mu-\varepsilon_i)/k_{\text{B}}T}\right]$$
$$= g_s \frac{|e|BV}{c(2\pi\hbar)^2} \int dp_3 \sum_n \ln\left\{1 + e^{[\mu-\varepsilon(p_3,n)]/k_{\text{B}}T}\right\} \qquad (7.206)$$

The distribution numbers also follow as in Eqs. (7.99), and the particle number is

$$N = \sum_i \frac{1}{e^{(\varepsilon_i-\mu)/k_{\text{B}}T}+1}$$
$$= g_s \frac{|e|BV}{c(2\pi\hbar)^2} \int dp_3 \sum_n \frac{1}{e^{[\varepsilon(p_3,n)-\mu]/k_{\text{B}}T}+1} \qquad (7.207)$$

This equation can, in principle, be inverted to obtain the chemical potential $\mu(N/V, T, B)$.

The magnetization of the assembly arising from the charged particle motion is then given by the general thermodynamic relations in Eqs. (7.167)–(7.168).[36] With the use of the second of Eqs. (7.205), one finds

$$\mathcal{M} = k_{\text{B}}T \frac{\partial}{\partial B} \ln(\text{G.P.F.})_{\text{B}}$$
$$= -g_s \frac{|e|BV}{c(2\pi\hbar)^2} \int dp_3 \sum_n \frac{1}{e^{[\varepsilon(p_3,n)-\mu]/k_{\text{B}}T}+1} \left[\frac{|e|\hbar}{mc}\left(n+\frac{1}{2}\right)\right]$$
$$+ \frac{k_{\text{B}}T}{B} \ln(\text{G.P.F.})_{\text{B}} \qquad (7.208)$$

Note, in particular, the sign of the first term and the presence of the second term arising from the density of states.

7.4.4.4 *High-Temperature Limit*

At high temperature, the chemical potential becomes very large and negative, and we recover Boltzmann statistics

$$\frac{\mu}{k_{\text{B}}T} \to -\infty \quad ; \; T \to \infty \qquad (7.209)$$

[36] The intrinsic magnetic moment of the particles is treated as in the last section.

In this limit

$$N = \ln{(\text{G.P.F.})_{\text{B}}} = \sum_i e^{(\mu-\varepsilon_i)/k_{\text{B}}T} \qquad ; T \to \infty$$

$$= e^{\mu/k_{\text{B}}T} g_s \frac{|e|BV}{c(2\pi\hbar)^2} \int_{-\infty}^{\infty} dp_3 \sum_{n=0}^{\infty} e^{-p_3^2/2mk_{\text{B}}T} \exp\left[-\frac{|e|\hbar B}{mck_{\text{B}}T}\left(n+\frac{1}{2}\right)\right] \tag{7.210}$$

Equation (6.20) provides the general relation

$$PV = k_{\text{B}}T \ln{(\text{G.P.F.})} \tag{7.211}$$

It follows from the first of Eqs. (7.210) that at high temperature we recover the equation of state of a perfect gas

$$PV = Nk_{\text{B}}T \qquad ; T \to \infty \tag{7.212}$$

The integral over p_3 and sum over n are now familiar from our previous analysis, and hence Eqs. (7.210) become

$$\ln{(\text{G.P.F.})_{\text{B}}} = e^{\mu/k_{\text{B}}T} g_s \frac{|e|BV}{c(2\pi\hbar)^2} (2\pi mk_{\text{B}}T)^{1/2} \frac{e^{-x}}{1-e^{-2x}}$$

$$x \equiv \frac{|e|\hbar B}{2mck_{\text{B}}T} \tag{7.213}$$

At high temperature the parameter $x \ll 1$, and the s.h.o. partition function can be expanded as

$$\frac{e^{-x}}{1-e^{-2x}} = \frac{1}{e^x - e^{-x}} = \frac{1}{2x + 2x^3/6 + \cdots} = \frac{1}{2x}\left(1 - \frac{x^2}{6} + \cdots\right) \tag{7.214}$$

This gives[37]

$$\ln{(\text{G.P.F.})_{\text{B}}} = N$$

$$= e^{\mu/k_{\text{B}}T} g_s V \left(\frac{2\pi mk_{\text{B}}T}{h^2}\right)^{3/2} \left[1 - \frac{1}{6}\left(\frac{e\hbar}{2mc}\right)^2 \left(\frac{B}{k_{\text{B}}T}\right)^2\right]$$

$$; B \to 0 \quad ; T \to \infty \tag{7.215}$$

The low-field magnetization then follows from the first of Eqs. (7.208)

$$\frac{\mathcal{M}}{V} = -\frac{1}{3}\left(\frac{e\hbar}{2mc}\right)^2 \left(\frac{N}{V}\right) \frac{B}{k_{\text{B}}T} \qquad ; B \to 0 \quad ; T \to \infty \tag{7.216}$$

[37]Notice the nice check that at $B = 0$, one recovers $N = \lambda(\text{p.f.})_{\text{free}}$.

The diamagnetic susceptibility is thus identified as

$$\kappa_{\text{Landau}} = -\frac{1}{3k_{\text{B}}T}\left(\frac{e\hbar}{2mc}\right)^2\left(\frac{N}{V}\right) \qquad ; T \to \infty$$

$$= -\frac{\mu_0^2}{3k_{\text{B}}T}\left(\frac{N}{V}\right) \qquad ; \mu_0 \equiv \frac{e\hbar}{2mc} \qquad (7.217)$$

Some comments:

- In contrast to the situation described by the Bohr-Van Leeuwan theorem, the magnetic susceptibility is now non-zero due to the quantization of the orbits;[38]
- Note the sign of κ;
- This is the high-T limit of *Landau diamagnetism*;
- Note also the $1/T$ behavior;
- This expression should be compared with the previous result for the paramagnetic susceptibility of an assembly of particles with spin J and intrinsic magnetic moment $\boldsymbol{\mu}^2 = \mu_0^2$ in Eq. (3.173).

7.4.4.5 *Low-Temperature Limit*

As the temperature goes to zero, the orbits are filled to the Fermi level $\mu = \varepsilon_{\text{F}}$ as illustrated in Fig. 7.9(a). Then

$$n^\star = 1 \qquad ; \varepsilon \le \mu \qquad ; T \to 0$$

$$n^\star = 0 \qquad ; \varepsilon > \mu \qquad (7.218)$$

With the use of the second of Eqs. (7.205), the condition $\varepsilon \le \mu$ becomes

$$\frac{p_3^2}{2m} + \frac{|e|\hbar B}{mc}\left(n + \frac{1}{2}\right) \le \mu \qquad ; \text{filled levels} \qquad (7.219)$$

Equation (7.219) puts limits on the various sums. The particle number in Eq. (7.207), for example, is now given by

$$N = g_s \frac{|e|BV}{c(2\pi\hbar)^2} \sum_n \int_{-p_{3\,\text{max}}}^{+p_{3\,\text{max}}} dp_3 \qquad ; T \to 0 \qquad (7.220)$$

[38] Note that the energy of motion along the field line, and that of the quantized circular orbits in the transverse plane, now enter in a very different way in the logarithm of $(\text{G.P.F.})_{\text{B}}$ in Eq. (7.206).

where

$$p_{3\,\max} \equiv \sqrt{2m\left[\mu - \frac{|e|\hbar B}{mc}\left(n+\frac{1}{2}\right)\right]} \tag{7.221}$$

At finite T, the logarithm of the grand partition function is given by Eq. (7.206) as

$$\ln{(\text{G.P.F.})_\text{B}} = g_s \frac{|e|BV}{c(2\pi\hbar)^2}\sum_n \int dp_3 \ln\left\{1 + e^{[\mu-\varepsilon(p_3,n)]/k_\text{B}T}\right\} \tag{7.222}$$

This expression can be partially integrated on p_3 using

$$\begin{aligned} dp_3 &= du \qquad ; \; v = \ln\left\{1 + e^{[\mu-\varepsilon(p_3,n)]/k_\text{B}T}\right\} \\ p_3 &= u \qquad ; \; dv = -\frac{p_3 dp_3}{mk_\text{B}T}\frac{1}{e^{[\varepsilon(p_3,n)-\mu]/k_\text{B}T}+1} \end{aligned} \tag{7.223}$$

Then[39]

$$\ln{(\text{G.P.F.})_\text{B}} = g_s \frac{|e|BV}{mc(2\pi\hbar)^2 k_\text{B}T}\sum_n \int p_3^2\, dp_3 \frac{1}{e^{[\varepsilon(p_3,n)-\mu]/k_\text{B}T}+1} \tag{7.224}$$

The $T \to 0$ limit of this expression can now be taken to give

$$k_\text{B}T\ln{(\text{G.P.F.})_\text{B}} = g_s \frac{|e|BV}{mc(2\pi\hbar)^2}\sum_n \int_{-p_{3\,\max}}^{+p_{3\,\max}} p_3^2\, dp_3 \qquad ; \; T \to 0 \tag{7.225}$$

A combination of these results allows us to express the $T \to 0$ limit of the magnetization in Eqs. (7.208) as

$$\begin{aligned} \mathcal{M} = &-g_s \frac{|e|BV}{c(2\pi\hbar)^2}\sum_n \frac{|e|\hbar}{mc}\left(n+\frac{1}{2}\right)\int_{-p_{3\,\max}}^{+p_{3\,\max}} dp_3 \\ &+g_s \frac{|e|V}{mc(2\pi\hbar)^2}\sum_n \int_{-p_{3\,\max}}^{+p_{3\,\max}} p_3^2\, dp_3 \qquad ; \; T \to 0 \end{aligned} \tag{7.226}$$

From Eq. (7.219), the limit on the final sum over n for a given μ is

$$n_\text{max} = \frac{mc}{|e|\hbar B}\mu - \frac{1}{2} \qquad ; \; n \le n_\text{max} \tag{7.227}$$

Evidently

$$\mu \ge \frac{|e|\hbar B}{2mc} \tag{7.228}$$

[39] Note that $[uv]_{-\infty}^{\infty} = 0$; remember, the partial integration is carried out at finite T.

Let us analyze these expressions in detail. Parameterize the chemical potential with a non-negative quantity x defined by

$$\mu \equiv \frac{|e|\hbar B}{2mc}(1+2x) \qquad ; \text{ parameterization} \qquad (7.229)$$

Then from Eq. (7.227)

$$n_{\text{max}} = x \qquad ; n \leq n_{\text{max}} \qquad (7.230)$$

If the integrals are now evaluated, and a little algebra employed, one has from Eqs. (7.220), (7.221), and (7.226)

$$N = g_s \frac{2|e|BV}{c(2\pi\hbar)^2}\left(\frac{2|e|\hbar B}{c}\right)^{1/2} \sum_n (x-n)^{1/2}$$

$$\mathcal{M} = -g_s \frac{2|e|BV}{c(2\pi\hbar)^2}\left(\frac{2|e|\hbar B}{c}\right)^{1/2} \frac{|e|\hbar}{2mc} \times \sum_n \left[(2n+1)(x-n)^{1/2} - \frac{4}{3}(x-n)^{3/2}\right] \qquad (7.231)$$

Now define a new field B_0 by

$$\frac{N}{B_0^{3/2}} \equiv g_s \frac{2|e|V}{c(2\pi\hbar)^2}\left(\frac{2|e|\hbar}{c}\right)^{1/2} = 2\pi g_s V \left(\frac{|e|}{\pi\hbar c}\right)^{3/2} \qquad (7.232)$$

Equations (7.231) then become

$$1 = \left(\frac{B}{B_0}\right)^{3/2} \sum_{n=0}^{n_{\text{max}}} (x-n)^{1/2}$$

$$\frac{\mathcal{M}}{V} = -\frac{|e|\hbar}{2mc}\left(\frac{B}{B_0}\right)^{3/2}\left(\frac{N}{V}\right)\sum_{n=0}^{n_{\text{max}}}\left[(2n+1)(x-n)^{1/2} - \frac{4}{3}(x-n)^{3/2}\right] \qquad (7.233)$$

Several comments:

- Equations (7.233) form our principal result for *Landau diamagnetism in a Fermi gas at $T = 0$ in a uniform magnetic field*;
- The field B_0 appearing in these relations is defined in Eq. (7.232) as

$$B_0 = \frac{\pi\hbar c}{|e|}\left(\frac{1}{2\pi g_s}\right)^{2/3}\left(\frac{N}{V}\right)^{2/3} \qquad (7.234)$$

Note that B_0 depends on the particle density as $(N/V)^{2/3}$;[40]

- The sums are both performed up to the largest integer satisfying Eqs. (7.230)[41]

$$n \leq n_{\max} = x \tag{7.235}$$

- We now have the solution to the problem in parametric form:

(1) The first of Eqs. (7.233) provides an implicit relation for $x(N/V, B)$, or equivalently from Eq. (7.229), the chemical potential $\mu(N/V, B)$

$$\mu = \frac{|e|\hbar B}{2mc}(1 + 2x) \tag{7.236}$$

(2) The second of Eqs. (7.233) then determines the magnetization $\mathcal{M}/V \equiv M(N/V, B)$;

- The large field limit of these results with $B \gg B_0$ is readily obtained. In this case $x \to 0$, and only the first term with $n = 0$ contributes in Eqs. (7.233). These equations become

$$\begin{aligned} 1 &= \left(\frac{B}{B_0}\right)^{3/2} x^{1/2} \qquad ; 0 < x < 1 \;;\; B \gg B_0 \\ \frac{\mathcal{M}}{V} &= -\frac{|e|\hbar}{2mc}\left(\frac{B}{B_0}\right)^{3/2}\left(\frac{N}{V}\right)x^{1/2}\left(1 - \frac{4}{3}x\right) \\ &= -\frac{|e|\hbar}{2mc}\left(\frac{N}{V}\right)\left(1 - \frac{4}{3}x\right) \end{aligned} \tag{7.237}$$

- Now decrease B until $B = B_0$, and $x = 1$. The term with $n = 1$ then starts to contribute to the sums, and

$$\begin{aligned} 1 &= \left(\frac{B}{B_0}\right)^{3/2}\left[x^{1/2} + (x-1)^{1/2}\right] \qquad ; 1 < x < 2 \\ \frac{\mathcal{M}}{V} &= -\frac{|e|\hbar}{2mc}\left(\frac{B}{B_0}\right)^{3/2}\left(\frac{N}{V}\right) \\ &\quad \times \left[x^{1/2} + 3(x-1)^{1/2} - \frac{4}{3}x^{3/2} - \frac{4}{3}(x-1)^{3/2}\right] \end{aligned} \tag{7.238}$$

- Continue this process. The chemical potential μ, magnetization $\mathcal{M}/V$, and susceptibility $\boldsymbol{\kappa} = \partial/\partial B(\mathcal{M}/V)$ all show *breaks* at $x = 1, 2, 3, \cdots$;

---

[40] See Prob. 7.18.

[41] In Mathcad11, this is called "floor(x)".

- To illustrate this, Fig. 7.15 shows the numerical solution for x as a function of B/B_0 arising from the first of Eqs. (7.233);

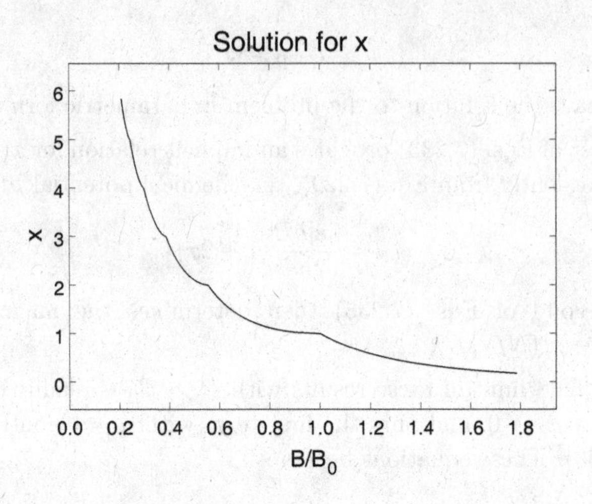

Fig. 7.15 Numerical solution of the first of Eqs. (7.233) for x as a function of B/B_0. Note the breaks in the curve at integer values of x.

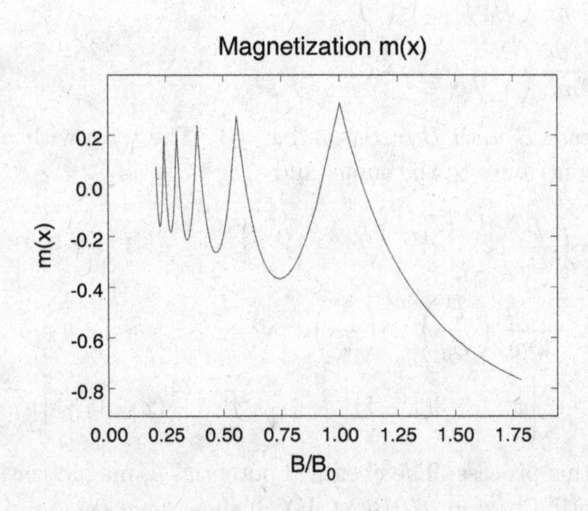

Fig. 7.16 Calculation of the magnetization $m(x)$ based on Fig. 7.15 and the second of Eqs. (7.233). Here $\mathcal{M}/V \equiv (|e|\hbar/2mc)(N/V)m(x)$.

- Figure 7.16 then shows the implied magnetization $m(x)$ from the second of Eqs. (7.233), where

$$\frac{\mathcal{M}}{V} \equiv \frac{|e|\hbar}{2mc}\left(\frac{N}{V}\right) m(x) \tag{7.239}$$

- The breaks in the magnetization and susceptibility are known as the *De Haas-Van Alphen effect*;
- The study of a two-dimensional Fermi gas of electrons in a magnetic field forms a central thrust of current condensed matter physics.[42]

[42]See, for example, "*2DEG*" and "*Fractional Quantum Hall Effect*" in [Wiki (2010)].

Chapter 8

Special Topics

There are many directions in which one can now go with statistical mechanics; here we discuss only a few special topics. We first consider the properties of a condensed phase with two components, that is, of a *solution*.

8.1 Solutions

8.1.1 *Perfect Solutions*

Start with the problem of two components in two phases, gas and condensed. Make the following two assumptions:

(1) The gas phase can be treated as a *perfect gas* of independent, non-localized systems;
(2) The *Einstein model* of independent, localized systems can be used for the condensed phase.

We work with the canonical ensemble and compute the canonical partition function for this assembly.

8.1.1.1 *Canonical Partition Function*

If there are (N_A, N_B) atoms of components (A, B), with (N_1, N_2) of each type in the gas phase and (M_1, M_2) of each type in the condensed phase, then

$$
\begin{aligned}
M_1 + N_1 &= N_A \qquad ; \text{ components} \\
M_2 + N_2 &= N_B
\end{aligned}
\tag{8.1}
$$

We assume a total of $M_1 + M_2$ localized sites, so that there are $(M_1 + M_2)!/M_1!M_2!$ ways to distribute the (M_1, M_2) systems on the condensed sites

$$g(M_1, M_2) = \frac{(M_1 + M_2)!}{M_1!M_2!} \quad ; \text{ number of rearrangements} \quad M_1 + M_2 \text{ localized sites} \tag{8.2}$$

Our previous analysis tells us that the canonical partition function for this assembly is given by

$$(\text{P.F.}) = \sum_{M_1}\sum_{N_1}\sum_{M_2}\sum_{N_2} \frac{(M_1 + M_2)!}{M_1!M_2!}(\text{P.F.})_A(\text{P.F.})_B \tag{8.3}$$

Here $(\text{P.F.})_A$ is the canonical partition function for component A with (M_1, N_1) localized and non-localized systems, and $(\text{P.F.})_B$ is the canonical partition function for component B with (M_2, N_2) localized and non-localized systems. The sums are to be performed subject to the two constraints in Eqs. (8.1). Equations (4.12) and (4.14) can now be employed for the localized and non-localized contributions to the canonical partition functions of the two components. Thus[1]

$$(\text{P.F.}) = \sum_{M_1}\sum_{N_1}\sum_{M_2}\sum_{N_2} \frac{(M_1 + M_2)!}{M_1!M_2!}[(\text{p.f.})_A^c]^{M_1} \frac{[(\text{p.f.})_A^g]^{N_1}}{N_1!} \times [(\text{p.f.})_B^c]^{M_2} \frac{[(\text{p.f.})_B^g]^{N_2}}{N_2!} \tag{8.4}$$

This is the same problem dealt with in our discussion of chemical equilibria, and it is handled in the same fashion:

- Replace the sums by the largest term, obtained by maximizing the logarithm of the summand subject to the two constraints;
- Incorporate the constraints with Lagrange multipliers (α_A, α_B);
- Use Stirling's approximation for (M_1, N_1, M_2, N_2) so that, for example,

$$\ln M! = M \ln M - M$$

$$\frac{\partial}{\partial M} \ln M! = \ln M \tag{8.5}$$

[1] Compare Eq. (3.233).

8.1.1.2 *Helmholtz Free Energy*

As before, the resulting Helmholtz free energy of the assembly is then

$$
\begin{aligned}
A &= -k_{\rm B}T\ln{(\rm P.F.)} = A_c + A_g \\
A_c &= -k_{\rm B}T\ln\left\{\frac{(M_1^\star + M_2^\star)!}{M_1^\star!M_2^\star!}\left[({\rm p.f.})_A^c\right]^{M_1^\star}\left[({\rm p.f.})_B^c\right]^{M_2^\star}\right\} \\
A_g &= -k_{\rm B}T\ln\left\{\frac{\left[({\rm p.f.})_A^g\right]^{N_1^\star}}{N_1^\star!}\frac{\left[({\rm p.f.})_B^g\right]^{N_2^\star}}{N_2^\star!}\right\}
\end{aligned}
\tag{8.6}
$$

The mean particle numbers are those obtained by maximizing the logarithm of the summand with the incorporated constraints. From the four derivatives of the logarithm with respect to (M_1, M_2, N_1, N_2) respectively, one has[2]

$$
\begin{aligned}
\ln{(M_1^\star + M_2^\star)} + \ln{({\rm p.f.})_A^c} - \ln M_1^\star + \alpha_A &= 0 \qquad ; M_1 \text{ eqn} \\
\ln{(M_1^\star + M_2^\star)} + \ln{({\rm p.f.})_B^c} - \ln M_2^\star + \alpha_B &= 0 \qquad ; M_2 \text{ eqn} \\
\ln{({\rm p.f.})_A^g} - \ln N_1^\star + \alpha_A &= 0 \qquad ; N_1 \text{ eqn} \\
\ln{({\rm p.f.})_B^g} - \ln N_2^\star + \alpha_B &= 0 \qquad ; N_2 \text{ eqn}
\end{aligned}
\tag{8.7}
$$

The solution to these equations takes the form

$$
\begin{aligned}
\frac{M_1^\star}{M_1^\star + M_2^\star} &= \lambda_A({\rm p.f.})_A^c \qquad ; \lambda_A = e^{\alpha_A} \\
\frac{M_2^\star}{M_1^\star + M_2^\star} &= \lambda_B({\rm p.f.})_B^c \qquad ; \lambda_B = e^{\alpha_B} \\
N_1^\star &= \lambda_A({\rm p.f.})_A^g \\
N_2^\star &= \lambda_B({\rm p.f.})_B^g
\end{aligned}
\tag{8.8}
$$

Hence

$$
\begin{aligned}
N_1^\star &= \frac{M_1^\star}{M_1^\star + M_2^\star}\frac{({\rm p.f.})_A^g}{({\rm p.f.})_A^c} \\
N_2^\star &= \frac{M_2^\star}{M_1^\star + M_2^\star}\frac{({\rm p.f.})_B^g}{({\rm p.f.})_B^c}
\end{aligned}
\tag{8.9}
$$

Now use the fact that $({\rm p.f.})^g \propto V$, and define

$$
p_1 \equiv k_{\rm B}T\frac{N_1^\star}{V} \qquad ; p_2 \equiv k_{\rm B}T\frac{N_2^\star}{V}
\tag{8.10}
$$

---

[2]Compare, for example, Eqs. (3.200).

Then

$$p_1 = \frac{M_1^\star}{M_1^\star + M_2^\star} p_1^0(T) = x_1 p_1^0(T)$$
$$p_2 = \frac{M_2^\star}{M_1^\star + M_2^\star} p_2^0(T) = x_2 p_2^0(T) \tag{8.11}$$

We make several comments:

- The pressures (p_1, p_2) are the individual *partial pressures* in the gas;
- The quantities (x_1, x_2) are the *mole fractions* in the condensed phase

$$x_1 \equiv \frac{M_1^\star}{M_1^\star + M_2^\star} \quad ; \; x_2 \equiv \frac{M_2^\star}{M_1^\star + M_2^\star} \tag{8.12}$$

  They satisfy

$$x_1 + x_2 = 1 \qquad ; \text{ mole fractions in condensed phase} \tag{8.13}$$

- The total gas pressure is

$$p = p_1 + p_2$$
$$= x_1 p_1^0(T) + x_2 p_2^0(T) \tag{8.14}$$

  The use of Eq. (8.13) then gives

$$p = x_1 p_1^0(T) + (1 - x_1) p_2^0(T) \qquad ; \text{ Raoult's law} \tag{8.15}$$

  This straight-line behavior as a function of x_1 is *Raoult's law*; It is illustrated in Fig. 8.1;

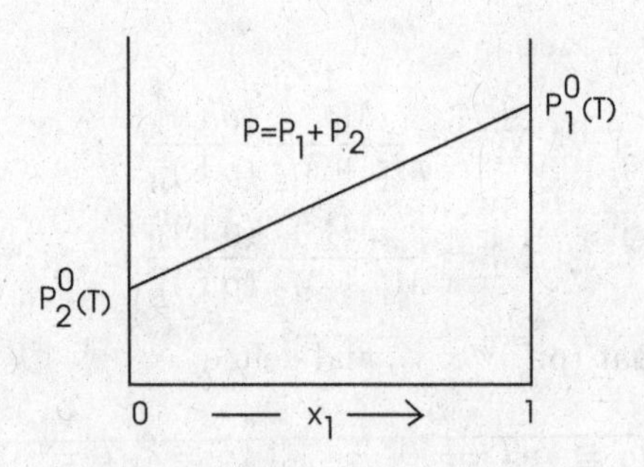

Fig. 8.1 Illustration of *Raoult's law*. Vapor pressure of a perfect solution as a function of the mole fraction of a constituent in the condensed phase.

- For ease of writing, we now again suppress the star on the mean numbers (M_1, N_1, M_2, N_2);
- For pure condensed phases of (A, B) with (M_1, M_2) constituents, the Helmholtz free energy of each is

$$A_1 = -k_B T \ln [(\text{p.f.})_A^c]^{M_1} \qquad ; \text{ pure condensed phases}$$
$$A_2 = -k_B T \ln [(\text{p.f.})_B^c]^{M_2} \qquad (8.16)$$

We can then compute the difference in Helmholtz free energy between the condensed-phase solution and pure separated condensed-phase constituents

$$\Delta A = A_c - A_1 - A_2 = -k_B T \ln \frac{(M_1 + M_2)!}{M_1! M_2!} \qquad (8.17)$$

With the aid of Stirling's formula this becomes

$$\begin{aligned}\Delta A &= -k_B T \left[(M_1 + M_2) \ln (M_1 + M_2) - M_1 \ln M_1 - M_2 \ln M_2\right] \\ &= k_B T \left[M_1 \ln \frac{M_1}{M_1 + M_2} + M_2 \ln \frac{M_2}{M_1 + M_2}\right] \\ &= k_B T (M_1 + M_2) \left[x_1 \ln x_1 + x_2 \ln x_2\right] \qquad (8.18)\end{aligned}$$

This can be re-written as

$$\frac{\Delta A}{k_B T (M_1 + M_2)} = x_1 \ln x_1 + (1 - x_1) \ln (1 - x_1) \qquad (8.19)$$

Now the change in Helmholtz free energy is related to the change in energy and change in entropy of the condensed phase by

$$\Delta A = \Delta E - T \Delta S \qquad (8.20)$$

Hence, from Eq. (8.17),

$$\Delta E = -T^2 \frac{\partial}{\partial T} \left(\frac{\Delta A}{T}\right) = 0$$
$$\Delta S = k_B \ln \frac{(M_1 + M_2)!}{M_1! M_2!} \qquad ; \text{ perfect solution} \qquad (8.21)$$

There is no *energy* of mixing, but only an *entropy* of mixing. These relations define a *perfect solution*;

- With the aid of Eq. (8.19), the entropy of mixing in a perfect solution takes the form

$$\frac{\Delta S}{k_B (M_1 + M_2)} = -\left[x_1 \ln x_1 + (1 - x_1) \ln (1 - x_1)\right] \qquad (8.22)$$

- The quantities $\Delta A/k_{\rm B}T(M_1+M_2)$ and $\Delta S/k_{\rm B}(M_1+M_2)$ for a perfect solution are sketched as a function of the mole fraction x_1 in Fig. 8.1.

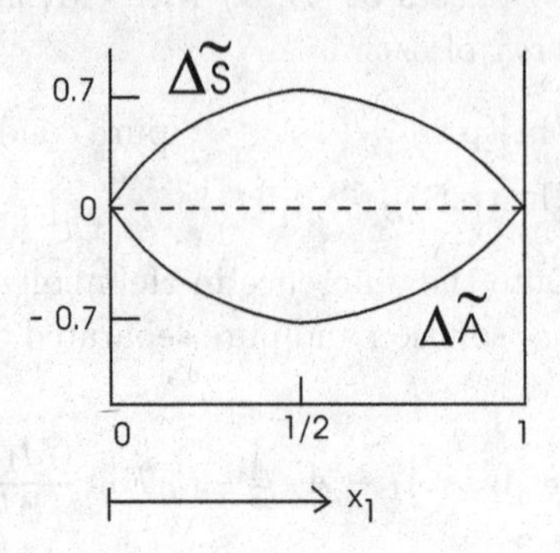

Fig. 8.2 The change in Helmholtz free energy $\Delta\tilde{A} = \Delta A/k_{\rm B}T(M_1 + M_2)$ and $\Delta\tilde{S} = \Delta S/k_{\rm B}(M_1 + M_2)$ for a perfect solution as a function of the mole fraction x_1. Note $\Delta\tilde{A}(1/2) = -\Delta\tilde{S}(1/2) = -\ln 2$.

- One concludes that the two components will always dissolve in one another at any x_1 in a perfect solution, since this *lowers the free energy.*[3]

Let us try to improve the treatment of the condensed phase. We still focus on two components; however, the discussion is applicable, for example, to assemblies and phenomena as diverse as

- Regular solutions
- Binary alloys
- Order-disorder transitions in crystals

8.1.2 *Regular Solutions*

Consider a more detailed description of the condensed phase, still within the framework of the *Einstein model* of independent localized systems. Assume the canonical partition function for a pure condensed phase with N systems has the general form[4]

$$\begin{aligned}
({\rm P.F.}) &= \left[\left(\frac{2\pi m k_{\rm B}T}{h^2}\right)^{3/2} J(T)\int e^{-\psi(r)/k_{\rm B}T}\, d\tau\right]^N \\
&\equiv [f(T,v)]^N
\end{aligned} \tag{8.23}$$

[3] See also Prob. 8.1.
[4] See [Rushbrooke (1949)].

Here

- The factor $J(T)$ is meant to describe the internal vibrations and rotations of an individual system;
- $\psi(r)$ is a *potential energy* for a system in the condensed phase, which varies as the system is displaced a distance r from its mean position;
- The integral $\int d\tau$ goes over one system, and v is the volume/system;
- The extension of this idea to a *perfect solution* with (N_A, N_B) constituents represents the canonical partition function of the condensed phase as

$$
\begin{aligned}
(\text{P.F.}) &= \left[\left(\frac{2\pi m_A k_B T}{h^2}\right)^{3/2} J_A(T) \int e^{-\psi_A(r)/k_B T}\, d\tau\right]^{N_A} \\
&\quad \times \left[\left(\frac{2\pi m_B k_B T}{h^2}\right)^{3/2} J_B(T) \int e^{-\psi_B(r)/k_B T}\, d\tau\right]^{N_B} g(N_A, N_B) \\
&= [f_A(T,v)]^{N_A}\, [f_B(T,v)]^{N_B}\, g(N_A, N_B)
\end{aligned} \tag{8.24}
$$

where

$$
g(N_A, N_B) \equiv \text{number of rearrangements of the } (A,B) \text{ systems} \tag{8.25}
$$

- Raoult's law is reproduced with the expression for a random distribution in Eq. (8.2)[5]

$$
g(N_A, N_B) = \frac{(N_A + N_B)!}{N_A! N_B!} \quad ; \text{ gives Raoult's law} \tag{8.26}
$$

8.1.2.1 *Improved Model of Localized Systems*

For the condensed phase, the Einstein model of independent localized systems can be improved in two ways:

(1) Assign a common *coordination number* Z for each system, corresponding to the number of nearest neighbors;[6]
(2) Include a constant *energy shift* depending on the nature of the neighbors

$$
\begin{aligned}
&\omega_{AA} \quad \text{for an AA pair} \\
&\omega_{BB} \quad \text{for a BB pair} \\
&\omega_{AB} \quad \text{for an AB pair}
\end{aligned} \tag{8.27}
$$

[5] See Prob. 8.2.
[6] This concept is, of course, precisely defined for a lattice.

This gives an additional *configuration energy* for the assembly

$$E_C = N_{AA}\omega_{AA} + N_{BB}\omega_{BB} + N_{AB}\omega_{AB} \quad \text{; configuration energy} \tag{8.28}$$

where (N_{AA}, N_{BB}, N_{AB}) are the total number of nearest neighbors of each type.

Consider the total number of nearest neighbors of the A systems. This number can be expressed in terms of the types of nearest neighbors according to

$$ZN_A = N_{AB} + 2N_{AA} \quad \text{; neighbors of the A systems} \tag{8.29}$$

The factor of 2 in the last term arises since all the AA neighbors are counted twice in this enumeration. Similarly, for the B systems

$$ZN_B = N_{AB} + 2N_{BB} \quad \text{; neighbors of the B systems} \tag{8.30}$$

The configuration energy in Eq. (8.28) can then be written as

$$\begin{aligned} E_C &= \frac{1}{2}ZN_A\,\omega_{AA} + \frac{1}{2}ZN_B\,\omega_{BB} + N_{AB}\left(\omega_{AB} - \frac{1}{2}\omega_{AA} - \frac{1}{2}\omega_{BB}\right) \\ &\equiv E_A^C + E_B^C + \omega N_{AB} \end{aligned} \tag{8.31}$$

In this expression:

- The first two terms are the configuration energies of the *pure condensed phases*;
- ω is the energy change on mixing and creation of an AB pair (see Fig. 8.3)

$$\omega = \left(\omega_{AB} - \frac{1}{2}\omega_{AA} - \frac{1}{2}\omega_{BB}\right) \tag{8.32}$$

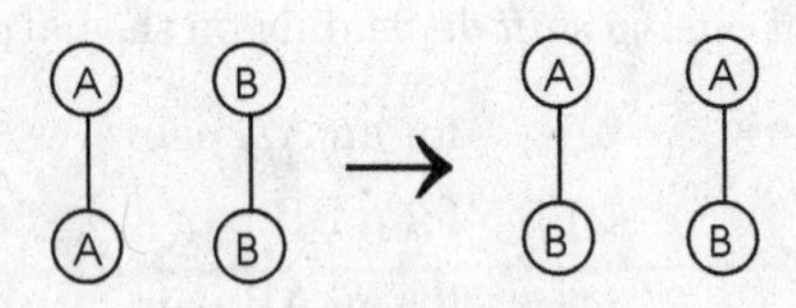

Fig. 8.3 The pure materials are mixed, and for each AA and BB interaction that is eliminated by the interchange of an A and and a B system, *two* AB interactions are introduced.

8.1.2.2 *Configuration Partition Function*

With the inclusion of a term e^{-E_C/k_BT}, which depndes on N_{AB}, the partition function in Eq. (8.24) gets extended in this improved model to read

$$(\text{P.F.}) = \left(f_A^{N_A} e^{-E_A^C/k_BT}\right)\left(f_B^{N_B} e^{-E_B^C/k_BT}\right) \times \sum_{N_{AB}} g(N_A, N_B, N_{AB}) e^{-\omega N_{AB}/k_BT} \quad \text{; regular solution} \tag{8.33}$$

The sum here is over all configurations with a given number (N_A, N_B) of component systems, and $g(N_A, N_B, N_{AB})$ counts those configurations with a number N_{AB} of AB nearest neighbors.

The free energy of the assembly is correspondingly modified to

$$A_{\text{assembly}} = -k_BT \ln (\text{P.F.}) = A_A + A_B + \Delta A \tag{8.34}$$

The change in free energy upon forming a solution from the pure materials is thus given by

$$\Delta A = -k_BT \ln (\text{P.F.})_C \tag{8.35}$$

$$(\text{P.F.})_C = \sum_{N_{AB}} g(N_A, N_B, N_{AB}) e^{-\omega N_{AB}/k_BT} \quad \text{; configuration (P.F.)}$$

Assemblies described by the canonical partition function in Eq. (8.33), with a free energy of mixing given by Eqs. (8.35), are said to form *regular solutions*. The evaluation of the configuration partition function $(\text{P.F.})_C$ in Eqs. (8.35) is one of the *fundamental problems in statistical mechanics*.

Note that the corresponding *energy* of mixing is given by

$$\Delta E = -T^2 \frac{\partial}{\partial T}\left(\frac{\Delta A}{T}\right) = \frac{\omega \sum_{N_{AB}} N_{AB}\, g(N_A, N_B, N_{AB}) e^{-\omega N_{AB}/k_BT}}{\sum_{N_{AB}} g(N_A, N_B, N_{AB}) e^{-\omega N_{AB}/k_BT}} \tag{8.36}$$

The configuration partition function has two useful general properties:

(1) If there is no energy change upon nearest-neighbor interchange and $\omega = 0$, then one simply has the total number of configurations with random mixing given in Eq. (8.2);

$$\sum_{N_{AB}} g(N_A, N_B, N_{AB}) = \frac{(N_A + N_B)!}{N_A! N_B!} \tag{8.37}$$

(2) If $\omega = 0$, the mean number of AB pairs, denoted by $\overset{o}{\overline{N}}_{AB}$,[7] is given by

$$\overset{o}{\overline{N}}_{AB} \equiv \frac{\sum_{N_{AB}} N_{AB}\, g(N_A, N_B, N_{AB})}{\sum_{N_{AB}} g(N_A, N_B, N_{AB})} = Z\frac{N_A N_B}{N_A + N_B} \tag{8.38}$$

This is demonstrated as follows: Since with $\omega = 0$ there is random mixing of the AB systems, one can use a simple probability argument. Consider two neighboring sites 1 and 2 (Fig. 8.4).

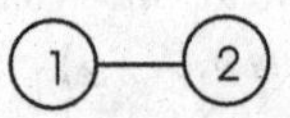

Fig. 8.4 Two neighboring sites used to compute the mean number of AB pairs $\overset{o}{\overline{N}}_{AB}$ in the case of random mixing.

Then

- The probability of finding an A on site 1 is $N_A/(N_A + N_B)$;
- The probability of finding a B on site 2 is $N_B/(N_A + N_B)$;
- The probabilty of finding an AB pair is then

$$P[A(1)B(2)] + P[A(2)B(1)] = \frac{2N_A N_B}{(N_A + N_B)^2} \tag{8.39}$$

- The total number of neighboring sites is $(Z/2)(N_A + N_B)$;
- Hence the mean number of AB pairs is

$$\overset{o}{\overline{N}}_{AB} = \frac{Z}{2}(N_A + N_B)\frac{2N_A N_B}{(N_A + N_B)^2} = Z\frac{N_A N_B}{N_A + N_B} \tag{8.40}$$

8.1.2.3 *Bragg-Williams Approximation*

The Bragg-Williams approximation replaces N_{AB} in the exponent in the configuration partition function in Eq. (8.35) by its mean value computed with $\omega = 0$; this is the expression in Eq. (8.38). The exponential factors out of the sum, and Eq. (8.37) can then be invoked. Thus

$$\begin{aligned}(\text{P.F.})_C &\approx \sum_{N_{AB}} g(N_A, N_B, N_{AB}) \exp\left[-\frac{\omega}{k_{\rm B}T}\overset{o}{\overline{N}}_{AB}\right] \quad ; \text{ Bragg-Williams} \\ &= \frac{(N_A + N_B)!}{N_A! N_B!} \exp\left[-\frac{\omega Z}{k_{\rm B}T}\frac{N_A N_B}{N_A + N_B}\right] \end{aligned} \tag{8.41}$$

[7]We use the notation of [Rushbrooke (1949)].

The free energy and energy of mixing in Eqs. (8.35)–(8.36) then follow as

$$\Delta A = \omega Z \frac{N_A N_B}{N_A + N_B} - k_{\rm B}T \ln \frac{(N_A + N_B)!}{N_A! N_B!}$$

$$\Delta E = \omega \overset{o}{\bar{N}}_{AB} = \omega Z \frac{N_A N_B}{N_A + N_B} \tag{8.42}$$

The entropy of mixing is thus given by

$$T\Delta S = \Delta E - \Delta A = k_{\rm B}T \ln \frac{(N_A + N_B)!}{N_A! N_B!} \tag{8.43}$$

With the aid of Eq. (8.22), these results can be re-written as

$$\frac{\Delta A}{k_{\rm B}T(N_A + N_B)} = \frac{\omega Z}{k_{\rm B}T} x(1-x) + [x \ln x + (1-x) \ln (1-x)]$$

$$\frac{\Delta S}{k_{\rm B}(N_A + N_B)} = -[x \ln x + (1-x) \ln (1-x)]$$

$$\frac{\Delta E}{(N_A + N_B)} = \omega Z x(1-x) \qquad ; \text{ Bragg-Williams} \tag{8.44}$$

Several comments:

- Since these expressions are symmetric in $x \leftrightarrow (1-x)$, x is now the mole fraction of either component in the solution;
- The mixing entropy ΔS is positive for $0 < x < 1$, and its contribution to the free energy $-T\Delta S$ is correspondingly negative;
- The most interesting case is a positive energy of mixing with $\omega > 0$, for then "*the energy fights the entropy in the free energy*";
- As $T \to \infty$, the entropy term dominates, The Helmholtz free energy will always be lowered upon mixing, and hence the two components will always go into solution at high enough temperature;
- As the temperature is lowered with $\omega > 0$, one arrives at the situation illustrated in Fig. 8.5;
- Consider the second derivative of the free energy with respect to x

$$\frac{\partial^2}{\partial x^2}\left[\frac{\Delta A}{k_{\rm B}T(N_A + N_B)}\right] = -\frac{2\omega Z}{k_{\rm B}T} + \left[\frac{1}{x} + \frac{1}{1-x}\right]$$

$$\geq -\frac{2\omega Z}{k_{\rm B}T} + 4 \tag{8.45}$$

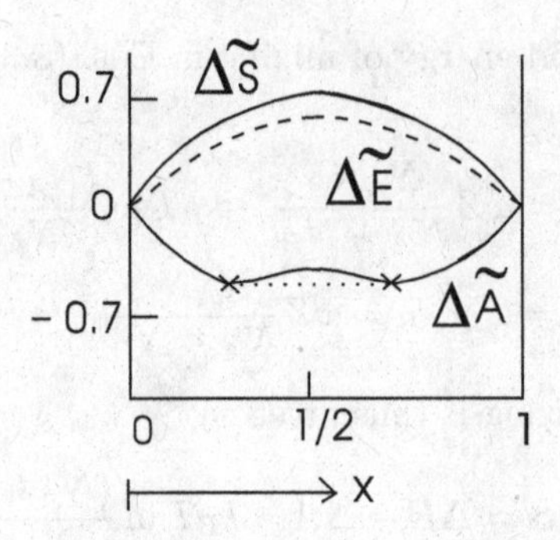

Fig. 8.5 Free energy and entropy of mixing at low-enough temperature in the Bragg-Williams approximation, with $\omega > 0$ in Eqs. (8.44). Here $\Delta\tilde{A} \equiv \Delta A/k_{\rm B}T(N_A + N_B)$ and $\Delta\tilde{S} \equiv \Delta S/k_{\rm B}(N_A + N_B)$. The shape of $\Delta\tilde{E} \equiv \Delta E/(N_A + N_B)$ is indicated, and the first term in $\Delta\tilde{A}$ is $\Delta\tilde{E}/k_{\rm B}T$. Compare Prob. 8.3.

This is positive for $T > T_C$, where T_C is the temperature at which this second derivative first vanishes

$$k_{\rm B}T_C = \frac{1}{2}\omega Z \tag{8.46}$$

It vanishes first at a mole fraction of $x = 1/2$.

- Below T_C, there will be a region centered on $x = 1/2$ for which the second derivative in Eq. (8.45) is negative, as illustrated in Fig. 8.5;
- The free energy of the assembly can then be *lowered* by separating into *two* solutions at the concentrations indicted with crosses in Fig. 8.5;[8]
- The Bragg-Williams approximation provides us with a simple model of a *phase transition in a regular solution as the temperature is lowered.*

8.1.2.4 *Quasi-Chemical Approximation*

A repetition of the arguments leading to Eq. (8.40) implies that in the Bragg-Williams approximation one has in addition

$$\overset{o}{N}_{AA} = \frac{Z}{2}\frac{N_A^2}{N_A + N_B} \quad ; \; \overset{o}{N}_{BB} = \frac{Z}{2}\frac{N_B^2}{N_A + N_B} \tag{8.47}$$

Hence

$$\frac{\overset{o}{N}_{AA}\overset{o}{N}_{BB}}{\left(\overset{o}{N}_{AB}\right)^2} = \frac{1}{4} \quad ; \text{ Bragg-Williams} \tag{8.48}$$

[8]See Prob. 8.4.

The quasi-chemical approximation, also known as the Guggenheim approximation, or Bethe-Peierls approximation, looks at this instead as a *chemical reaction* for nearest-neighbors

$$AA + BB \rightleftharpoons 2AB \tag{8.49}$$

There will be an equilibrium constant for this reaction

$$\frac{N_{AA}N_{BB}}{N_{AB}^2} = K_{\rm eq}(T)$$

$$K_{\rm eq}(T) = \frac{({\rm p.f.})_{\rm AA}({\rm p.f.})_{\rm BB}}{({\rm p.f.})_{\rm AB}^2} \tag{8.50}$$

Two features are then incorporated in the partition functions:

(1) There is a bond dissociation energy of 2ω;
(2) There is a symmetry factor of $\sigma = 2$ for each of the AA and BB bonds.

It follows that

$$K_{\rm eq}(T) = \frac{1}{4}e^{2\omega/k_{\rm B}T} \tag{8.51}$$

Equation (8.48) thus gets generalized to

$$\frac{N_{AA}N_{BB}}{N_{AB}^2} = \frac{1}{4}e^{2\omega/k_{\rm B}T} \qquad ; \text{quasi-chemical approximation} \tag{8.52}$$

One consequence of the quasi-chemical approximation is that the transition temperature now becomes (see [Rushbrooke (1949)])

$$k_{\rm B}T_C = \frac{\omega}{\ln\left[Z/(Z-2)\right]} \tag{8.53}$$

8.2 Order-Disorder Transitions in Crystals

We turn next to the topic of order-disorder transitions in crystals.

8.2.1 λ-*Point Transitions*

A phase transition of the *first kind*, such as an ordinary gas-liquid transition, has a discontinuity in the heat capacity C_V. In this case, two forms of the material, such as a liquid and its vapor, can co-exist in equilibrium at the same temperature (see Fig. 5.18). A phase transition of the *second kind* is one in which the heat capacity C_V is continuous, but where there is a discontinuity in the *slope* of C_V at the transition temperature (see Fig. 7.7).

A λ-point transition is a phase transition of the second kind in which the heat capacity becomes infinite at the transition temperature, as illustrated in Fig. 8.6.

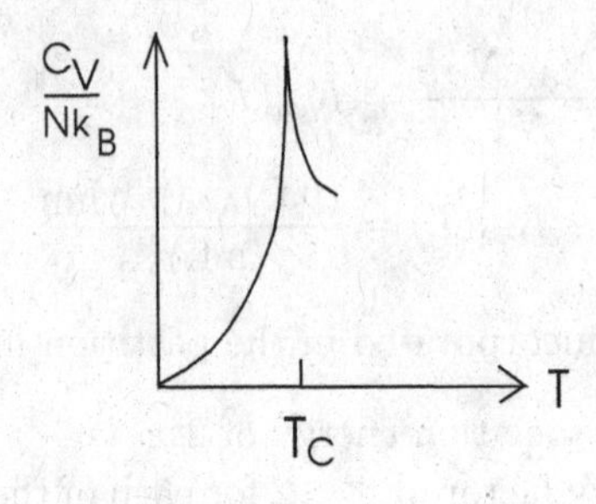

Fig. 8.6 Heat capacity C_V/Nk_B for a λ-point transition.

There are many examples of λ-point transitions:

- The ortho-para transition in solid H_2;
- The phase transition in β-brass, formed from (Cu, Zn), which goes into a body-centered cubic lattice with distinct, ordered centers below $T_C = 742\,^{\circ}$K (Fig. 8.7);

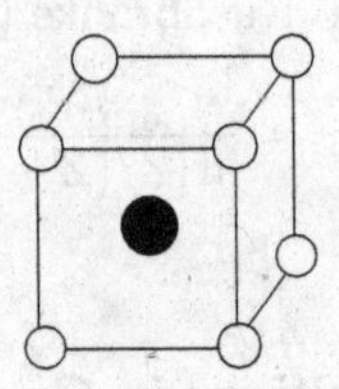

Fig. 8.7 Body-centered cubic lattice, with distinct, ordered centers.

- Paramagnetic-ferromagnetic transitions at the Curie point;
- The transition ${}^4\text{He}(I) \rightarrow {}^4\text{He}(II)$ at $T_\lambda = 2.17\,^{\circ}$K;
- The phase transition in crystalline ammonium chloride NH_4Cl, which shall form the paradigm for our discussion of order-disorder transitions in crystals. Here the ammonium ions $(NH_4)^+$ sit at the center of a body-centered cubic lattice of Cl^- ions. The ammonium ion has the

shape of a tetrahedron, and this tetrahedron can have different orientations between cells, as illustrated in Fig. 8.8. Below the transition temperature $T_C = 243\,^{\circ}\mathrm{K}$, the tetrahedra are completely ordered.

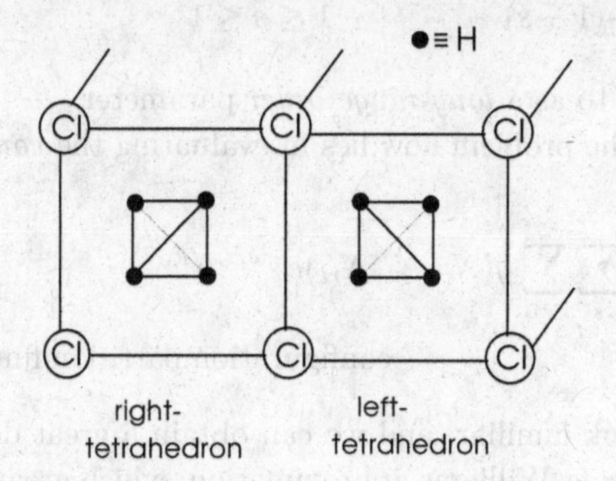

Fig. 8.8 Crystalline ammonium chloride. The ammonium ion $(NH_4)^+$ forms a tetrahedron at the center of a body-centered cubic lattice of chloride ions Cl^-, and the tetrahedra can have different left- and right-orientations between cells.

8.2.2 *Configuration Partition Function*

In NH_4Cl, a nearest neighbor rl pair of $(NH_4)^+$ ions has a slightly higher energy than a rr or ll pair. As before, we define the unlike-pair creation energy as ω. The coordination number for the $(NH_4)^+$ ions in the lattice is

$$Z = 6 \qquad ; \text{ coordination number for } (NH_4)^+ \text{ ions} \tag{8.54}$$

We work in the canonical ensemble formed from the set of $(NH_4)^+$ ions in the crystal, and write the canonical partition function as[9]

$$(\mathrm{P.F.}) = (\mathrm{P.F.})_{\mathrm{vib}}(\mathrm{P.F.})_C \tag{8.55}$$

The partition function $(\mathrm{P.F.})_{\mathrm{vib}}$ describes the internal structure of the individual $(NH_4)^+$ ions, and we assume it to be the *same for all*

[9]The lattice of Cl^- ions, although it can conduct heat, is assumed to be otherwise inert here.

configurations. Write the number of right- and left-tetrahedrons as

$$N_r \equiv \frac{N}{2}(1+s) \qquad ; \text{ s is long-range order parameter}$$
$$N_l \equiv \frac{N}{2}(1-s) \qquad -1 \leq s \leq 1 \tag{8.56}$$

Here s is referred to as a *long-range order* parameter.

The meat of the problem now lies in evaluating the *configuration partition function*

$$(\text{P.F.})_C = \sum_s \sum_{N_{rl}} g(N_r, N_l, N_{rl}) e^{-\omega N_{rl}/k_B T}$$
$$; \text{ configuration partition function} \tag{8.57}$$

This begins to look familiar, and we can obtain a great deal of insight by employing the Bragg-Williams approximation, which greatly simplifies the problem.

8.2.2.1 *Bragg-Williams Approximation*

The Bragg-Williams approximation replaces the number of rl pairs N_{rl} in the exponent of the configuration partition function by the mean number calculated with a random distribution of r and l tetrahedra. From Eq. (8.38), this is given by

$$\bar{N}_{rl} = Z\frac{N_r N_l}{N_r + N_l} \tag{8.58}$$

The exponential factors from the second sum in Eq. (8.57), and Eq. (8.37) can then be used on the sum over N_{rl}. Thus

$$(\text{P.F.})_C = \sum_s \frac{N!}{[N(1+s)/2]![N(1-s)/2]!} \exp\left[-\frac{NZ\omega}{4k_B T}(1-s^2)\right]$$
$$; \text{ Bragg-Williams} \tag{8.59}$$

Once again, we can just use the largest term in the sum, obtained by maximizing the logarithm of the summand $t(s)$ with respect to s. With the aid of Stirling's formula, one has

$$\ln t(s) = N\ln N - N - \frac{N}{2}(1+s)\ln\frac{N}{2}(1+s) + \frac{N}{2}(1+s)$$
$$-\frac{N}{2}(1-s)\ln\frac{N}{2}(1-s) + \frac{N}{2}(1-s) - \frac{NZ\omega}{4k_B T}(1-s^2) \tag{8.60}$$

This simplifies to

$$\ln t(s) = N \ln N - N \ln \frac{N}{2} - \frac{N}{2} [(1+s) \ln (1+s) + (1-s) \ln (1-s)] - \frac{NZ\omega}{4k_{\rm B}T}(1-s^2) \tag{8.61}$$

The maximum is obtained by setting the derivative with respect to s equal to zero

$$\frac{1}{N}\frac{\partial}{\partial s} \ln t(s) = -\frac{1}{2} [\ln (1+s) - \ln (1-s)] + \frac{Z\omega s}{2k_{\rm B}T} = 0 \tag{8.62}$$

Hence

$$\ln \frac{1+s}{1-s} = \frac{Z\omega s}{k_{\rm B}T}$$

$$\text{or ;} \qquad (1+s) = (1-s)e^{Z\omega s/k_{\rm B}T}$$

$$e^{Z\omega s/k_{\rm B}T} - 1 = s\left(e^{Z\omega s/k_{\rm B}T} + 1\right) \tag{8.63}$$

Thus, finally,

$$s = \tanh\left(\frac{Z\omega s}{2k_{\rm B}T}\right) \qquad ; \text{ determines } s \tag{8.64}$$

This is a transcendental equation for s, which can be solved graphically. The two curves $f(s) = \tanh(Z\omega s/2k_{\rm B}T)$ and $f(s) = s$ are shown in Fig. 8.9. Their points of intersection provide the solution for s. There are two possibilities:

(1) At high T, the only solution is at $s = 0$. This corresponds to complete *disorder* in Eqs. (8.56);
(2) At low-enough T, there will be three solutions: one at $s = 0$, and a pair that as $T \to 0$ occur at $s = \pm 1$. The latter correspond to a *complete ordering* in Eqs. (8.56). We leave it as a problem to show that the ordered pair has the lower free energy, and therefore represents the stable configuration. (Prob. 8.8).

It is clear from Fig. 8.9 that the transition between the two cases occurs when the slope of the hyperbolic tangent at the origin is equal to one. Thus the transition temperature is given by

$$\left[\frac{d}{ds}\tanh\left(\frac{Z\omega s}{2k_{\rm B}T}\right)\right]_{s=0} = \frac{Z\omega}{2k_{\rm B}T_C \cosh^2(0)} = \frac{Z\omega}{2k_{\rm B}T_C} = 1 \tag{8.65}$$

Hence[10]

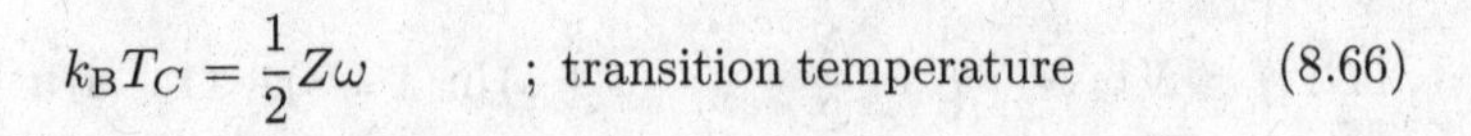

$$k_{\rm B}T_C = \frac{1}{2}Z\omega \qquad ; \text{ transition temperature} \tag{8.66}$$

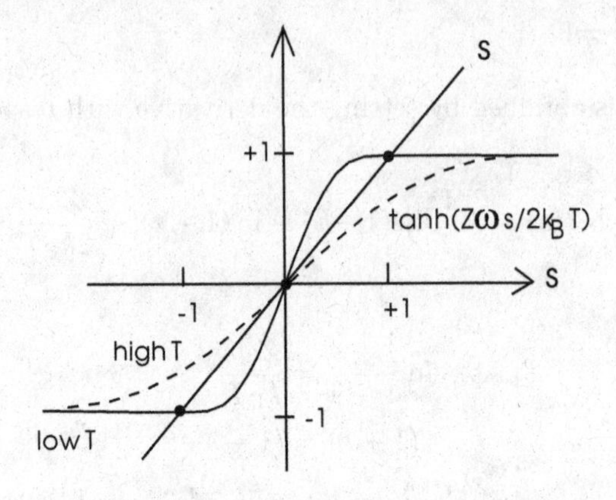

Fig. 8.9 Graphical solution of Eq. (8.64). The curves $f(s) = \tanh(Z\omega s/2k_{\rm B}T)$ and $f(s) = s$ are shown, and their points of intersection provide the solutions for s. The former curve is sketched for two values of T. At high T, the only point of intersection is at $s = 0$; at low enough T, there will be three points of intersection, indicated by the three dots.

The corresponding configuration Helmholtz free energy is obtained from Eq. (8.61) as

$$\begin{aligned} A_C &= -k_{\rm B}T \ln({\rm P.F.})_C \\ &= -Nk_{\rm B}T\left\{\ln 2 - \frac{1}{2}\left[(1+s)\ln(1+s) + (1-s)\ln(1-s)\right]\right\} \\ &\quad + \frac{1}{4}NZ\omega(1-s^2) \end{aligned} \tag{8.67}$$

We write this as

$$\begin{aligned} A_C &\equiv -TS_C + E_C \\ \frac{S_C}{Nk_{\rm B}} &= \ln 2 - \frac{1}{2}\left[(1+s)\ln(1+s) + (1-s)\ln(1-s)\right] \\ \frac{E_C}{N} &= \frac{Z\omega}{4}(1-s^2) \end{aligned} \tag{8.68}$$

The molar configuration entropy S_C/R is sketched in Fig. (8.10).

---

[10] Compare Eq. (8.46).

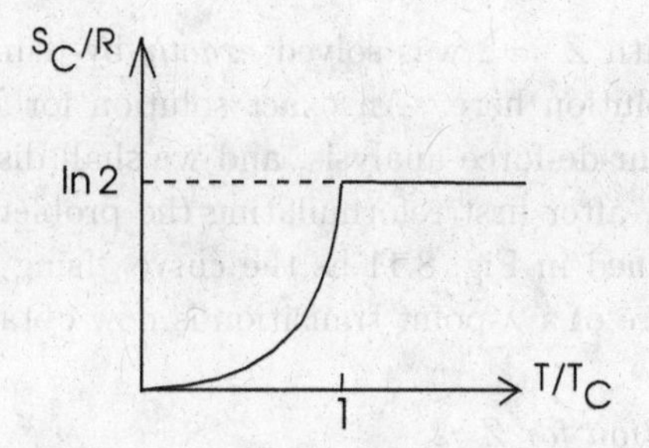

Fig. 8.10 Sketch of the molar configuration entropy S_C/R obtained from Eq. (8.64) and the second of Eqs. (8.68). Compare Prob. 8.9.

The corresponding configuration specific heat can be obtained by

$$C_V = T\frac{\partial S_C}{\partial T} \tag{8.69}$$

This is sketched in Fig. 8.11. The result is *qualitatively* correct.

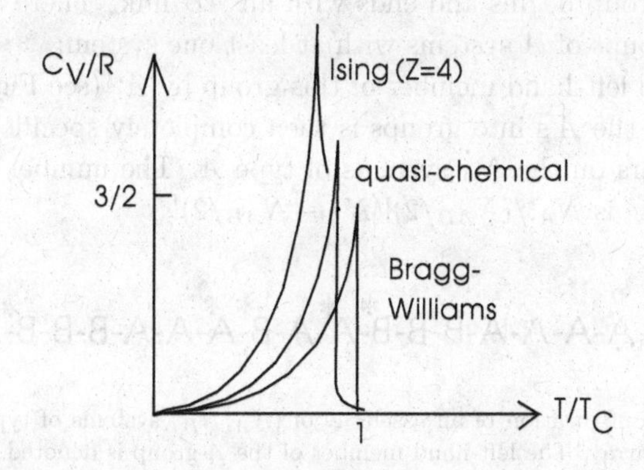

Fig. 8.11 Sketch of the molar heat capacity in the Bragg-Williams approximation obtained from Eq. (8.69) and the results in Fig. 8.10. (Compare Prob. 8.9.) Two other results are also shown (see text).

An improvement over Bragg-Williams is obtained with the quasi-chemical approximation of the last section. The improved transition temperature is given by Eq. (8.53)

$$k_{\rm B}T_C = \frac{\omega}{\ln\left[Z/(Z-2)\right]} \quad ; \text{ quasi-chemical} \tag{8.70}$$

The improved heat capacity is also sketched in Fig. 8.11.

The problem with $Z = 2$ was solved *exactly* by [Ising (1925)], and we go through that solution here. An exact solution for Z=4 was obtained by Onsager in a tour-de-force analysis, and we shall discuss that solution in the next section, after first reformulating the problem. That result for $Z = 4$ is also sketched in Fig. 8.11 as the curve "Ising (Z=4)". Note the characteristic feature of a λ-point transition is now obtained.

8.2.2.2 *Ising Solution for Z=2*

Consider a one-dimensional assembly composed of of N_A systems of type A and N_B systems of type B. The goal is to evaluate the configuration partition function

$$(\text{P.F.})_C = \sum_{N_{AB}} g(N_A, N_B, N_{AB}) e^{-\omega N_{AB}/k_B T} \tag{8.71}$$

We make a series of observations:

- Every A group begins and ends with an AB link. There are therefore $N_{AB}/2$ groups of A systems with at least one system;
- Denote the left-hand member of this group by $A^\star$ (see Fig. 8.12). The division of the A's into groups is then completely specified by placing $N_{AB}/2$ stars on the N_A systems of type A. The number of ways this can be done is $N_A!/(N_{AB}/2)!(N_A - N_{AB}/2)!$;

B*-A*-A-A-A-A-B-B-B*-A*-A-B*-A*-A-A-B-B-B*-A*-

Fig. 8.12 One configuration of an assembly of (N_A, N_B) systems of types (A, B) in a one-dimensional array. The left-hand member of the A-group is denoted with a star, as is the right-hand member of the B-group.

- Denote the right-hand member of the $N_{AB}/2$ groups of B systems by $B^\star$. The number of ways this can be done is $N_B!/(N_{AB}/2)!(N_B - N_{AB}/2)!$;
- Now arrange the systems so the ith $A^\star$ system immediately follows the ith $B^\star$ system. This uniquely specifies the arrangement of the A's and B's in a straight line;
- The configuration in Fig. 8.12 starts with a B on the left, and one could just as well have started with an A. This is an end effect, and end effects will only contribute to the final result in $O(1/N)$ where $N = N_A + N_B$.

- For large N, the total number of configurations with a given N_{AB} is therefore

$$g(N_A, N_B, N_{AB}) = \frac{N_A!}{(N_A - N_{AB}/2)!(N_{AB}/2)!} \frac{N_B!}{(N_B - N_{AB}/2)!(N_{AB}/2)!} \quad (8.72)$$

The configuration partition function for this problem therefore takes the form

$$(\text{P.F.})_C = \sum_{N_{AB}} \frac{N_A!}{(N_A - N_{AB}/2)!(N_{AB}/2)!} \frac{N_B!}{(N_B - N_{AB}/2)!(N_{AB}/2)!} \times e^{-\omega N_{AB}/k_B T} \quad (8.73)$$

- One can now simply pick out the largest term in the sum using Stirling's formula, since all of the N's are large.[11] If the summand is denoted by $t(N_{AB})$, then the derivative of $\ln t(N_{AB})$ is[12]

$$\left[\frac{\partial \ln t(N_{AB})}{\partial N_{AB}}\right]_{N_{AB}^\star} = \frac{1}{2}\ln\left(N_A - \frac{1}{2}N_{AB}^\star\right) + \frac{1}{2}\ln\left(N_B - \frac{1}{2}N_{AB}^\star\right) - \ln\frac{1}{2}N_{AB}^\star - \frac{\omega}{k_B T} = 0 \quad (8.74)$$

Thus

$$\left(N_A - \frac{1}{2}N_{AB}^\star\right)\left(N_B - \frac{1}{2}N_{AB}^\star\right) = \left(\frac{N_{AB}^\star}{2}\right)^2 e^{2\omega/k_B T} \quad (8.75)$$

From Eqs. (8.29)–(8.30), we have the following relations for $Z = 2$

$$2N_A = 2N_{AA}^\star + N_{AB}^\star$$
$$2N_B = 2N_{BB}^\star + N_{AB}^\star \quad (8.76)$$

Hence Eq. (8.75) can be written as

$$\frac{N_{AA}^\star N_{BB}^\star}{(N_{AB}^\star)^2} = \frac{1}{4}e^{2\omega/k_B T} \quad (8.77)$$

The quasi-chemical result in Eq. (8.52) is exact here!

[11] The sum can be done exactly with the (G.P.F.)—see [Rushbrooke (1949)].
[12] Use $d(x\ln x - x)/dN_{AB} = [d(x\ln x - x)/dx](dx/dN_{AB}) = \ln x\,(dx/dN_{AB})$.

• Equation (8.70) then implies

$$k_{\rm B}T_C = 0 \qquad ; \text{ for } Z = 2 \tag{8.78}$$

There is no phase transition with a one-dimensional array. Entropy always wins.

8.3 The Ising Model

Before continuing, we reformulate the problem we have been studying.

8.3.1 *Heisenberg Hamiltonian*

The Heisenberg hamiltonian was originally introduced in the study of ferromagnetism arising from the electron spins. Consider electrons on neighboring atoms in a solid, and for concreteness, go back to the electron wave functions for the simplest diatomic molecule in Fig. 3.5, as sketched again in Fig. 8.13.

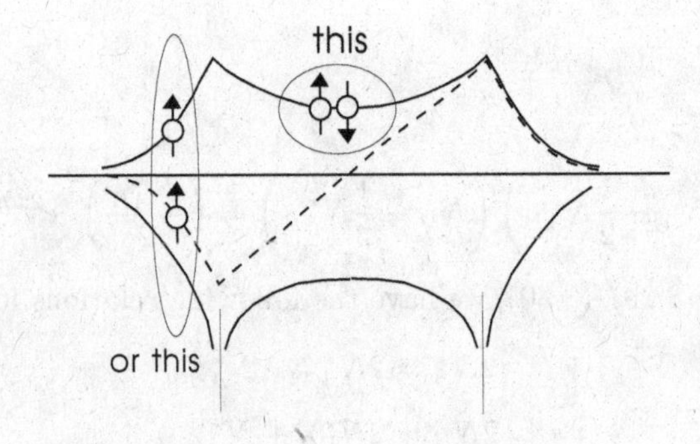

Fig. 8.13 Electron spin states in a diatomic molecule.

The Pauli principle prohibits two identical electrons from going into the same state, and hence two electrons in the same orbital must go into opposite spin states. If the two electrons have the same spin, then one of them must go into a distinct orbital. Thus the energy of the electrons depends on their *relative spins.* Although it is ultimately a reflection of the Coulomb interaction between the electrons, one can consider this physics to produce an *effective interaction between the electron spins.*[13]

Recall that

$$\begin{aligned} \boldsymbol{\sigma}_1 \cdot \boldsymbol{\sigma}_2 &= +1 \qquad ; \text{ triplet state} \\ &= -3 \qquad ; \text{ singlet state} \end{aligned} \tag{8.79}$$

The Heisenberg hamiltonian is an effective interaction between the electron spins of the form

$$H_{\text{Heisenberg}} = -J \sum_{ij} \boldsymbol{\sigma}(i) \cdot \boldsymbol{\sigma}(j) \qquad ; \text{ Heisenberg hamiltonian} \quad \text{nearest-neighbor sum} \tag{8.80}$$

Here J is a constant and the sum goes over nearest neighbors. For a ferromagnet, $J > 0$. The presence of the non-commuting components (σ_x, σ_y) complicates this interaction, and it was further simplified by Ising.

8.3.2 *One-Dimensional Ising Model*

While retaining much of the physics, the Heisenberg hamiltonian can be simplified by keeping just the z-components $\sigma_z(i)\sigma_z(j)$ in the spin-spin interaction in Eq. (8.80). In one-dimension, the hamiltonian of the Ising model is then

$$\begin{aligned} H_{\text{Ising}} &= -J \sum_{\kappa=1}^{N} \sigma_z(\kappa)\sigma_z(\kappa+1) \\ &\equiv -\frac{\omega}{2} \sum_{\kappa=1}^{N} S_\kappa S_{\kappa+1} \qquad ; \text{ Ising model} \end{aligned} \tag{8.81}$$

Here $S_\kappa = \pm 1$ denotes (twice) the z-component of the spin of system κ, and the second line explicitly exhibits the spin matrix elements of the first.

Now observe that this is *exactly the problem we have been studying!* Consider two nearest neighbors. The interaction energy is

$$\begin{aligned} H_{\uparrow\uparrow} &= -\frac{\omega}{2} \equiv \omega_{AA} \\ H_{\downarrow\downarrow} &= -\frac{\omega}{2} \equiv \omega_{BB} \\ H_{\uparrow\downarrow} = H_{\downarrow\uparrow} &= +\frac{\omega}{2} \equiv \omega_{AB} \end{aligned} \tag{8.82}$$

[13] For electrons with the same spin, the energy also involves both a direct and an exchange integral over the Coulomb repulsion between them; it is the exchange interaction that is responsible for the binding of metals (see [Fetter and Walecka (2003)]).

Hence, just as in Eq. (8.32), the unlike pair creation energy is

$$\omega_{AB} - \frac{1}{2}\omega_{AA} - \frac{1}{2}\omega_{BB} = \omega \tag{8.83}$$

For a ferromagnet, $\omega > 0$.

8.3.2.1 *Canonical Partition Function*

The spin array for the one-dimensional Ising model is illustrated in Fig. 8.14. To simplify the problem, connect the two ends of the chain and apply periodic boundary conditions

$$S_{N+1} = S_1 \qquad ; \text{ p.b.c.} \tag{8.84}$$

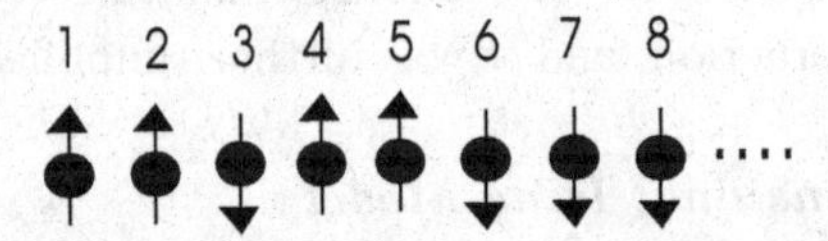

Fig. 8.14 Electron spins $S_\kappa = \pm 1$, with $\kappa = 1, 2, \cdots, N$, in the one-dimensional Ising model. We connect the ends and apply periodic boundary conditions so that $S_{N+1} = S_1$.

The canonical partition function for this spin assembly is given by[14]

$$(\text{P.F.}) = \sum_{S_1} \cdots \sum_{S_N} \exp\left[\frac{\omega}{2k_BT}\sum_{\kappa=1}^{N} S_\kappa S_{\kappa+1}\right] \qquad ; S_\kappa = \pm 1 \tag{8.85}$$

This problem is now identical to that studied previously, where the individual numbers $(N_A, N_B) = (N_\uparrow, N_\downarrow)$ are allowed to vary while their sum is kept fixed

$$N_\uparrow + N_\downarrow = N \qquad ; \text{ fixed} \tag{8.86}$$

8.3.2.2 *Matrix Solution*

The *matrix solution* to this problem is due to [Kramers and Wannier (1941)] (see [Wannier (1987)]). Since it is very instructive, we go through that analysis here. Define a 2×2 matrix $\underline{P}$ through its matrix elements

$$\langle S|P|S'\rangle \equiv e^{\omega SS'/2k_BT} \qquad ; (S, S') = \pm 1 \tag{8.87}$$

[14] We are here discussing the canonical ensemble of these spin assemblies.

Thus

$$\begin{aligned}\langle +1|P|+1\rangle = \langle -1|P|-1\rangle = e^{\omega/2k_BT} \\ \langle +1|P|-1\rangle = \langle -1|P|+1\rangle = e^{-\omega/2k_BT}\end{aligned} \qquad (8.88)$$

Explicitly, in matrix notation,

$$\underline{P} = \begin{pmatrix} e^{\omega/2k_BT} & e^{-\omega/2k_BT} \\ e^{-\omega/2k_BT} & e^{\omega/2k_BT} \end{pmatrix} \qquad (8.89)$$

The canonical partition function in Eq. (8.85) can then be written as

$$\begin{aligned}(\text{P.F.}) &= \sum_{S_1} \cdots \sum_{S_N} \prod_{\kappa=1}^{N} \exp\left[\frac{\omega}{2k_BT} S_\kappa S_{\kappa+1}\right] \\ &= \sum_{S_1} \cdots \sum_{S_N} \langle S_1|P|S_2\rangle\langle S_2|P|S_3\rangle \cdots \langle S_{N-1}|P|S_N\rangle\langle S_N|P|S_{N+1}\rangle\end{aligned} \qquad (8.90)$$

From the p.b.c in Eq. (8.84), the last factor is

$$\langle S_N|P|S_{N+1}\rangle = \langle S_N|P|S_1\rangle \qquad ;\ \text{p.b.c.} \qquad (8.91)$$

This ties the first and last matrix indices together, and the partition function becomes the *trace* of the matrix product

$$\begin{aligned}(\text{P.F.}) &= \text{trace}\,[\underline{P}\,\underline{P}\,\underline{P}\cdots\underline{P}] \qquad ;\ N \text{ factors} \\ &= \text{trace}\,[\underline{P}^N]\end{aligned} \qquad (8.92)$$

This expression is analyzed as follows:

- The matrix $\underline{P}$ is a real symmetric matrix. It can be diagonalized with an orthogonal transformation[15]

$$\underline{U}\,\underline{P}\,\underline{U}^{-1} = \underline{P}_D = \begin{pmatrix} \lambda_+ & 0 \\ 0 & \lambda_- \end{pmatrix} \qquad (8.93)$$

- The eigenvalues (λ_+, λ_-) of the real symmetric matrix $\underline{P}$ are real. Let λ_+ be the *largest* eigenvalue;
- Insert $\underline{U}^{-1}\underline{U}$ everywhere in the matrix product in Eq. (8.92). Since the trace is invariant under cyclic permutations, this becomes

$$\text{trace}\,[\underline{P}^N] = \text{trace}\left[\left(\underline{U}\,\underline{P}\,\underline{U}^{-1}\right)^N\right] = \text{trace}\,[\underline{P}_D^N] \qquad (8.94)$$

[15]See Prob. 8.14.

- Then

$$\text{trace}\left[\underline{P}_D^N\right] = \text{trace}\begin{pmatrix} \lambda_+^N & 0 \\ 0 & \lambda_-^N \end{pmatrix} = \lambda_+^N + \lambda_-^N$$

$$\to \lambda_+^N \qquad ; \, N \to \infty \tag{8.95}$$

As $N \to \infty$, the canonical partition function for the one-dimensional Ising model is given entirely by the largest eigenvalue of the matrix $\underline{P}$ in Eq. (8.89)!

$$(\text{P.F.}) = \lambda_+^N \qquad ; \, N \to \infty \tag{8.96}$$

- The eigenvalues of $\underline{P}$ are obtained by setting the following determinant equal to zero

$$\det\begin{vmatrix} e^{\omega/2k_BT} - \lambda & e^{-\omega/2k_BT} \\ e^{-\omega/2k_BT} & e^{\omega/2k_BT} - \lambda \end{vmatrix} = 0 \tag{8.97}$$

This gives

$$\left(e^{\omega/2k_BT} - \lambda\right)^2 - e^{-\omega/k_BT} = 0$$

$$\text{or ;} \qquad \lambda^2 - 2\lambda e^{\omega/2k_BT} + e^{\omega/k_BT} - e^{-\omega/k_BT} = 0 \tag{8.98}$$

Hence

$$\lambda_\pm = \frac{1}{2}\left\{2e^{\omega/2k_BT} \pm \left[4e^{\omega/k_BT} - 4\left(e^{\omega/k_BT} - e^{-\omega/k_BT}\right)\right]^{1/2}\right\}$$
$$= e^{\omega/2k_BT} \pm e^{-\omega/2k_BT} \tag{8.99}$$

The largest eigenvalue is therefore given by[16]

$$\lambda_+ = 2\cosh\left(\frac{\omega}{2k_BT}\right) \tag{8.100}$$

The canonical partition function and Helmholtz free energy then follow from Eq. (8.96)

$$(\text{P.F.}) = \left[2\cosh\left(\frac{\omega}{2k_BT}\right)\right]^N$$

$$A = -k_BT\ln(\text{P.F.}) = -Nk_BT\ln\left[2\cosh\left(\frac{\omega}{2k_BT}\right)\right] \tag{8.101}$$

[16] Note that both eigenvalues here are real and positive.

The energy of the assembly is obtained as

$$E = -T^2 \frac{\partial}{\partial T}\left(\frac{A}{T}\right)$$
$$= -T^2\left[-\frac{Nk_{\rm B}}{2\cosh(\omega/2k_{\rm B}T)} 2\sinh\left(\frac{\omega}{2k_{\rm B}T}\right)\right]\left(-\frac{\omega}{2k_{\rm B}T^2}\right) \tag{8.102}$$

Thus we arrive at the following expression for the energy

$$E = -\frac{N\omega}{2}\tanh\left(\frac{\omega}{2k_{\rm B}T}\right) \tag{8.103}$$

This result can be written as

$$E = \frac{N\omega}{2}e(x)$$
$$e(x) = -\tanh\left(\frac{1}{x}\right) \qquad ; \; x = \frac{2k_{\rm B}T}{\omega} \tag{8.104}$$

The function $e(x)$ is plotted as a function of x in Fig. 8.15.

The heat capacity is obtained from

$$C_V = \frac{\partial E}{\partial T}$$
$$= -\frac{N\omega}{2}\left[\text{sech}^2\left(\frac{\omega}{2k_{\rm B}T}\right)\right]\left(-\frac{\omega}{2k_{\rm B}T^2}\right) \tag{8.105}$$

Hence

$$\frac{C_V}{Nk_{\rm B}} = \left(\frac{\omega}{2k_{\rm B}T}\right)^2 \text{sech}^2\left(\frac{\omega}{2k_{\rm B}T}\right) \tag{8.106}$$

This result can similarly be written as

$$C_V = Nk_{\rm B}\, c(x)$$
$$c(x) = \left(\frac{1}{x^2}\right)\text{sech}^2\left(\frac{1}{x}\right) \qquad ; \; x = \frac{2k_{\rm B}T}{\omega} \tag{8.107}$$

The function $c(x)$ is plotted as a function of x in Fig. 8.16.[17]

Several comments:

- *These expressions provide an exact solution for the one-dimensional Ising model in the limit of large N!*

[17] Note $c(x) = de(x)/dx$.

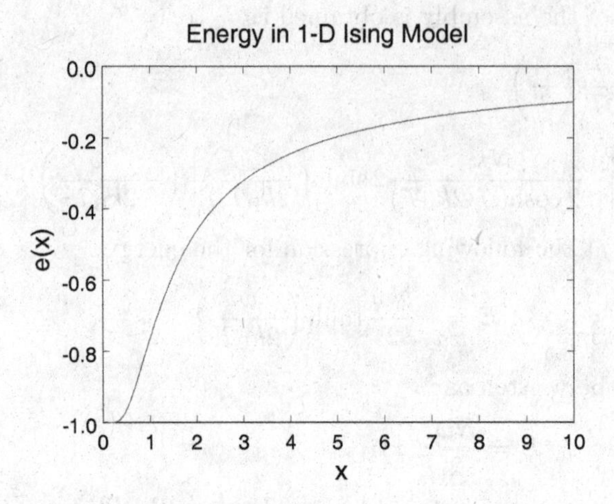

Fig. 8.15 Energy $E = (N\omega/2)e(x)$ as a function of $x = 2k_\mathrm{B}T/\omega$ in the one-dimensional Ising model. Here $e(x) = -\tanh(1/x)$.

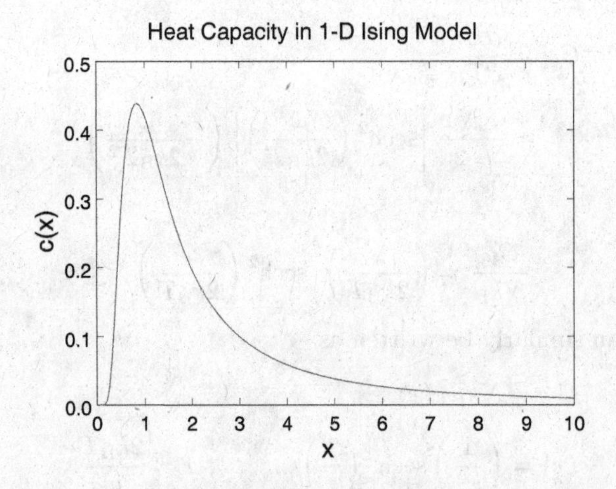

Fig. 8.16 Heat capacity $C_V = Nk_\mathrm{B}c(x)$ as a function of $x = 2k_\mathrm{B}T/\omega$ in the one-dimensional Ising model. Here $c(x) = (1/x^2)\,\mathrm{sech}^2(1/x)$.

- At low temperature (small x), the energy is $E = -N\omega/2$ and the spins are all aligned; at high temperature (large x), $E = 0$, and the spins are randomly oriented;
- The heat capacity C_V is a smooth, continuous curve. *There is no phase*

transition;

- The one-dimensional Ising model can again be solved exactly in the presence of a magnetic field $B\mathbf{e}_z$, where there is an additional interaction $H' = -\mu_0 B \sum_{\kappa=1}^{N} S_\kappa$.[18] *There is no ferromagnetism*;
- A general proof that the linear chain in Fig. 8.14 cannot be ferromagnetic is obtained as follows:
  - Suppose the N spins are all aligned. Now break the chain somewhere (Fig. 8.17)

Fig. 8.17 Break in a chain of N aligned spins.

  - The increase in energy of the chain is $\Delta E = \omega$;
  - Since the break can occur at any of N positions, the increase in entropy of the chain of spins is

$$\Delta S = k_B \ln N \tag{8.108}$$

  - The change in free energy of the chain is therefore

$$\begin{aligned} \Delta A &= \Delta E - T\Delta S \\ &= \omega - k_B T \ln N \end{aligned} \tag{8.109}$$

  - For $N \to \infty$, the entropy always wins. At any temperature, one can always lower the free energy of a one-dimensional assembly of aligned spins by breaking the chain;[19]
- To get a ferromagnetic spin system, one needs *more nearest neighbors*.

8.3.3 *Two-Dimensional Ising Model (Z=4)*

The two-dimensional Ising model, with a coordination number $Z = 4$, was solved by Onsager in a truly impressive calculation using complicated matrix techniques [Onsager (1944)]. The proof is given in more advanced texts such as [Huang (1987); Wannier (1987)]. We leave that for a future course. Here we simply summarize some of the principal results:

[18]See Prob. 8.12.

[19]The physics here is that although it costs slightly more energy to create one of the broken configurations, there are overwhelmingly many more of them. Note that the mean value of the magnetization in the broken chain vanishes.

(1) There is a phase transition;
(2) The transition temperature is given by

$$\sinh\left(\frac{\omega}{k_{\rm B}T_C}\right) = 1 \tag{8.110}$$

(3) The energy and heat capacity in the vicinity of T_C exhibit the characteristics of a λ-point transition

$$\begin{aligned} E &\sim (T - T_C)\ln|T - T_C| \\ C_V &\sim \ln|T - T_C| \end{aligned} \tag{8.111}$$

8.3.4 *Mean Field Theory*

While not exhibiting the detailed behavior of the exact solution, one can obtain a great deal of insight into a many-body problem such as this by employing *mean field theory* (MFT), which reduces it to a solvable one-body problem. Here the Ising model is studied within the following MFT framework:

- We assume the *dimension* d of the problem satisfies

$$d \geq 2 \qquad ; \text{ dimension} \tag{8.112}$$

- Periodic boundary conditions are assumed in all directions. The configuration for the two-dimensional Ising model is illustrated in Fig. 8.18;
- The following notation is used

$$\sum_{ij} \equiv \sum_{\text{nearest neighbors}} \tag{8.113}$$

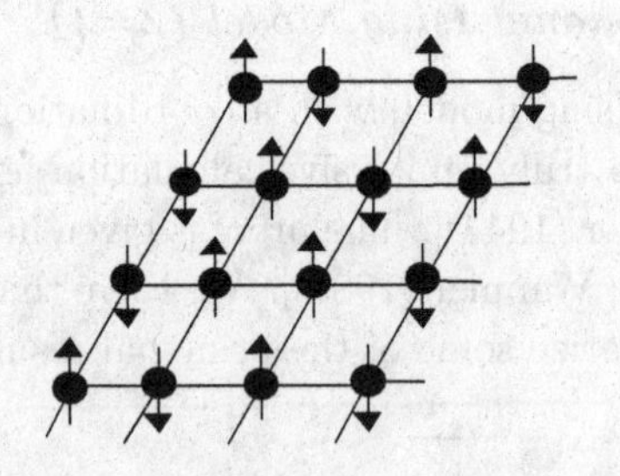

Fig. 8.18 Two-dimensional Ising model. Periodic boundary conditions are assumed in both directions (see Prob. 8.13).

- The total spin S of the assembly is given by

$$S = \sum_{j=1}^{N} S_j \qquad ; \; S_j = \pm 1 \tag{8.114}$$

The expectation value of the total spin in the Ising model is then

$$\begin{aligned} \langle S \rangle &= \frac{\sum_{S_1} \cdots \sum_{S_N} \left(\sum_i S_i\right) \exp\left(J \sum_{jk} S_j S_k / k_\mathrm{B} T\right)}{\sum_{S_1} \cdots \sum_{S_N} \exp\left(J \sum_{jk} S_j S_k / k_\mathrm{B} T\right)} \\ &\equiv Nm \end{aligned} \tag{8.115}$$

Here the last line defines the magnetization per particle m.[20] This expression can be analyzed as follows:

- With periodic boundary conditions, all spins are equivalent. Hence Eqs. (8.115) reduce to the following, and one can focus on the ith spin,

$$m = \frac{\sum_{S_1} \cdots \sum_{S_N} S_i \exp\left(J \sum_{jk} S_j S_k / k_\mathrm{B} T\right)}{\sum_{S_1} \cdots \sum_{S_N} \exp\left(J \sum_{jk} S_j S_k / k_\mathrm{B} T\right)} \tag{8.116}$$

- Mean-field theory (MFT) keeps just an *average coupling* of the ith spin to its neighbors, and approximates the sum in the exponential by

$$\begin{aligned} \sum_{jk} S_j S_k &\to z \langle S_j \rangle S_i + \sum_{jk \neq i} S_j S_k \\ &= zm S_i + \sum_{jk \neq i} S_j S_k \end{aligned} \tag{8.117}$$

  Here m is again the magnetization per particle, and z is an *effective coordination number*, discussed below;
- Since the ith spin no longer appears in the remaining sum, all the remaining spin sums *factor and cancel in the ratio* in Eq. (8.116). Thus

$$m = \frac{\sum_S S\, e^{JzmS/k_\mathrm{B} T}}{\sum_S e^{JzmS/k_\mathrm{B} T}} \qquad ; \; S = \pm 1 \tag{8.118}$$

  where we have simply relabeled $S_i \to S$;
- We have now accomplished our goal of reducing the calculation of the magnetization in the Ising model to a one-body problem.

[20] The magnetization m is now in units of the intrinsic moment μ_0.

The sums in Eq. (8.118) are immediately performed. Define

$$\alpha \equiv \frac{Jz}{k_{\rm B}T} \qquad ; J \geq 0 \tag{8.119}$$

where we assume an attractive interaction with $J \geq 0$. Equation (8.118) then becomes

$$m = \frac{e^{\alpha m} - e^{-\alpha m}}{e^{\alpha m} + e^{-\alpha m}} = \tanh(\alpha m) \tag{8.120}$$

The result is a *self-consistency equation for* m, the mean magnetization per particle

$$m = \tanh\left(\frac{Jzm}{k_{\rm B}T}\right) \qquad ; \text{ self-consistency equation for } m \tag{8.121}$$

This is exactly Eq. (8.64)! An identical analysis then leads to the existence of a critical temperature, below which there is a pair of non-zero solutions for m, and above which the only solution is $m = 0$,

$$k_{\rm B}T_C = Jz \qquad ; \text{ critical temperature} \tag{8.122}$$

We now have to face the issue of identifying the effective coordination number z. At first blush, one would simply take the number of nearest neighbors Z. A pondering of Eq. (8.117), however, makes one realize that the mean field contribution in the sum in the exponent must be *apportioned* between the ith spin and its jth nearest neighbor. The simplest way to do this is to form the spin lattice from *unit cells*, out of which the lattice can be built by repetition, and then simply assign the unit cell to each spin to determine its appropriate mean-field contribution.[21] For the two-dimensional Ising model, the unit cell is that pictured in Fig. 8.19 and $z = 2$. Similarly, for the three-dimensional Ising model one has $z = 3$, and so on. Evidently $z = d$, where d is the *dimension* of the problem

$$\begin{aligned} z &= d \qquad ; \text{ effective coordination number} \\ k_{\rm B}T_C &= Jd \end{aligned} \tag{8.123}$$

Several comments:

- With $J > 0$, mean field theory predicts the existence of a phase transition to an ordered state of the spins for the Ising model in any number of dimensions d;

[21]The proper way to do this is to write the mean-field contribution of the entire lattice, and then minimize the free energy—this goes beyond the present development.

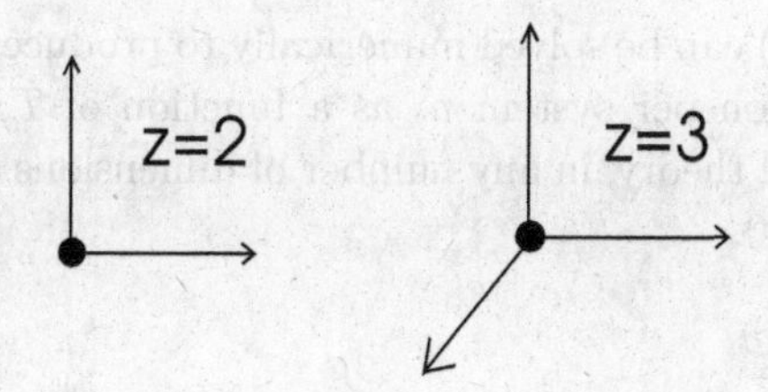

Fig. 8.19 Unit cells and effective coordination numbers in two- and three-dimensional Ising models.

- The critical temperature is given by

$$k_B T_C = Jd \tag{8.124}$$

where J is the coupling constant in $H_{\text{Ising}} = -J\sum_{ij} S_i S_j$, with $S_i = \pm 1$ in the nearest-neighbor sum, and d is the dimension of the problem;
- Below T_C, the magnetization per particle is given by the solution to the equation

$$m = \tanh\left(\frac{T_C}{T}m\right) \tag{8.125}$$

- Above T_C, one has $m = 0$;
- We know there is no phase transition in one dimension, so mean field theory *fails* for $d = 1$;
- A comparison with Onsager's exact result for the two-dimensional Ising model in Eq. (8.110) is shown in Table 8.1.[22] This is not bad for such a simple calculation;

Table 8.1 Transition temperature for Ising model in mean field theory.

Dimension d	$k_B T_C$	Exact Result
1	J	no phase transition
2	2J	2.2692J
3	3J	??

- The result for $d = 3$ is also shown in Table 8.1; the three-dimensional problem has never been solved analytically;
- *Thus mean field theory becomes more reliable in just that regime where analytic solution becomes more prohibitive!*

[22]See Prob. 8.10.

Equation (8.125) can be solved numerically to produce a universal curve for the magnetization per system m as a function of T/T_C for the Ising model, in mean field theory, in any number of dimensions $d \geq 2$. The result is shown in Fig. 8.20.

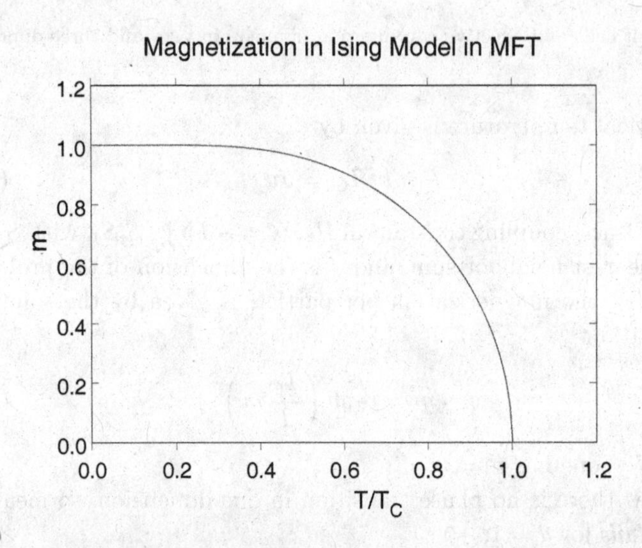

Fig. 8.20 Magnetization per system m for the Ising model in MFT in any number of dimensions $d \geq 2$. (Compare Prob. 8.11.)

8.3.5 *Numerical Methods*

Problems in statistical mechanics lend themselves to numerical analysis, particularly with the easy availablity of powerful desktop computers. The author has found this to be especially fun for students. Here, Probs. 8.15–8.18 take the reader through two widely employed numerical methods, Monte Carlo calculations and the use of the Metropolis algorithm, as applied to the one-dimensional Ising model where an exact analytic solution is available for comparison.

Our final special topic is lattice gauge theory.

8.4 Lattice Gauge Theory

8.4.1 *The Standard Model*

The marvelously successful standard model of the strong, electromagnetic, and weak interactions is a Yang-Mills local gauge theory built on a symmetry structure of $SU(3)_C \bigotimes SU(2)_W \bigotimes U(1)_W$, where the mechanism of spontaneous symmetry breaking gives rise to the particle masses.[23] The theory of the strong interactions, *quantum chromodynamics* (QCD), is based on an $SU(3)_C$ color symmetry of underlying quark and gluon fields. This theory has two remarkable properties arising from the gauge couplings:

(1) *Confinement*: The underlying degrees of freedom, the quarks and gluons, are confined to the interior of the observed strongly interacting hadrons (baryons and mesons); they are never oberved in free, asymptotic scattering states in the laboratory;
(2) *Asymptotic Freedom*: At very large momentum transfer, or short distance, the coupling constant of QCD becomes small, and one can do perturbation theory in this regime.

In the low-energy, low-momentum-transfer regime, QCD is a strong-coupling theory giving rise to the observed hadrons and their interactions.

A procedure for solving the theory in this regime, lattice gauge theory (LGT), was developed by Wilson in a paper that has had a profound influence on nuclear and particle physics [Wilson (1974)]. The dynamical variables are the fields, defined throughout space-time. In LGT, the theory is placed on a finite space-time lattice. Asymptotic freedom then allows one to go smoothly to the continuum limit. Exact local gauge invariance is maintained at all steps. While the non-abelian nature of the group elements significantly complicates the analysis, large-scale numerical calculations are currently available which produce striking results. Such calculations will play a major role in nuclear and particle physics for the foreseeable future.

While a detailed discussion of lattice gauge theory goes well beyond the confines of this text, it is appropriate in an introductory book on statistical mechanics to illustrate the basic approach in a simplified version of the theory.

[23]The standard model and lattice gauge theory are discussed in detail in [Walecka (2004)]. A thorough list of references can be found there. For Yang-Mills theories and spontaneous symmetry breaking, see also [Walecka (2010)].

8.4.1.1 *Quantum Electrodynamics (QED)*

Consider *quantum electrodynamics* (QED), which is a Yang-Mills local gauge theory built on an underlying $U(1)$ symmetry group of phase transformations of the charged fields. This set of one-dimensional phase transformations is abelian, with commuting elements, which simplifies the analysis.

The lagrangian density for the pure electromagnetic gauge field in QED is

$$\mathcal{L} = -\frac{1}{4} F_{\mu\nu} F_{\mu\nu}$$

$$F_{\mu\nu} = \frac{\partial A_\nu}{\partial x_\mu} - \frac{\partial A_\mu}{\partial x_\nu} \tag{8.126}$$

Here the four-vectors are $x_\mu = (\mathbf{x}, it)$ and $A_\mu = (\mathbf{A}, i\Phi)$.[24] This theory is invariant under the following local gauge transformation

$$A_\mu \to A_\mu + \frac{1}{e_0} \frac{\partial \Lambda}{\partial x_\mu} \tag{8.127}$$

where $\Lambda(x)$ is a scalar function of position in space-time. The *action* for this field theory is given by

$$S = \int d^4x \, \mathcal{L}(x) \tag{8.128}$$

8.4.2 *Partition Function in Field Theory*

First, as *motivation*, observe the following relation

$$e^{-iHt} \to e^{-\beta H} \qquad ; \; t \to -i\beta \tag{8.129}$$

The l.h.s. of this relation is the quantum mechanical propagator, while the r.h.s. is the statistical operator. Hence we observe that the the action and amplitude for propagation for imaginary time will be related to a statistical average.

Suppose one has a neutral scalar field theory with a non-linear self-coupling

$$\mathcal{L} = -\frac{1}{2} \left[\left(\frac{\partial \phi}{\partial x_\mu} \right)^2 + m^2 \phi^2 \right] - \frac{\lambda}{4!} \phi^4 \tag{8.130}$$

[24] Repeated Greek indicies are summed from 1 to 4. In this last section, we work with units where $\hbar = c = 1$, and for ease with signs we use $\beta \equiv +1/k_{\rm B}T$, as in [Fetter and Walecka (2003)].

It is shown in [Walecka (2004)] that the quantum partition function for this theory is given by the following expression

$$Z = \int \mathcal{D}(\phi)\, e^{-S(\beta,0)} \tag{8.131}$$

Here

- $\int \mathcal{D}(\phi)$ is a *path integral* over all values of the field at each point in space-time;
- The analysis is carried out with imaginary time, so that

$$t \to -i\tau \qquad ;\ \text{imaginary time} \tag{8.132}$$

- The exponent is the action with imaginary time evaluated over the interval $[\beta, 0]$ [25]

$$S(\beta, 0) = -\int_0^\beta d\tau \int d^3x\, \mathcal{L}(\mathbf{x}, \tau) \tag{8.133}$$

- Periodic boundary conditions are imposed in the τ-direction, with period $\beta = 1/k_\mathrm{B}T$.

This analysis will be applied to the $U(1)$ local gauge theory of QED, with the lagrangian density in Eq. (8.126).

8.4.3 *U*(1) *Lattice Gauge Theory*

The $U(1)$ theory is *discretized* through the following series of steps:

(1) One works with imaginary time $t \to -i\tau$;

(2) With a contour rotation so that $\Phi \to -iA_4$, a path integral over the field variables can be re-expressed as

$$i\int \cdots \int dA_1 dA_2 dA_3 d\Phi \to \int \cdots \int dA_1 dA_2 dA_3 dA_4 \qquad ;\ \text{contour rotation} \tag{8.134}$$

All four vectors are now expressed in the *euclidian metric*, for example

$$\begin{aligned} A_\mu &= (A_1, A_2, A_3, A_4) \qquad &; A_\mu^2 = A_1^2 + A_2^2 + A_3^2 + A_4^2 \\ x_\mu &= (x_1, x_2, x_3, \tau) \qquad &; x_\mu^2 = x_1^2 + x_2^2 + x_3^2 + \tau^2 \end{aligned} \tag{8.135}$$

[25] Note that for imaginary time, $\mathcal{L}(\mathbf{x}, \tau) \equiv [\mathcal{L}(x_\mu)]_{t=-i\tau} = -\mathcal{H}(\mathbf{x}, \tau)$, where $\mathcal{H}$ is the hamiltonian density (see Prob. 8.19).

(3) The theory is placed on a finite space-time lattice, as illustrated in Fig. (8.21). The sites denote the points in space-time, and the length of the directional links between sites is a

$$a \equiv \text{length of link} \tag{8.136}$$

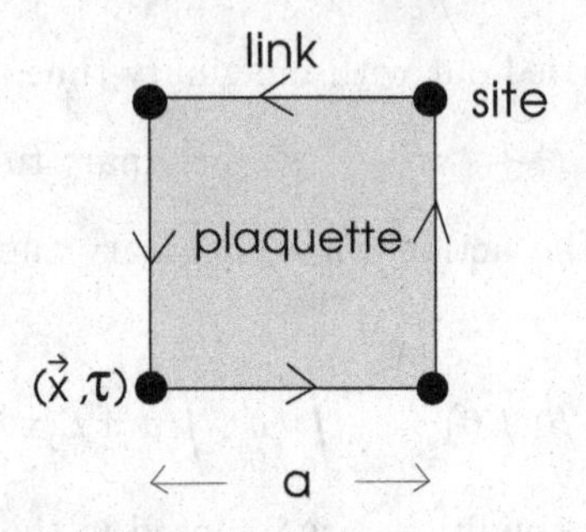

Fig. 8.21 Definitions of site, link, and plaquette for lattice gauge theory. The sites denote discrete points in space-time $(\mathbf{x}, \tau)$, and the directional link length is a. The plaquette is denoted by $\Box$.

Here we assume a square lattice, but it is easy to use a different lattice spacing in the τ-direction, where the total lattice length is β;

(4) Periodic boundary conditions are assumed in all directions;

(5) The field variables are associated with the *links*, in the following fashion

$$\begin{aligned} U_{\text{link}} &\equiv e^{ie_0 A_\mu(x)[x(j)-x(i)]_\mu} \\ &\equiv e^{i\phi_l} \end{aligned} \tag{8.137}$$

(6) The contribution to the action from these link variables is taken as the *product around a plaquette*

$$\begin{aligned} U_\Box &\equiv U_1 U_2 U_3 U_4 \\ S_\Box &\equiv 2\sigma\,(1 - \text{Re}\, U_\Box) \end{aligned} \tag{8.138}$$

where 2σ is a constant, which can depend on the charge e_0 and lattice spacing a. One has to keep track of the *direction* of the link, with the phase positive in the direction of the positive coordinate axes (see Fig. 8.22).

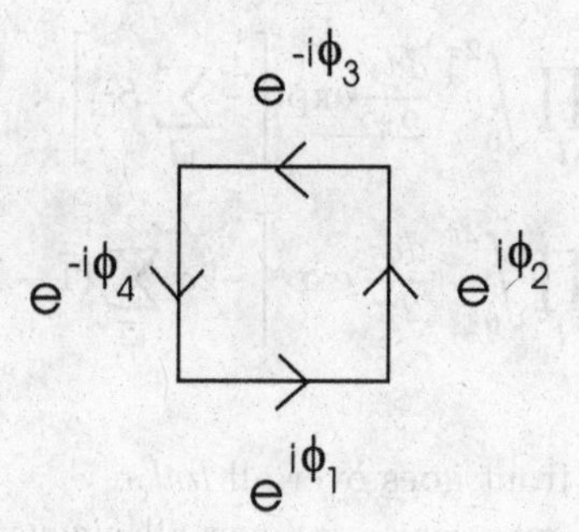

Fig. 8.22 Calculation of $U_\Box$ from the link variables.

Thus, in detail,

$$U_\Box = e^{i\phi_1}e^{i\phi_2}e^{-i\phi_3}e^{-i\phi_4} \tag{8.139}$$

(7) Since $(\partial\Lambda/\partial x_\mu)dx_\mu = d\Lambda$ is an exact differential, the change in a link variable under the gauge transformation in Eq. (8.127) is given for small a by

$$U \rightarrow e^{i\Lambda(j)}Ue^{-i\Lambda(i)} \tag{8.140}$$

The product of link variables around a plaquette is *unchanged* under this transformation, and $U_\Box$ and $S_\Box$ are therefore *gauge invariant*;

(8) The total action is now obtained from the sum over all plaquettes

$$S = \sum_\Box S_\Box \tag{8.141}$$

(9) Since the link variable in Eq. (8.137) is periodic in the phase ϕ_l, a gauge-invariant integral over all field configurations is obtained by simply integrating ϕ_l over the interval $[0, 2\pi]$[26]

$$\int [dA_\mu] \rightarrow \int_0^{2\pi} \frac{d\phi_l}{2\pi} \tag{8.142}$$

Note that as $a \rightarrow 0$, this encompasses an infinite range of field values;

(10) The partition function for this $U(1)$ lattice gauge theory is then given by

[26]See Prob. 8.28.

$$Z \equiv \prod_l \int_0^{2\pi} \frac{d\phi_l}{2\pi} \exp\left[-\sum_\Box S_\Box\right]$$
$$= \prod_l \int_0^{2\pi} \frac{d\phi_l}{2\pi} \exp\left[-2\sigma \sum_\Box (1 - \mathrm{Re}\, U_\Box)\right] \qquad (8.143)$$

Here

- The product in front goes over all *links*;
- The sum in the exponent goes over all *plaquettes*;
- The coupling occurs because a given link variable will appear in several plaquettes.

(11) One now reproduces the correct *continuum limit* of the theory by choosing the constant $2\sigma(e_0, a)$ so that, using the first of Eqs. (8.137), the exponent in Eq. (8.143) reduces to the correct continuum action in the euclidian metric (see Probs. 8.20-8.21)

$$\sum_\Box S_\Box \to \int d\tau\, d\mathbf{x}\, \left[\tfrac{1}{4} F_{\mu\nu} F_{\mu\nu}\right] \qquad ; \; a \to 0 \qquad (8.144)$$

Several comments:

- The theory for Z is now well-formulated;[27]
- It is gauge invariant;
- It has the correct continuum limit;[28]
- *When the dust clears, the identification of Z in Eq. (8.143) with the canonical partition function*

$$Z \equiv (\text{P.F.}) \qquad ; \; 2\sigma \equiv \beta_{\text{eff}} \qquad (8.145)$$

presents a well-defined problem in statistical mechanics, where $2\sigma \equiv \beta_{\text{eff}}$ plays the role of an effective inverse temperature!

- We can now use all the weapons in our arsenal, for example:

[27] The relation of Z to observables is discussed in chap. 33 of [Walecka (2004)].

[28] With the employment of the measure in Eq. (8.142), all the dependence on the coupling constant e_0 is transfered to $\sigma(e_0, a)$, the one remaining parameter in the theory (see Probs. 8.20-8.21). One then fails to recover the fact that the free, continuum theory in Eq. (8.126) is *independent* of this coupling constant. It is only in asymptotically-free theories that one can pass smoothly from the discrete to the continuum limit; this is accomplished by including the dependence of the coupling constant on distance scale (compare Prob. 8.30). Whatever is done on the lattice with QED is thus just a model; however, as it does illustrate the approach, it provides an interesting study.

– Expansion of the exponential and analytic evaluation as $\beta_{\text{eff}} \to 0$ ("high-temperature");
– Mean field theory to reduce it to a one-body problem as $\beta_{\text{eff}} \to \infty$ ("low-temperature");
– Numerical Monte Carlo evaluation for all β_{eff}.

Let us start with mean field theory, where the discussion closely parallels that in the Ising model.[29]

8.4.3.1 *Mean Field Theory (MFT)*

The plaquettes in the exponent in Eq. (8.143) are coupled through their common links $U_l = e^{i\phi_l}$, where the phase is integrated over. Assume there is a state of the assembly in which a common, real, "magnetization" develops

$$\langle e^{i\phi_l} \rangle = \langle \cos\phi_l \rangle \equiv m \qquad ; \text{ effective magnetization} \qquad (8.146)$$

In MFT, one dynamical link is retained, and all the coupled neighbors are replaced by their mean value. This is illustrated in two dimensions in Fig. 8.23.

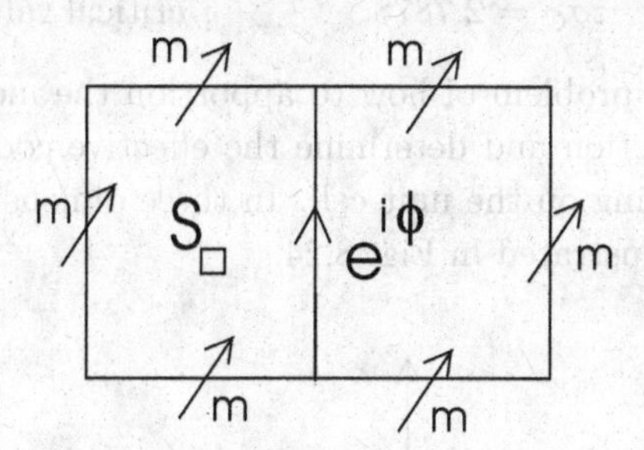

Fig. 8.23 Mean field theory for a link for $U(1)$ lattice gauge theory in two dimensions.

In MFT, the contribution to the action from the plaquettes coupled to this link is then replaced by

$$\begin{aligned} S_\Box &= 2\sigma(1 - \operatorname{Re} U_\Box) \\ &\to 2\sigma(1 - m^3\cos\phi) \qquad ; \text{ MFT} \end{aligned} \qquad (8.147)$$

Although the discussion is framed in terms of a link variable, we ultimately deal with $S_\Box$, which is a *gauge-invariant* quantity.

[29] Wilson's original paper introduced mean field theory in LGT [Wilson (1974)].

In the exponent, one then replaces

$$\sum_{\Box} S_{\Box} \to 2z\sigma(1 - m^3 \cos\phi) \tag{8.148}$$

where z is the *effective coordination number* for the plaquettes coupled to that link. The expectation value of $S_{\Box}$ in MFT is then given by

$$\begin{aligned}\langle S_{\Box}\rangle &= 2\sigma(1-m^4) \\ &= \frac{\int_0^{2\pi} d\phi\, 2\sigma(1-m^3\cos\phi)e^{-2z\sigma(1-m^3\cos\phi)}}{\int_0^{2\pi} d\phi\, e^{-2z\sigma(1-m^3\cos\phi)}}\end{aligned} \tag{8.149}$$

With the cancellation of common factors, this can be written as

$$m = \frac{\int_0^{2\pi} d\phi\, \cos\phi\, e^{2z\sigma m^3 \cos\phi}}{\int_0^{2\pi} d\phi\, e^{2z\sigma m^3\cos\phi}} \tag{8.150}$$

This is a transcendental equation for the magnetization m. Numerical analysis indicates there will be a solution for $\sigma > \sigma_C$ where the critical value is[30]

$$z\sigma_C = 2.7878 \qquad ; \text{ critical value} \tag{8.151}$$

Once again, the problem of how to apportion the mean-field contributions in the total action and determine the effective coordination number z is solved by focusing on the unit cell. In three dimensions, the situation in the unit cell is illustrated in Fig. 8.24.

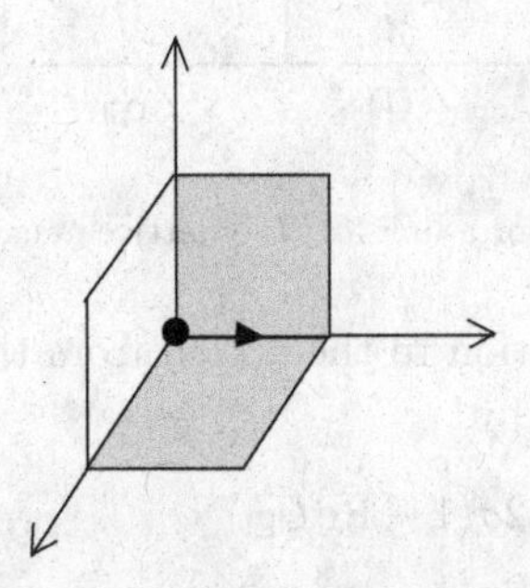

Fig. 8.24 Site, link, and coupling plaquettes in the unit cell for $U(1)$ lattice gauge theory in three dimensions. Here the effective coordination number is z = plaquettes/link = 2.

[30]See Prob. 8.23.

The number of plaquettes coupled to a link along the direction of one of the positive coordinate axes is the number of planes that can be passed through that axis. This is just one less than the number of dimensions d of the problem. Thus the effective coordination number in MFT is

$$z = d - 1 \qquad ; \text{ effective coordination number} \tag{8.152}$$

Equation (8.151) can thus be recast as

$$(d-1)\sigma_C = 2.7878 \qquad ; \text{ MFT} \tag{8.153}$$

This is compared with numerical results in Table 8.2.

Table 8.2 "Transition temperature" $\sigma_C = \beta_{\text{eff}}^C/2$ for $U(1)$ lattice gauge theory in mean-field theory. The exact result for $d = 4$ is from [Dubach (2004)].

Dimension d	σ_C(MFT)	Exact Result
2	2.7878	no phase transition
4	0.9293	0.4975

Not as accurate as in the Ising model, but at least there is a phase transition for $d = 4$.

8.4.3.2 *Numerical Monte Carlo*

The "exact result" for $d = 4$ in Table 8.2 is a numerical Monte Carlo done in $3 + 1$ dimensions on a 5^4 lattice by [Dubach (2004)]. His result for the magnetization m^4 as a function of σ_C/σ is compared with the MFT result obtained from the solution to Eq. (8.150) in Fig. 8.25.[31]

8.4.3.3 *Strong-Coupling Limit*

For high effective temperature, or small $\beta_{\text{eff}} = 2\sigma$, the exponential in Eq. (8.143) can be expanded, and the integrations done term by term using

$$\frac{1}{2\pi}\int_0^{2\pi} d\phi\, e^{in\phi} = \delta_{n,0} \tag{8.154}$$

[31] We speak of an "effective temperature", but what one is really doing here by varying $\sigma(e_0, a)$ is studying the phase diagram as a function of coupling constant e_0 for fixed lattice spacing a [compare Probs. 8.20–8.21].

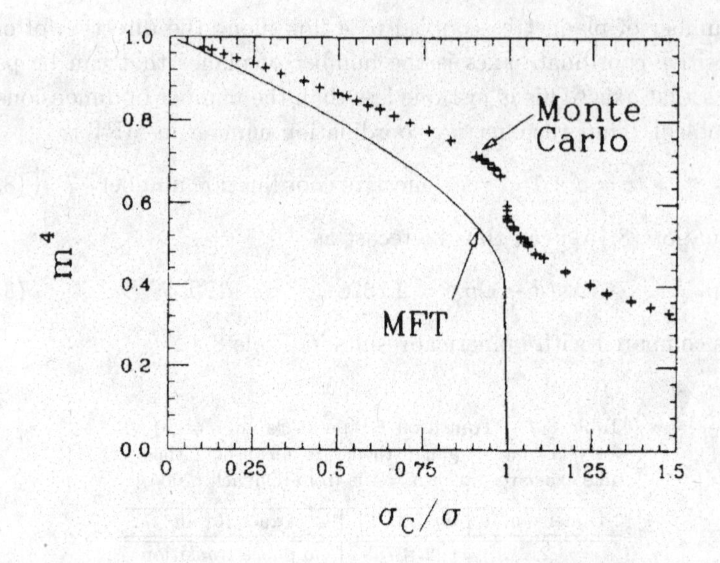

Fig. 8.25 The magnetization m^4 for $U(1)$ lattice gauge theory in 3 + 1 dimensions ($d = 4$) in MFT, as a function of $\sigma_C/\sigma = \beta_{\rm eff}^C/\beta_{\rm eff}$, compared with the essentially exact Monte Carlo calculation on a 5^4 lattice carried out by Dubach [Dubach (2004)]. Here $S_\Box \equiv 2\sigma(1-m^4)$.

One can then calculate, for example, the expectation value of a product of link variables around a closed curve (a "Wilson loop") by using

$$\begin{aligned} S_\Box &= \sigma\left\{[1-(U_\Box)_{\hookrightarrow}] + [1-(U_\Box)_{\hookleftarrow}]\right\} \\ &= \sigma\left\{[1-(U_\Box)_{\hookrightarrow}] + [1-(U_\Box)^\star_{\hookrightarrow}]\right\} \end{aligned} \tag{8.155}$$

In this fashion, one can obtain a power-series expansion of such expectation values.[32] For example, it is shown in Prob. 8.22 that the strong-coupling limit of $\langle S_\Box\rangle$ in $U(1)$ lattice gauge theory is

$$\langle S_\Box\rangle = 2\sigma - \frac{1}{2}(2\sigma)^2 \qquad ; \ \sigma \to 0 \tag{8.156}$$

8.4.3.4 *Improved Analytic Approximations*

It is of interest to investigate how far one can go with improved analytic approximations in this simple model. This serves both to provide insight in

[32] Recall Eq. (8.139).

LGT, where extensive numerical calculations can often obscure the physics, and as a check on the numerical calculations themselves.

Mean field theory, valid at large $\beta_{\text{eff}} = 2\sigma$, can be improved as follows:

- The validity of MFT increases with increasing effective coordination number z. It is clear from Eq. (8.152) that this is accomplished by working in a *higher dimension* d;
- One can formulate MFT through a variational principle, and in this means determine an optimal z;
- *Gauge fixing* allows one to factor spurious contributions, which then cancel in an expectation-value ratio.

The *strong-coupling expansion*, valid at small $\beta_{\text{eff}} = 2\sigma$, can be improved as follows:

- More terms can be kept in the series expansion in β_{eff};[33]
- A *Padé approximant* can be constructed from the series expansion, which increases the convergence and allows for the possibility of singularities in the amplitude. In this approach, one constructs

$$\sum_{i=0}^{n} c_i \beta^i = \frac{\sum_{j=0}^{p} a_j \beta^j}{\sum_{k=0}^{q} b_k \beta^k} \quad ; \text{ Padé approximant } [p,q]$$

$$p + q = n \tag{8.157}$$

The coefficients $\{a_j, b_k\}$ are then determined by expanding the r.h.s. in a power series in β and matching the known coefficients c_j on the l.h.s. (see Prob. 8.26).

The analysis described above has been carried out by [Barmore (1999)], and his results are shown in Fig. 8.26. We make several comments:

- This is for *six dimensions* ($d = 6$);
- Here the effective plaquette energy $E_\Box$ is defined by

$$E_\Box \equiv \frac{1}{2\sigma}\langle S_\Box \rangle = (1 - m^4) \tag{8.158}$$

- The effective inverse temperature β_{eff} is[34]

$$\beta_{\text{eff}} \equiv 2\sigma \tag{8.159}$$

[33] An amazing number of terms in this expansion exists in the literature!

[34] Note that the ordinate here differs from that in Fig. 8.25, and the abcissa is inverted.

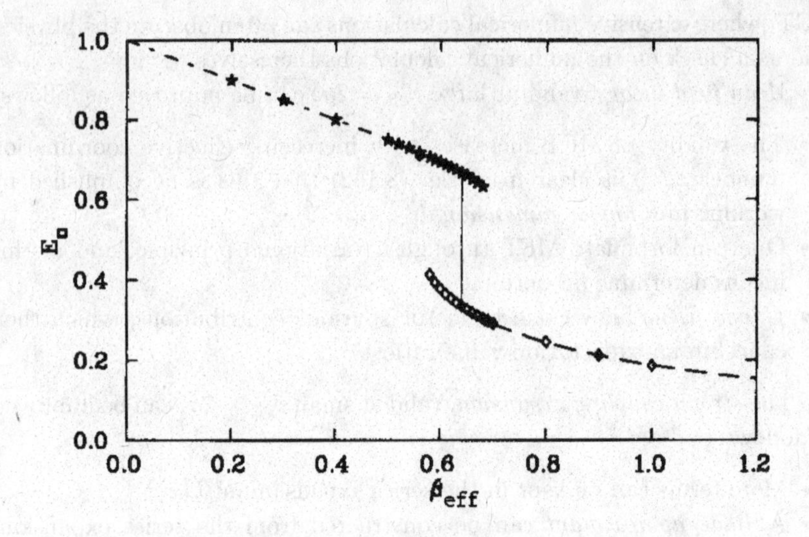

Fig. 8.26 Analytic phase diagram for $U(1)$ lattice gauge theory in six dimensions. Here $E_\square \equiv \langle S_\square \rangle / 2\sigma$ and $\beta_{\text{eff}} \equiv 2\sigma$. Dash-dot line is a [7, 8] Padé approximant for the strong-coupling expansion up to the transition point, the solid vertical line marks the transition point, and the long-dashed line is axial-gauge MFT beyond the transition point. Also shown are the Monte Carlo data on a 5^6 lattice. From [Barmore (1999)]. (Compare Prob. 8.27.)

- The phase transition is located in the following manner:

(1) Recall the Helmholtz free energy and energy of the assembly are given in thermodynamics by[35]

$$A = -\frac{1}{\beta} \ln (\text{P.F.})$$

$$E = \frac{\partial}{\partial \beta}(\beta A) \tag{8.160}$$

(2) This last relation can be integrated to obtain

$$\Delta(\beta A) = \int^{\beta} E \, d\beta \tag{8.161}$$

[35]Here $\beta = 1/k_{\text{B}}T$; see Prob. 8.25.

We will apply this thermodynamic relation in Fig. 8.26.

(3) The analytic results in Fig. 8.26 are smoothly joined with an S-shaped curve, as with the Van der Waal's equation of state in the region of the phase transition in Fig. (5.18). The above integration is then carried out along the resulting curve until $\Delta(\beta_{\text{eff}}A)$ crosses itself. At this point β_{eff}^C, one has two phases with a common $(\beta_{\text{eff}}A)$.

- Also shown in Fig. 8.26 are Barmore's numerical Monte Carlo data on a 5^6 lattice, which lie right on the analytic results;
- *In this example, one has analytic results which agree with the numerical calculations and hold well through the region of the phase transition!*

8.4.4 Non-Abelian Theory $SU(n)$

In $U(1)$ lattice gauge theory, the link variables are just phases

$$U = e^{i\phi} \qquad ; \ U(1) \text{ theory} \tag{8.162}$$

In the non-abelian $SU(2)$ lattice gauge theory, the link elements are 2×2 *matrices*

$$\underline{U} = \exp\left\{\frac{i}{2}\boldsymbol{\omega}\cdot\boldsymbol{\tau}\right\} \qquad ; \ SU(2) \text{ theory} \tag{8.163}$$

where $\boldsymbol{\tau} = (\tau_1, \tau_2, \tau_3)$ are the Pauli matrices, and $\boldsymbol{\omega} = (\omega_1, \omega_2, \omega_3)$ denotes a set of three real parameters. These $SU(2)$ matrices $\underline{U}$ do not commute, which complicates matters, but the same basic LGT approach again applies.[36]

In quantum chromodynamics (QCD), one has the corresponding 3×3 $SU(3)$ matrices. The current state of lattice gauge theory calculations in QCD is truly impressive (see, for example, [Lattice (2002)]). As stated at the outset, the extension of such calculations will provide a base for nuclear and particle physics for the foreseeable future.

---

[36] See [Walecka (2004)]; a variational MFT calculation for $SU(n)$ is also detailed there.

Chapter 9

Problems

1.1 Prove from Eq. (1.1) that the integral in Eq. (1.3) is independent of path.

1.2 Start from either statement of the second law, and see how far you can get in verifying the statements leading to Eqs. (1.5) and (1.6); then compare with [Zemansky (1968)].

1.3 (a) Why does each point on the dotted curve in Fig. 1.2 correspond to a given T?[1]

(b) How could one carry out the Carnot cycles shown in Fig. 1.2 in a continuous manner with each segment being covered only once?

(c) Why is it unnecessary for the construction of the entropy to actually traverse opposing segments of the adiabats in (b)?

(d) Show that the total heat input and total work output in (b) satisfy $Q = W$.

1.4 A perfect gas obeys the equation of state $PV = nRT = 2E(T)/3$, where n is the number of moles and R is the gas constant. Such a gas, in contact with a heat bath at temperature T, is initially confined by a thin membrane to one-half of a box of volume V. A hole is punched in the membrane so that the gas now fills the entire box.

(a) Find a reversible path, and show that the entropy change of the gas is $\Delta S = nR\ln 2$;

(b) What is the heat flow from the bath when the hole is punched?[2] What is the entropy change of the heat bath? Of the combined system of sample and heat bath?

---

[1] *Hint*: What is the equation of state of a perfect gas?

[2] *Hint*: Use the first law. Note that in this problem the reversible and irreversible processes connect two points on the *same isotherm* for the gas, $PV = nRT$.

1.5 N objects of spin 1/2 sit on distinct, localized sites, and the assembly is unpolarized.

(a) Show the spin entropy of this system is $S = Nk_{\rm B} \ln 2$;

(b) At a temperature T the spins are observed to align. Show that an amount of heat $Q = Nk_{\rm B}T \ln 2$ must have been extracted to produce this configuration;

(c) What is the corresponding change in internal spin energy?

1.6 (a) A piston under a pressure P expands quasistatically. Show that the reversible work done in the surroundings is $dW = PdV$;

(b) Show that the second law implies that the reversible (quasistatic) heat flow to a sample at an absolute temperature T is $dQ = TdS$.

1.7 Use an argument similar to that given in the text for the Helmholtz free energy to derive Gibbs criterion in Eq. (1.34) for equilibrium in a sample at fixed (P, T).

1.8 The *enthalpy* is a state function that is useful at fixed pressure. It is defined by making the following Legendre transformation

$$\mathcal{H} \equiv E + PV \qquad ; \text{ enthalpy}$$

(a) Show that with pressure-volume work, the first law of thermodynamics becomes

$$d\mathcal{H} = đQ + VdP \ ; \text{ first law}$$

(b) Show that for reversible (quasistatic) processes, the first and second laws become

$$d\mathcal{H} = TdS + VdP \qquad ; \text{ first and second law}$$

(c) Show that Gibbs criterion for equilibrium takes the form

$$(\delta\mathcal{H})_{S,P} \geq 0 \qquad ; \text{ Gibbs criterion}$$

A sample in equilibrium at fixed (S, P) will minimize its *enthalpy*.

1.9 (a) Consider a sample of volume V, surface temperature T, and surface pressure P. Divide it into tiny subunits. Introduce a heat flow dQ_R distributed over the surface, and show $\delta S = dQ_R/T$;[3]

---

[3] It is assumed here that any subsequent reversible heat flow across the surface of *interior* subunits cancels in this argument (compare the discussion of entropy conservation in Chapter 60 of [Fetter and Walecka (2003a)]).

(b) Now introduce *any* heat flow $đQ$ distributed over the surface, and derive Eq. (1.30);

(c) Derive the first of Eqs. (1.31).

1.10 Given an assembly of N localized systems with two energy levels separated by ε. Suppose the systems are initially all in the excited state and the temperature satisfies $k_B T \ll \varepsilon$.

(a) What is the initial entropy of the assembly? What is the entropy change ΔS to the state of the assembly where the systems are all in the ground state? What is the energy change ΔE?

(b) Use the finite form of the stability criterion in Eq. (1.29), $(\Delta E)_{S,V} \geq 0$, to show the initial state of this assembly is *unstable*;

(c) What is the heat flow in the transition to the final state? Show that the inequality $\Delta Q/T - \Delta S \leq 0$ is satisfied.

2.1 Construct the complexions for 3 excited systems on 5 sites and show the total number of complexions is $\Omega = 5!/3!\,2! = 10$;

2.2 Separate an analytic function into its real and imaginary parts $f(z) = u(x,y) + iv(x,y)$. An analytic function has a derivative that is independent of the direction in which it is taken in the complex plane. Evaluate the derivative first in the x-direction and then in the y-direction to show

$$\frac{df}{dz} = \frac{\partial u}{\partial x} + i\frac{\partial v}{\partial x} = \frac{\partial u}{\partial iy} + i\frac{\partial v}{\partial iy}$$

Hence deduce the *Cauchy-Riemann equations*

$$\frac{\partial u}{\partial x} = \frac{\partial v}{\partial y} \quad ; \quad \frac{\partial u}{\partial y} = -\frac{\partial v}{\partial x}$$

2.3 The modulus of the integrand in Eq. (2.71) is defined in terms of its real and imaginary parts by $|I(z)| = \sqrt{u^2 + v^2}$. Along the real axis, $I(z)$ is real so that $|I| = u$ (positive) and $v = 0$.

(a) The condition that the integrand has a minimum in the x-direction at $x = x_0$ is then $[\partial u/\partial x]_{x_0} = [\partial v/\partial x]_{x_0} = 0$. Show this implies $[\partial |I|/\partial x]_{x_0} = 0$;

(b) Use the Cauchy-Riemann equations of Prob. 2.2 to show that $[\partial v/\partial y]_{x_0} = [\partial u/\partial y]_{x_0} = 0$. Hence show that $[\partial |I|/\partial y]_{x_0} = 0$;

(c) Use these results to show

$$\left[\frac{\partial^2 |I|}{\partial x^2}\right]_{x_0} = \left[\frac{\partial^2 u}{\partial x^2}\right]_{x_0} \quad ; \quad \left[\frac{\partial^2 |I|}{\partial y^2}\right]_{x_0} = \left[\frac{\partial^2 u}{\partial y^2}\right]_{x_0}$$

Hence conclude that

$$\left[\left(\frac{\partial^2}{\partial x^2}+\frac{\partial^2}{\partial y^2}\right)|I|\right]_{x_0} = \left[\left(\frac{\partial^2}{\partial x^2}+\frac{\partial^2}{\partial y^2}\right)u\right]_{x_0}$$

(d) Show $|[f(z)]^N/z^E| = |f(z)|^N/|z|^E$.

2.4 (a) Write $z = x_0 + iy$, expand in y, and verify the statement that the $1/z$ in the integrand in Eq. (2.91) can be evaluated at the saddle point as $1/x_0$ to $O(1/N)$;

(b) Verify that the error incurred in extending the limits on the y-integral to $\pm\infty$ in Eq. (2.82) is covered by the stated error in Eq. (2.91);

(c) Explain the appropriate limiting process used in arriving at the error estimate in Eq. (2.93).

2.5 Show from Eq. (2.100) that the method of steepest descent reproduces the familiar Boltzmann distribution $n_i^\star/N = e^{\beta\varepsilon_i}/\sum_i e^{\beta\varepsilon_i}$ for the occupation numbers.

2.6 Use the microcanonical ensemble in equilibrium statistical mechaniocs to derive the following results:

(a) The Maxwell-Boltzmann distribution of velocities in an ideal gas

$$\frac{\Delta n}{n} = \frac{4\pi v^2 \Delta v}{(2\pi k_{\rm B}T/m)^{3/2}} \exp\left(-\frac{mv^2}{2k_{\rm B}T}\right)$$

(b) Halley's formula for the density distribution in an isothermal atmosphere

$$\rho = \rho_0 \exp\left(-\frac{mgh}{k_{\rm B}T}\right)$$

(c) The concentration distribution of macro-molecules in solution in an ultracentrifuge

$$c = c_0 \exp\left(\frac{mr^2\omega^2}{2k_{\rm B}T}\right)$$

2.7 Verify that the molar constant-volume heat capacity of a perfect gas is $C_V = (3/2)R$. Compare with the law of Dulong and Petit in Eq. (2.54).

2.8 (a) Assume the transition from a classical to a quantum gas occurs at a transition temperature $T_{\rm C}$ where $n^\star \approx 1$. Show from Eq. (2.144) that this criterion can be written as

$$\frac{\hbar^2}{2m}n^{2/3} \approx \frac{1}{4\pi}k_{\rm B}T_{\rm C}$$

where $n = N/V$ is the density.

(b) Use $\rho = 0.145\,\text{gm/cm}^3$ and $m = 6.64 \times 10^{-24}\,\text{gm}$ for (liquid) ^{4}He. Compute T_C. Compare with the measured λ-point temperature of $T_\lambda = 2.17\,^\circ$K for the transition from the normal to the superfluid phase of ^{4}He (see [Fetter and Walecka (2003)]).

2.9 A liquid is in equilibrium with its vapor in a container at fixed (P, T). The temperature is sufficiently high, and pressure sufficiently low, that the vapor behaves as a perfect gas, Show that the chemical potential of a system in the liquid, no matter how complicated the liquid structure, is given by Eq. (2.139).

2.10 Explain why the classical partition function for a perfect gas in Eq. (2.158) reproduces the result in Eq. (2.130) for all T.

2.11 Show that the Dirichlet integral in Eq. (2.175) gives

$$I_1(R) = 2R \qquad ; \; I_2(R) = \pi R^2 \qquad ; I_3(R) = \frac{4\pi}{3} R^3$$

Interpret these results.

2.12 (a) Introduce polar coordinates in two dimensions with $(x_1 = r\cos\phi,\, x_2 = r\sin\phi)$. Show $x_1^2 + x_2^2 = r^2$ and $dx_1 dx_2 = r dr d\phi$;

(b) Introduce polar-spherical coordinates in three dimensions

$$x_1 = r\sin\theta\cos\phi \qquad ; \; x_2 = r\sin\theta\sin\phi \qquad ; \; x_3 = r\cos\theta$$

Show $x_1^2 + x_2^2 + x_3^2 = r^2$ and $dx_1 dx_2 dx_3 = r^2 \sin\theta\, dr d\theta d\phi$;

(c) Introduce polar-spherical coordinates in four dimensions

$$\begin{aligned} x_1 &= r\sin\beta\sin\theta\cos\phi \qquad &; \; x_2 = r\sin\beta\sin\theta\sin\phi \\ x_3 &= r\sin\beta\cos\theta \qquad &; \; x_4 = r\cos\beta \end{aligned}$$

Show $x_1^2 + x_2^2 + x_3^2 + x_4^2 = r^2$ and $dx_1 dx_2 dx_3 dx_4 = r^3 \sin^2\beta \sin\theta\, dr d\beta d\theta d\phi$;

(d) Do the integral over angles in each case with $0 \leq \phi \leq 2\pi$, $0 \leq \theta \leq \pi$, and $0 \leq \beta \leq \pi$. Reproduce Eq. (2.181), with C_n given by Eq. (2.184), for $n = 2, 3, 4$.

2.13 (a) The energy of the one-dimensional simple harmonic oscillator is given in Eq. (2.159). Make a phase-space plot where the ordinate is p and the abcissa is x. Show the constant-energy phase-space orbit is an ellipse with semi-major axis $a = (2\varepsilon/m\omega^2)^{1/2}$ and semi-minor axis $b = (2m\varepsilon)^{1/2}$;

(b) Verify that the area of the ellipse is $\pi ab = 2\pi\varepsilon/\omega$;

(c) In quantum mechanics, the energy is quantized with $\varepsilon_n = \hbar\omega(n + 1/2)$. Show the area between the states with $n+1$ and n is exactly that given in Eq. (2.153).

2.14 (a) Prove that if the classical expression for the energy of a system can be written as

$$\varepsilon(q_1, \cdots, q_n; p_1, \cdots, p_n) = \sum_{i=1}^{n} a_i q_i^2 + \sum_{i=1}^{n} b_i p_i^2$$

where all the a's and b's are non-zero and positive, then the classical internal energy of the assembly is nk_BT per system.

(b) Show that if only l of the a's and m of the b's are different from zero, the classical value of the internal energy is $[(l+m)/2]k_BT$ per system. This is one statement of the *classical equipartition theorem.*

2.15 Suppose, in contrast to statistical assumption I in Eq. (1.41), that the summand in Eq. (2.2) were to be weighted with some probability $P(m)$, where $0 \leq P(m) \leq 1$, with $P(1/2) = 1$ and $P'(1/2) = 0$. Show the entropy is still given by Eq. (2.12), and the argument in Eqs. (2.13)–(2.14) is unchanged. Discuss the implications of this observation.

3.1 Show that the second of Eqs. (3.4) follows from the first.

3.2 Use the analysis in Eqs. (2.119)–(2.124) to convert the sum to an integral, and verify Eq. (3.39).

3.3 Use Eq. (3.43) to show that the electronic partition function in Eq. (3.41) contributes only a constant to the energy, and thus makes a vanishing contribution to the constant-volume heat capacity.

3.4 Show from Eqs. (3.43)–(3.44) that every mode in the internal partition function that has $(\text{p.f.})_{\text{mode}} \propto T^\nu$ in the high-temperature limit, will contribute νR to the molar constant-volume heat capacity.

3.5 Pick representative values of (θ_R, θ_V), and make a good numerical calculation of the molar heat capacity of a diatomic gas C_V/R in Fig. 3.4.[4]

3.6 (a) Make a good numerical calculation of the molar heat capacity of molecular H_2 obtained from Eq. (3.52) with $A = 1$ and $\rho = 2$. Express the result as a function of T/θ_R;

(b) Now repeat the calculation for a metastable assembly that retains the high-temperature ratio of $N_{\text{ortho-H}_2}/N_{\text{para-H}_2} = 3/1$ down to low temperatures [see Eq. (3.58)];

[4] Recall the arguments in Eqs. (2.51)–(2.53).

(c) Compare these results with each other and with experiment as sketched in Fig. 3.7.

3.7 Verify the high-temperature relations satisfied by the rotational partition function in Eq. (3.54).

3.8 The high-temperature limit of the rotational partition function for the homonuclear diatomic molecule in Eq. (3.55) contains a factor of ρ^2 for the nuclear spin degeneracy, where $\rho = 2\mathcal{I} + 1$. Work through the corresponding Helmholtz free energy.

(a) What is the effect of this degeneracy factor on the energy E? On the heat capacity C_V?

(b) What is the effect of this factor on the entropy S? On the chemical potential μ?

3.9 (a) Suppose the vibrational state changes, while the electronic state remains unchanged. Derive the first selection rule in Eq. (3.63) for the operator in Eq. (3.62) taken between simple-harmonic-oscillator vibrational eigenstates $\psi_n^{\rm vib}(x)$ with the energy eigenvalues given in Eq. (3.33);

(b) Use the properties of the matrix element of the spherical harmonic in Eq. (3.62) taken between the rotational eigenstates in Eq. (3.36) to show that $\Delta l = 0, \pm 1$;

(c) Use the parity of the operator $\mathbf{r}$ to show that l must change, and hence arrive at the second selection rule in Eq. (3.63).[5]

3.10 The electric dipole moment for a collection of charges is defined by $|e|\mathbf{d} = \sum_p q_p \mathbf{r}_p$ where the sum goes over all the charges. The electric dipole moment for the diatomic molecule in Fig. 3.1 is therefore

$$\mathbf{d} = Z_A\mathbf{r}_A + Z_B\mathbf{r}_B - \sum_i \mathbf{r}_i$$

(a) Show that the electric dipole moment of two neutral, spherically symmetric atoms placed at $\mathbf{r}_A$ and $\mathbf{r}_B$ vanishes;

(b) The ground-state electronic wave function in the molecule is of the form $\Psi_{\rm el}(\mathbf{r}_i;\mathbf{r})$. Define the expectation value of the last term in $\mathbf{d}$ by

$$\langle\Psi_{\rm el}|\sum_i \mathbf{r}_i\,|\Psi_{\rm el}\rangle \equiv Z_A\mathbf{r}_A + Z_B\mathbf{r}_B - \zeta\mathbf{r}$$

Hence conclude that the effective dipole moment arises from the redistri-

[5] [Schiff (1968)] and [Edmonds (1974)] are good resources for this problem.

bution of electronic charge in the molecule[6]

$$|e|\mathbf{d} = |e|\zeta\mathbf{r}$$

(c) What is the electric dipole moment of a homonuclear diatomic molecule? Give your argument.

3.11 A more accurate treatment of the internal motion of a diatomic molecule is obtained by writing $r = r_0 + x$ in the second term on the r.h.s. of Eqs. (3.25), and then expanding in x. Discuss the effects of the additional *rotation-vibration coupling*.

3.12 One way of arriving at the Born-Oppenheimer approximation for the diatomic molecule in Eq. (3.14) is to substitute a wave function ansatz of the form in Eq. (3.46) into the separated internal stationary-state Schrödinger equation, use Eq. (3.13), and then take electronic matrix elements with the wave function $\psi_{\text{el}}(\mathbf{r}_i; \mathbf{r})$. Discuss the approximations made in this approach, and indicate how one would go about estimating the size of the correction terms.

3.13 The dependence on γ, the angle of rotation about the figure axis, of the wave function for a symmetric top in Eq. (3.78) is $e^{i\kappa\gamma}$ (see [Edmonds (1974)]). In general, the quantum number κ describing the angular momentum along the figure axis is restricted by the internal symmetry of the molecule.

(a) Suppose that the molecule is unchanged under rotations of $\gamma \to \gamma + 2\pi/\sigma$ about the figure axis, where σ is a positive integer, and the wave function is required to be *periodic* under such rotations. Show this implies $\kappa = p\sigma$ where $p = 0, \pm 1, \pm 2, \cdots$;

(b) Show the modification of the partition function in Eq. (3.88) is then

$$(\text{p.f.})_{\text{top}} = \frac{\sqrt{\pi}}{\sigma}\left(\frac{8\pi^2 k_{\text{B}} T}{h^2}\right)^{3/2} (I^2 I_3)^{1/2} \qquad ; T \to \infty$$

Interpret this result.

3.14 (a) Show that the enthalpy of a perfect gas is $\mathcal{H} = (5/2)nRT$ (recall Prob. 1.8);

(b) Show that the molar constant-pressure heat capacity of a perfect gas is given by $\mathcal{C}_P = (5/2)R = \mathcal{C}_V + R$.

[6] There is a theorem that either parity or time-reversal invariance implies the vanishing of an electric dipole moment; however, here there is an external vector $\mathbf{r}$ in the electron problem arising from the fact that the two nuclei are heavy and fixed. Indeed, diatomic molecules do exhibit electric dipole moments.

3.15 (a) Directly obtain the result for the induced magnetic moment for spin-1/2 in the first of Eqs. (3.146) and Eq. (3.149) by explicitly carrying out the sums in Eqs. (3.136) and (3.142);

(b) The experimental value of the magnetic moment μ in Eq. (3.135) is defined with respect to J, and it can have either sign. Show that $\langle\mu_F\rangle$ in Eq. (3.146) is even in μ, and hence one can employ $|\mu|$ in discussing it.

3.16 Include a contribution of $-\alpha F^2/2$ in the hamiltonian in Eq. (3.117), assume a uniform polarizability α, with no angle or momentum dependence, and show that the analysis of the induced moment arising from the orientation of the dipoles in Eqs. (3.123)–(3.131) is unaffected.

3.17 Consider two charges $\pm q$ connected by a spring and originally unseparated. Now apply an electric field $\mathbf{E}$, which will stretch the spring and create a dipole $\mathbf{d}$. The polarizability α of this system is defined by $\mathbf{d} = \alpha\mathbf{E}$ where α is a constant (Fig. 9.1).

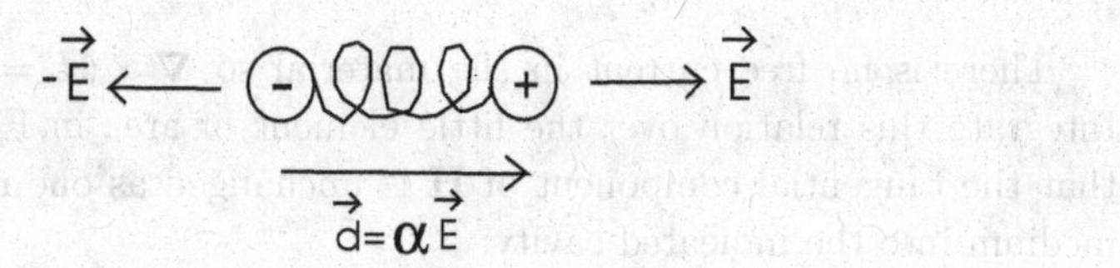

Fig. 9.1 Induced dipole in external electric field modeled with two charges on a spring.

(a) Show that the work done as the electric field is increased by an amount dE is

$$dW = Fds = (E + dE)\alpha[(E + dE) - E] \approx \alpha E dE$$

(b) Integrate this result to show the work required to create the dipole $\mathbf{d}$ is $W = \alpha\mathbf{E}^2/2$;

(c) Show the energy of the induced dipole in the external field is therefore

$$E_{\text{in}} = -\mathbf{d}_{\text{ind}} \cdot \mathbf{E} + \frac{1}{2}\alpha\mathbf{E}^2 = -\frac{1}{2}\alpha\mathbf{E}^2$$

(d) Show the polarizability of the spring is $\alpha = q^2/k$ where k is the spring constant.

3.18 Consider a paramagnetic sample placed in a solenoid, which produces a uniform magnetic field $\mathbf{H}$. A uniform magnetization per unit volume $\mathbf{M}$ is induced in the sample (Fig. 9.2).

(a) The magnetic field in the material is

$$\mathbf{B} = \mathbf{H} + 4\pi\mathbf{M}$$

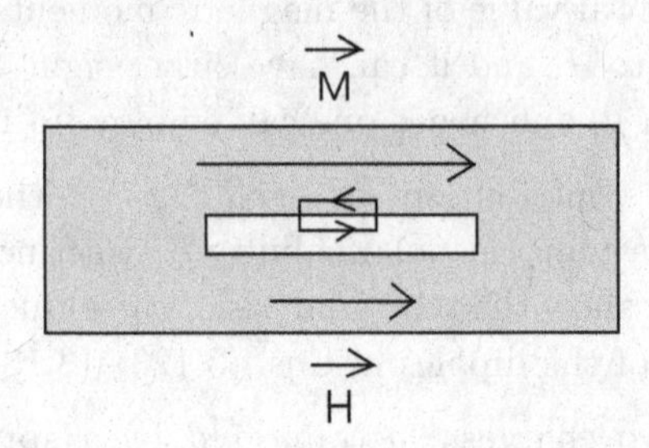

Fig. 9.2 Paramagnetic sample in a solenoid, which produces a uniform magnetic field **H**. A uniform magnetic moment per unit volume **M** is induced in the sample. Also shown is a needle-shaped cavity and a small element of area used to derive a boundary condition..

There is no free current in the material so $\nabla \times \mathbf{H} = 0$ everywhere. Integrate this relation over the little element of area in Fig. 9.2 to show that the tangential component of **H** is unchanged as one moves from the medium into the indicated cavity;

(b) The magnetic field satisfies $\nabla \cdot \mathbf{B} = 0$ everywhere. Take the divergence of the expression in (a), and define a magnetic charge by

$$\rho_m \equiv -\nabla \cdot \mathbf{M} \qquad ; \text{ magnetic charge}$$

Hence show

$$\nabla \cdot \mathbf{H} = 4\pi\rho_m \qquad ; \nabla \times \mathbf{H} = 0$$

Use the above results to establish a strict analogy between ($\mathbf{H} \rightleftharpoons \mathbf{E}$), ($\mathbf{M} \rightleftharpoons \mathbf{P}$), and ($\mathbf{B} \rightleftharpoons \mathbf{D}$). Where is ρ_m non-zero in Fig. 9.2?

(c) Now repeat the argument on the polarization in a dielectric medium to obtain the effective magnetic field at the center of a spherical cavity in the paramagnetic material

$$\mathbf{H}_{\text{eff}} = \mathbf{H} + \frac{4\pi}{3}\mathbf{M}$$

(d) Show $\mathbf{B}_{\text{eff}} = \mathbf{H}_{\text{eff}}$ in the cavity.

3.19 Consider the following chemical reaction between perfect gases

$$A_2 + B_2 \rightleftharpoons 2AB$$

(a) Show the law of mass action and equilibrium constant in this case are given by

$$\frac{N_{AA}N_{BB}}{[N_{AB}]^2} = \frac{(\text{p.f.})_{\text{AA}}(\text{p.f.})_{\text{BB}}}{[(\text{p.f.})_{\text{AB}}]^2} \equiv K_{\text{eq}}(T)$$

where $K_{\text{eq}}(T)$ is independent of V.

(b) Show the chemical potentials satisfy

$$\mu_1 + \mu_2 = 2\mu_{12}$$

3.20 (a) Use the result in Prob. 3.19, and find an expression for the equilibrium constant $K_{\text{eq}}(T)$ for the following reaction[7]

$$\text{H}_2 + \text{I}_2 \rightleftharpoons 2\text{HI}$$

in terms of the partition functions for these diatomic molecules.

(b) Compute $K_{\text{eq}}(T)$ at 500°K from the following table of constants

Table 9.1 Molecular constants.

	H_2	I_2	HI
D	38,436	12,625	24,944
ω	4,395	214.6	2,309
r	0.7417	2.667	1.604

The quantities appearing in this table are

$$D = \text{dissociation energy in cm}^{-1}$$

$$\omega = \text{oscillator angular frequency in cm}^{-1}$$

$$r = \text{internuclear spacing in Å}$$

(c) Compare with the experimental value of the equilibrium constant at that temperature

$$\log_{10}[K_{\text{eq}}]^{-1} = 2.078 \qquad ; \text{ experiment}$$

3.21 Consider all the participants in a chemical reaction to be perfect gases. Show that the volume dependence in the equilibrium constant $K(T, V)$ is just such as to allow one to re-express the law of mass action in terms of the particle densities $n_i = N_i/V$.[8]

---

[7]This problem takes a little more work, but it may be the most instructive of all.

[8]These are often referred to as the "concentrations".

3.22 (a) Assume the model of a solid used in Einstein's theory of the specific heat, and treat the vapor as a perfect gas. Derive an explicit expression for the vapor pressure $g(T)^{-1}$ in Eq. (3.232);

(b) Sketch the T-dependence of $g(T)^{-1}$. Discuss.

3.23 Go through the arguments in the method of steepest descent in detail, and verify Eqs. (3.187) and (3.190).[9]

3.24 (a) Make a model of a system on a surface site as a particle bound in a two-dimensional harmonic oscillator in the transverse directions, and in a square-well potential in the direction perpendicular to the surface. Compute the ratio of partition functions $g(T)$ in Eq. (3.246), and obtain an explicit expression for the Langmuir adsorption isotherm in Eq. (3.248);

(b) Discuss the validity of this model. How would you improve it?

3.25 Add explicit subscripts to indicate the variables that are to be held fixed in computing each of the partial derivatives in Eqs. (3.215)–(3.217).

4.1 (a) Show that the error in Eq. (4.40) is indeed of $O(1/n)$;

(b) Show that Eq. (4.41) is then Stirling's formula;

(c) Show that with the neglect of terms of $O(1/n)$, Eq. (4.41) provides an analytic continuation of Stirling's formula to non-integer n.

4.2 Define the mean-square deviation from 1 of x in Fig. 4.3 as

$$\langle (x-1)^2 \rangle \equiv \frac{\int_0^\infty (x-1)^2 I(x)dx}{\int_0^\infty I(x)dx}$$

where $I(x)$ is defined in Eq. (4.36). Take the root-mean-square deviation as a measure of the width of $I(x)$, and show that for large n.

$$\sqrt{\langle (x-1)^2 \rangle} = \frac{1}{\sqrt{n}} \quad ; \; n \to \infty$$

4.3 Suppose the members of a given set of n_k systems in Fig. 2.1 are in an excited state with energy ε_k and degeneracy ω_k. Show the number of complexions from this configuration is $\omega_k^{n_k}$.

5.1 Consider the triple sum over modes $\sum_{n_x} \sum_{n_y} \sum_{n_z} f(\mathbf{k}^2)$ where $k_i = (\pi/L)n_i$ with $i = (x, y, z)$, and $n_i = 1, 2, 3, \cdots$. Here f is an arbitrary function of $\mathbf{k}^2 = k_x^2 + k_y^2 + k_z^2$.

---

[9] *Hint*: Leave the steepest-descent result in the form of Eq. (2.85).

(a) Convert the sum to

$$\sum_{n_x} \Delta n_x \sum_{n_y} \Delta n_y \sum_{n_z} \Delta n_z \, f(\mathbf{k}^2)$$

where $\Delta n_i = 1$;

(b) Introduce $\Delta k_i \equiv (\pi/L)\Delta n_i$ and convert this sum to

$$\left(\frac{L}{\pi}\right)^3 \sum_{k_x} \Delta k_x \sum_{k_y} \Delta k_y \sum_{k_z} \Delta k_z \, f(\mathbf{k}^2)$$

(c) Now take the limit $L \to \infty$, and use the definition of the integral, to show

$$\sum_{n_x}\sum_{n_y}\sum_{n_z} f(\mathbf{k}^2) \to \left(\frac{L}{\pi}\right)^3 \int_{\text{first octant}} f(\mathbf{k}^2)\, d^3k \quad ; \; L \to \infty$$

5.2 (a) Start from Prob. 5.1(c), and show that

$$\sum_{n_x}\sum_{n_y}\sum_{n_z} f(\mathbf{k}^2) \to \left(\frac{L}{2\pi}\right)^3 \int_{\text{all k}} f(\mathbf{k}^2)\, d^3k \quad ; \; L \to \infty$$

(b) Show it follows that

$$\sum_{n_x}\sum_{n_y}\sum_{n_z} f(\mathbf{k}^2) \to \left(\frac{L}{2\pi}\right)^3 \int_0^\infty f(k^2)\, 4\pi k^2 dk \quad ; \; L \to \infty$$

5.3 A Debye temperature of $\theta_D = 1890°\text{K}$ is found to give an excellent fit to the molar specific heat of diamond (Fig. 5.5). Diamond is pure carbon with an atomic weight of 12 (so that $12\,\text{gm} = 1\,\text{mole}$), and the measured mass density of diamond is approximately $3.25\,\text{gm/cm}^3$.

(a) Express c_{av} in terms of θ_D in the Debye model, and deduce the speed of sound in diamond;

(b) Compare the value of c_{av} found in part (a) with the measured speed of sound in diamond.[10]

5.4 (a) Show that the minimum wavelength in the Debye model is

$$\lambda_{\text{min}} = \frac{c_{\text{av}}}{\nu_m} = \left[\frac{3}{4\pi}\left(\frac{N}{V}\right)\right]^{-1/3}$$

[10] One goal of this part of the problem is to get the reader to locate such measured values. (*Hint*: Try the *Handbook of Chemistry and Physics*, or the *Web*.)

(b) Evaluate this quantity for diamond (see Prob. 5.3). Discuss.

5.5 Start from the spectral weight in the lattice model in Eq. (5.65), and verify that the total number of normal modes is given by $\int_0^{\nu_M} g(\nu)d\nu = N$.

5.6 (a) Do the integral numerically, and make a good plot of the molar heat capacity in the Debye model in Eq. (5.33). Plot the result as a function of T/θ_D;

(b) Plot the molar heat capacity in the Einstein model in Eq. (2.53) as a function of T/θ_E, where $\theta_E \equiv h\nu_0/k_B$ (here ν_0 denotes the single oscillator frequency). Compare with the result in (a).

5.7 Why is it inappropriate to use $\kappa\{\varepsilon[x + u(x) + \Delta x] - \varepsilon[x + u(x)]\}$ in Eq. (5.39)?

5.8 Consider the transverse planar oscillations of a string with displacement $q(x,t)$ and fixed endpoints. The normal-mode amplitudes are $q(x) = A \sin kx$, with $k = n\pi/L$, $n = 1, 2, 3, \cdots$, and $\omega = kc$. Let $L \to \infty$.

(a) Show the spectral weight is $(1/L)g(\nu) = 2/c$;

(b) Show the normal-mode amplitude is the sum of two waves, one running to the right and one to the left;

(c) Shift the phase of one of the running waves by $\delta(k)$. Show the values of q on the far-away boundaries are altered, while the result in part (a) remains unchanged.

5.9 Consider a mass m attached to one end of a massless spring, with the other end fixed, and take $x = 0$ to mark the unextended length of the spring. Apply a constant force F to the mass which stretches the spring. The hamiltonian for this one-dimensional system is $H = p^2/2m + \kappa x^2/2 - Fx$.[11]

(a) Complete the square, and show the classical partition function is $(\text{p.f.}) = (\text{p.f.})_{\text{osc}}\, e^{F^2/2\kappa k_B T}$ where $(\text{p.f.})_{\text{osc}}$ is the result for a free oscillator with $F = 0$;

(b) Show the Helmholtz free energy of an assembly of N such springs is $A(N,T,V) = A(N,T,V)_{\text{osc}} - NF^2/2\kappa$;

(c) Show the change in the internal energy from the assembly at $F = 0$ is $\Delta E = -NF^2/2\kappa$; Derive this result by balancing forces in a system. What is the corresponding entropy change ΔS?

(d) Increase F slighty so that $F \to F + dF$. Show that the work done *by* a system is $dW = F\, dF/\kappa$. Explain the sign. Hence show that the work done

[11] An example is provided by gravity, with the mass hanging down on the spring and $F = mg$; one could then change F by changing g. It helps in visualizing this problem to then put an imaginary box around the whole system, including the mass.

by N systems when the force is increased from 0 to F is $\Delta W = NF^2/2\kappa$;

(f) Show that the isothermal heat flow when the force is increased reversibly from 0 to F, is given by $\Delta Q = T\Delta S$, where ΔS is the entropy change calculated in (c).

5.10 (a) Start from the partition function (p.f.) in Prob. 5.9(a). Show the mean displacement per system is $\langle x \rangle = k_B T(\partial/\partial F)\ln(\text{p.f.}) = F/\kappa$;

(b) Similarly, show $\langle \varepsilon \rangle = k_B T^2(\partial/\partial T)\ln(\text{p.f.}) = \langle \varepsilon \rangle_{\text{osc}} - F^2/2\kappa$;

(c) Write the Helmholtz free energy per system as $a(T, V, N, F) = -k_B T \ln(\text{p.f.})$ and work per system as dw. Show

$$\left(\frac{\partial a}{\partial F}\right)_{T,V,N} dF = -\left(\frac{F}{\kappa}\right) dF = -\langle x \rangle\, dF = -dw$$

where dw follows from Prob. 5.9(d). Hence write the analog of Eq. (3.180).

5.11 The Lennard-Jones (or "6-12") interatomic two-body potential is a useful empirical potential of the form $U(r) = 4\epsilon[(\sigma/r)^{12} - (\sigma/r)^6]$ where (ϵ, σ) are positive constants (see Fig. 5.19). Calculate the second virial coefficient $B(T)$ at high temperature with this potential.

5.12 The model interatomic two-body potential in Eq. (5.89) has a hard core at a distance σ in the relative coordinate r, surrounded by a $-c/r^6$ Van der Waal's attraction.

(a) Identify Van de Waal's parameters (a, b) in terms of the parameters (σ, c) by matching the second virial coefficients in Eqs. (5.92) and (5.93);

(b) Discuss the role of the parameters (a, b) in Van de Waal's equation of state in Eq. (5.73) in terms of the properties of the potential.

5.13 (a) Start from Eq. (5.97) and explicitly derive the linked-cluster decomposition in Eqs. (5.108) for $N = 4$;

(b) Show that the expression for the cluster of 3 systems in Eq. (5.101) is symmetric under the interchange of any pair of indices (i, j, k).

5.14 This problem uses Euler's theorem on homogeneous functions to prove that the Gibb's free energy can be expressed as $G = N\mu$ where μ is the chemical potential.

(a) Give the argument that $G(T, P, \alpha N) = \alpha G(T, P, N)$;

(b) Differentiate with respect to α, and then set $\alpha = 1$, to show

$$N\left(\frac{\partial G}{\partial N}\right)_{T,P} = N\mu = G(T, P, N)$$

This is an *important result.*

5.15 Consider the hard-sphere gas where there is a hard-core in the two-body potential that extends out to a distance σ in the relative coordinate. Show the second virial coefficient is

$$B(T) = \frac{2\pi\sigma^3}{3} = 4v_0 \qquad ; \text{ hard sphere}$$

where v_0 is the volume of each hard sphere.

5.16 Show the third virial coefficient with the Van der Waal's equation of state is

$$C(T) = b^2 \qquad ; \text{ Van der Waal's}$$

5.17 (a) Show that the third virial coefficient of a real gas is given by

$$C(T) = \frac{4g_2^2 - 2g_3g_1}{g_1^4}$$

(b) Show this reduces to

$$C(T) = B_2^2 - \frac{1}{3}B_3 = -\frac{1}{3V}\int\int\int d\tau_1 d\tau_2 d\tau_3\, f_{12}f_{23}f_{31}$$

5.18 Show the third virial coefficient for the hard-sphere gas in Prob. 5.15 is[12]

$$C(T) = \frac{5\pi^2\sigma^6}{18} \qquad ; \text{ hard sphere}$$

5.19 Verify the reduced form of Van der Waal's equation of state in Eq. (5.143).

6.1 Use the first and second laws of thermodynamics to show that the thermodynamic potential satisfies the conditions in Eqs. (6.21) and (6.22).

6.2 The pressure P of an assembly of non-localized systems, can be obtained either from Eq. (6.20) or the last of Eqs. (6.28). Equate these two expressions, and show the logarithm of the grand partition function must be of the form

$$\ln(\text{G.P.F.}) = Vf(T, \lambda) \qquad ; \text{ non-localized systems}$$

Compare with the results in Eqs. (6.30) and (6.43).

[12]The integrals here are more challenging, but do-able, and fun! (Use some solid geometry.)

6.3 If E_i represents the energy of the state i in an assembly of N systems, the grand partition function can be written

$$(\text{G.P.F.}) = \sum_N \sum_i e^{-E_i/k_\text{B}T} e^{N\mu/k_\text{B}T}$$

(a) In quantum mechanics, the trace of an operator represents the sum of the diagonal elements taken between a complete set of states $|\Psi_i\rangle$ in the appropriate Hilbert space. Show that if one uses simultaneous eigenstates of $(\hat{H}, \hat{N})$, then

$$(\text{G.P.F.}) = \text{Trace}\left\{e^{-(\hat{H}-\mu\hat{N})/k_\text{B}T}\right\}$$

(b) Show that the trace is invariant under a unitary transformation to any other complete set of states.

The expression in (a) forms the general definition of the grand partition function used as the starting point in [Fetter and Walecka (2003)].

6.4 In analogy to the grand partition function (G.P.F.) defined at externally fixed (T, μ, V), one can define a $(\mathcal{G}.\mathcal{P}.\mathcal{F})$ at fixed (T, N, P)

$$(\mathcal{G}.\mathcal{P}.\mathcal{F}) \equiv \sum_V \left(\sum_i e^{E_i/k_\text{B}T}\right) e^{-PV/k_\text{B}T} = \sum_V (\text{P.F.})_V e^{-PV/k_\text{B}T}$$
$$= \sum_V \left(\sum_E \Omega(E, V, N) e^{-E/k_\text{B}T}\right) e^{-PV/k_\text{B}T}$$

where everything is computed for a given N.

Pick out the largest term in the sum, and make the identification

$$G(T, N, P) = -k_\text{B}T \ln (\mathcal{G}.\mathcal{P}.\mathcal{F})$$

where $G(T, N, P)$ is the *Gibbs free energy* satisfying

$$dG = -SdT + VdP + \mu dN$$

6.5 Specialize the results in Prob. 6.4 to the one-dimensional case through the identification $V \to L$(length), and $P \to \tau$(tension). Use Boltzmann statistics, and consider the following problem:

N monomeric units are arranged in a straight line to form a chain molecule. Each monomeric unit is assumed to be capable of being either in an α state or a β state. In the former state, the length is a and the energy is ε_a, while in the latter the length is b and energy ε_b (see Fig. 9.3).

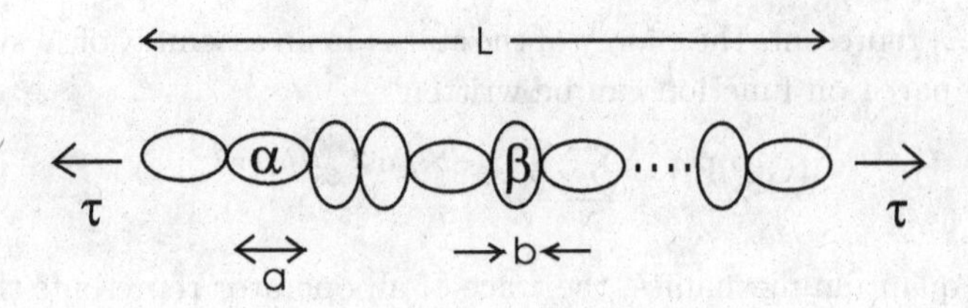

Fig. 9.3 N monomeric units arranged along a straight line to form a chain molecule. The systems have two states (α, β) with lengths (a, b) and energies $(\varepsilon_a, \varepsilon_b)$ respectively. The chain is placed under a tension τ.

(a) Compute the $(\mathcal{G}.\mathcal{P}.\mathcal{F})$ by summing over (N_α, N_β). Show

$$(\mathcal{G}.\mathcal{P}.\mathcal{F}) = \left[e^{-(\varepsilon_a - \tau a)/k_{\rm B}T} + e^{-(\varepsilon_b - \tau b)/k_{\rm B}T}\right]^N$$

$$G(T, N, \tau) = -Nk_{\rm B}T \ln\left[e^{-(\varepsilon_a - \tau a)/k_{\rm B}T} + e^{-(\varepsilon_b - \tau b)/k_{\rm B}T}\right]$$

(b) Show the equilibrium length of the chain is

$$\frac{L}{N} = \frac{ae^{-(\varepsilon_a - \tau a)/k_{\rm B}T} + be^{-(\varepsilon_b - \tau b)/k_{\rm B}T}}{e^{-(\varepsilon_a - \tau a)/k_{\rm B}T} + e^{-(\varepsilon_b - \tau b)/k_{\rm B}T}}$$

(c) Define the elasticity by $\chi \equiv [\partial(L/N)/\partial\tau]_{\tau=0}$. Show

$$\chi = \frac{(a-b)^2}{k_{\rm B}T}\frac{e^{-(\varepsilon_a + \varepsilon_b)/k_{\rm B}T}}{(e^{-\varepsilon_a/k_{\rm B}T} + e^{-\varepsilon_b/k_{\rm B}T})^2}$$

This is a model for the keratin molecules in wool.

7.1 (a) Show that with the periodic boundary conditions of Eqs. (5.24)–(5.25), the normal-mode solutions of Eq. (5.23) satisfy

$$\frac{1}{V}\int_{\rm box} e^{i(\mathbf{k}-\mathbf{k}')\cdot\mathbf{x}}\, d^3x = \delta_{\mathbf{k},\mathbf{k}'}$$

where $\delta_{\mathbf{k},\mathbf{k}'}$ is a Kronecker delta;

(b) Substitute Eqs. (7.22)–(7.23) into Eq. (7.25), do the integrals of the plane waves over the box, use the orthonormality of the unit vectors, and obtain the normal-mode expansion for electromagnetic radiation in a cavity in Eq. (7.26).[13]

7.2 Show that the total energy E can be written in the form in Eq. (7.20) for all three cases of Boltzmann, Bose-Einstein, and Fermi-Dirac statistics.

---

[13]See, for example, [Walecka (2008)].

Recall that both (V, λ) are to be kept constant in taking the partial derivative in Eq. (7.19).

7.3 (a) Compare with the Debye theory of the heat capacity of a crystal, and explain why one is always in the T^3-regime for the heat capacity of a photon gas;

(b) Explain the limiting procedure required to get from Eq. (5.35) to Eq. (7.44).

7.4 (a) Show the pressure $P(T)$ exerted by a photon gas is

$$P(T) = \left[\frac{\pi^2 k_{\rm B}^4}{45(\hbar c)^3}\right] T^4$$

(b) Put in numbers and compute P/T^4 in dynes/cm^2 $^\circ$K^4, and also in atm/$^\circ$K^4. What is P at 300 $^\circ$K? At 3000°K? At 3×10^6 $^\circ$K?

7.5 Use Fig. 5.6 to explain the exponential fall-off in Eq. (7.42).

7.6 Show that there is no Bose-Einstein condensation at any finite temperture for a two-dimensional ideal Bose gas.[14]

7.7 (a) Start from Eqs. (7.60), and explain why the curve in Fig. 7.7 goes to 3/2 at high T;

(b) Repeat for Eqs. (7.108) and Fig. 7.11.

7.8 (a) Use the density in Eq. (7.97) appropriate to the spin-zero boson ^{4}He with mass $m_{\rm He} = 6.64 \times 10^{-24}$ gm. Invert the last of Eqs. (7.60) numerically to find $\mu(N/V, T)$;

(b) Use the results from part (a), and make a good plot of the two curves in Fig. 7.6.

7.9 Define the mean value of k^n in the ground state of a Fermi gas by $\langle k^n \rangle_{\rm FG} \equiv \int_0^{k_{\rm F}} k^n d^3k / \int_0^{k_{\rm F}} d^3k$. Compute this mean value for $n = 2, 4, 6$.

7.10 Consider a non-interacting, non-relativistic Fermi gas of spin-1/2 particles in its ground state at a given particle density $n = N/V$ (electrons in a metal, neutrons in a neutron star, *etc.*). The particles have mass m and magnetic moment μ_0. A uniform magnetic field $\mathbf{B}$ is applied. Treat the assembly as two separate Fermi gases, one with magnetic moment aligned with the field, and one with magnetic moment opposed. Parameterize the number of particles of each type as

$$N_\uparrow = \frac{N}{2}(1 + \delta) \qquad ; N_\downarrow = \frac{N}{2}(1 - \delta)$$

[14]Compare Probs. 7.22–7.23.

(a) What is the contribution to the energy of the assembly coming from the interaction of the spin magnetic moments with the magnetic field for a given δ?

(b) The kinetic energy of the assembly must increase to achieve a configuration with finite δ. Let ε_F^0 be the value of the Fermi energy with $\delta = 0$. Show the increase in kinetic energy is given to order δ^2 by[15]

$$\Delta E = \frac{\delta^2}{3}\varepsilon_F^0 N$$

(c) Construct the total change in the energy as a sum of the two contributions in (a) and (b). Minimize with respect to δ to find the new ground state. Show

$$\delta = \frac{3\mu_0 B}{2\varepsilon_F^0}$$

(d) The magnetic spin susceptibility is defined in terms of the magnetic dipole moment per unit volume $\mathbf{M}$ according to $\mathbf{M} = \kappa_m \mathbf{B}$. Hence rederive the expression for the *Pauli paramagnetic spin susceptibility at zero temperature*

$$\kappa_{\text{Pauli}}(0) = \frac{3\mu_0^2}{2\varepsilon_F^0}\left(\frac{N}{V}\right)$$

7.11 Nuclear matter is a hypothetical material of uniform density filling a big box with periodic boundary conditions. It consists of four types of nucleons ($n\uparrow, n\downarrow, p\uparrow, p\downarrow$), with a baryon number of $B \equiv N + Z$ where N is the total number of neutrons and Z is the number of protons. The Coulomb interaction is turned off and nuclear matter is assumed to form a degenerate non-relativistic Fermi gas.[16]

(a) Show the baryon density of nuclear matter is related to the Fermi momentum $\hbar k_F$ by

$$\frac{B}{V} = \frac{2k_F^3}{3\pi^2} \quad ; \text{ nuclear matter}$$

(b) The observed Fermi wavenumber of nuclear matter, inferred from measurements of proton densities through electron scattering, is

$$k_F \approx 1.42\,\text{F}^{-1} \qquad ; \; 1\,\text{F} \equiv 10^{-13}\,\text{cm}$$

[15] Use the Taylor series expansion $(1+x)^n = 1 + nx + n(n-1)x^2/2! + \cdots$, which holds for $|x| < 1$ and any n (integer or non-integer).

[16] To a first approximation, it is the substance at the center of the Pb nucleus.

Compute the Fermi energy $\varepsilon_F = \hbar^2 k_F^2/2m_p$ of nuclear matter in MeV;
(c) Compute the baryon density of nuclear matter.

7.12 Assume that the neutron and proton densities in nuclear matter are now driven apart by the Coulomb interaction of the protons. Write

$$N = \frac{B}{2}(1+\delta) \qquad ; \; Z = \frac{B}{2}(1-\delta)$$

(a) Follow the analysis in Prob. 7.10, and show that the kinetic energy of nuclear matter at fixed baryon density, to leading order in δ, is increased by an amount

$$\frac{\Delta E}{B} = \frac{\varepsilon_{F0}}{3}\delta^2$$

Here ε_{F0} is the Fermi energy for symmetric nuclear matter.
(b) The semi-empirical mass formula for the ground-state energy of nuclei contains a term referred to as the *symmetry energy*[17]

$$E_{\rm sym} = a_4 \frac{(N-Z)^2}{B}$$

Show the result in part (a) implies a symmetry energy coefficient of

$$a_4 = \frac{\varepsilon_{F0}}{3}$$

(c) Use the numerical results from Prob. 7.11 to compute a_4. Compare with a measured value of $a_4 = 23.7\,\text{MeV}$. Discuss.

7.13 Consider an ultra-relativistic Fermi gas at zero temperature. The particle energy is now given by

$$\varepsilon(k) = pc = \hbar kc \qquad ; \; k \equiv |\mathbf{k}|$$

(a) Show the energy per particle in the assembly is

$$\frac{E}{N} = \frac{3}{4}\hbar k_F c = \frac{3}{4}\hbar c \left(\frac{6\pi^2}{g_s}\right)^{1/3} \left(\frac{N}{V}\right)^{1/3}$$

(b) Show the pressure is

$$P = -\left(\frac{\partial E}{\partial V}\right)_N = \frac{1}{4}\hbar c \left(\frac{6\pi^2}{g_s}\right)^{1/3} \left(\frac{N}{V}\right)^{4/3}$$

[17] See [Walecka (2008)].

(c) Hence conclude the equation of state is

$$PV = \frac{1}{3}E$$

7.14 As a model of a *white-dwarf star*, consider an electically neutral gas composed of a uniform background of inert, fully-ionized He (α particles) and a degenerate (zero-temperature) Fermi gas of electrons.[18]

(a) Show the equation of *local hydrostatic equilibrium* is

$$\frac{1}{\rho}\frac{dP}{dr} = -\frac{4\pi G}{r^2}\int_0^r \rho(r')r'^2 dr'$$

where $P(r)$ is the pressure, $\rho(r)$ is the mass density, and G is Newton's gravitational constant. What are the boundary conditions?

(b) Write this equation in the low-density (non-relativistic electron gas) and high-density (relativistic electron gas) limits assuming an ideal Fermi assembly at zero temperature (note Prob. 7.13);

(c) Find expressions for the mass density $\rho(r)$ and the relation between the total mass M and the radius R of the star;

(d) Show that there exists a maximum mass M_{max} comparable with the solar mass $M_\odot$. Explain the physics of why this is so;

(e) Check the initial model using the typical parameters of a white dwarf: $\rho \approx 10^7\,\text{g/cm}^3 \approx 10^7 \rho_\odot$; $M \approx 10^{33}\,\text{g} \approx M_\odot$. What is the Fermi energy? Note the following results obtained by numerical integration [Landau and Lifshitz (1980)]:

$$\frac{1}{\xi^2}\frac{d}{d\xi}\left(\xi^2 \frac{df}{d\xi}\right) = -f^{3/2} \qquad ;\ f'(0) = 0 \quad ;\ f(1) = 0$$

$$\Longrightarrow \qquad f(0) = 178.2 \qquad ;\ f'(1) = -132.4$$

and

$$\frac{1}{\xi^2}\frac{d}{d\xi}\left(\xi^2 \frac{df}{d\xi}\right) = -f^{3} \qquad ;\ f'(0) = 0 \quad ;\ f(1) = 0$$

$$\Longrightarrow \qquad f(0) = 6.897 \qquad ;\ f'(1) = -2.018$$

7.15 Liquid hydrogen has a mass density of $\rho_{\text{H}_2} = .07\,\text{g/cm}^3$. Suppose that at a density of $\sim 1\,\text{g/cm}^3$ it were to go into a metallic state. Compute the Fermi energy of that metal in eV.

---

[18]See, for example, [Walecka (2008)]. This problem is a little longer, but it is well worth it. The electron gas here is treated with Thomas-Fermi theory.

7.16 When a metal is heated to a sufficiently high temperature, electrons are emitted from the metal surface and can be collected as thermionic current. Assume the electrons form a non-interacting Fermi gas, and derive the Richardson-Dushman equation for the current

$$i = \frac{4\pi e m (k_B T)^2}{h^3} e^{-W/k_B T}$$

where W is the work function for the metal (that is, the energy necessary to remove the electrons).

7.17 Prove the *Bohr-Van Leeuwen theorem*, which states that the magnetic susceptibility of an assembly of charged point particles obeying classical mechanics and classical statistics *vanishes identically*. Introduce the magnetic field by means of a vector potential so that only the kinetic energy contains the magnetic field in the form

$$T = \frac{1}{2m}\left[\mathbf{p} - \frac{e}{c}\mathbf{A}(\mathbf{x})\right]^2$$

(a) First prove the result with $H = T$;

(b) Then show that the result holds even in the presence of two-body interactions, where $H = T + \sum_{i<j\leq N} V(ij)$.

7.18 (a) Show that B_0 in Eq. (7.234) has the dimensions of a magnetic field;

(b) If ε_F^0 is the free Fermi energy and μ_0 the appropriate Bohr magneton, show the following ratio is a pure number

$$\frac{|\mu_0| B_0}{\varepsilon_F^0} = \frac{1}{3}\left(\frac{3}{2}\right)^{1/3}$$

7.19 (a) Reproduce the numerical results in Figs. 7.15 and 7.16;

(b) Extend these results in both directions.

7.20 Show by direct differentiation that the magnetization $\mathcal{M}$ for Pauli spin paramagnetism in Eq. (7.153) is given by

$$\mathcal{M} = k_B T \frac{\partial}{\partial B} \ln(\text{G.P.F.})$$

where $(\text{G.P.F.})(\mu, V, T, B)$ is given by Eq. (7.151).

7.21 The expressions for the simple harmonic oscillator wave functions

$\psi_n(\xi)$ can be found in [Schiff (1968)]

$$\psi_n(\xi) = \left(\frac{1}{\sqrt{\pi}\,2^n n!}\right)^{1/2}\left[(-1)^n e^{\xi^2}\frac{\partial^n}{\partial\xi^n}e^{-\xi^2}\right]e^{-\xi^2/2}$$

$$\int_{-\infty}^{\infty}|\psi_n(\xi)|^2 d\xi = 1$$

Plot the probability distribution in the relative coordinate $y - y_0$ [see Eq. (7.196)] for the orbits with several values of n for a charged particle in a uniform magnetic field.

7.22 Consider a two-dimensional non-relativistic Fermi gas in a large square with side L and periodic boundary conditions.

(a) Show that as $L \to \infty$, the sum over states of a function of $|\mathbf{k}|$ becomes

$$\sum_i \to g_s\left(\frac{L}{2\pi}\right)^2\int_{\text{all k}} d^2k = g_s\frac{L^2}{2\pi}\int_0^\infty k\,dk = g_s\frac{\mathcal{A}}{4\pi}\left(\frac{2m}{\hbar^2}\right)\int_0^\infty d\varepsilon$$

where $\mathcal{A} = L^2$ is the area, and $\varepsilon(k) = \hbar^2k^2/2m$;

(b) Assume the Fermi gas is at temperature $T = 0$. Show

$$\frac{N}{\mathcal{A}} = \frac{g_s k_{\rm F}^2}{4\pi} \qquad ; \quad \frac{E}{N} = \frac{1}{2}\varepsilon_{\rm F}$$

where the Fermi energy is $\varepsilon_{\rm F} = \hbar^2k_{\rm F}^2/2m$.

7.23 In a two-dimensional sample, in analogy to the surface tension,[19] the pressure P becomes the *normal force per unit length*.

(a) Show all the thermodynamic arguments in the text go through with the substitution $V \to \mathcal{A}$, where $\mathcal{A}$ is the area of the sample;

(b) Show the pressure in Prob. 7.22(b) is

$$P = -\left(\frac{\partial E}{\partial \mathcal{A}}\right)_N = \frac{2\pi}{g_s}\frac{\hbar^2}{2m}\left(\frac{N}{\mathcal{A}}\right)^2$$

(c) Show the finite temperature analogs of Eqs. (7.108) are

$$\frac{N}{\mathcal{A}} = \frac{g_s}{4\pi}\left(\frac{2m}{\hbar^2}\right)\int_0^\infty \frac{d\varepsilon}{e^{(\varepsilon-\mu)/k_{\rm B}T}+1}$$

$$P = \frac{E}{\mathcal{A}} = \frac{g_s}{4\pi}\left(\frac{2m}{\hbar^2}\right)\int_0^\infty \frac{\varepsilon\,d\varepsilon}{e^{(\varepsilon-\mu)/k_{\rm B}T}+1}$$

[19] See, for example, [Fetter and Walecka (2003a)].

7.24 A positive point charge Ze_p placed into a uniform electron gas (imposed on a uniform, positive fixed background of charge density $e_p n_0$ that makes the unperturbed system neutral) will be *screened.* The Thomas-Fermi theory of this screening is achieved as follows:[20]

(a) Show the condition of local hydrostatic equilibrium for the electrons is

$$e_p \mathbf{E} = -\frac{1}{n} \boldsymbol{\nabla} P$$

where P is the pressure, $\mathbf{E}$ is the electric field, and $n = N/V$ is the electron density;

(b) Show Poisson's equation for the electric field gives

$$\boldsymbol{\nabla} \cdot \mathbf{E} = -\boldsymbol{\nabla}^2 \Phi = 4\pi \left[Ze_p \delta^{(3)}(\mathbf{x}) - e_p(n - n_0) \right]$$

where Φ is the electrostatic potential, with $\mathbf{E} = -\boldsymbol{\nabla}\Phi$;

(c) Write $n - n_0 \equiv \delta n$, and use the last of Eqs. (7.122) for the zero-temperature pressure of the Fermi gas of electrons to show from (a) that

$$\frac{2}{3} \frac{\hbar^2}{2m} \left(3\pi^2\right)^{2/3} \frac{1}{n^{1/3}} \boldsymbol{\nabla} \delta n = e_p \boldsymbol{\nabla} \Phi$$

Since the l.h.s. is already linear in small quantities, use the first of Eqs. (7.122) to write this as

$$\frac{2}{3} \frac{\hbar^2 k_{\mathrm{F}}^2}{2m} \frac{1}{n_0} \boldsymbol{\nabla} \delta n = e_p \boldsymbol{\nabla} \Phi$$

(d) Take the divergence of this result, and use part (b) to show

$$\left(\boldsymbol{\nabla}^2 - q_{\mathrm{TF}}^2\right) \delta n(\mathbf{x}) = -Z q_{\mathrm{TF}}^2 \delta^{(3)}(\mathbf{x}) \qquad ; \; q_{\mathrm{TF}} \equiv \left(\frac{6\pi e^2 n_0}{\varepsilon_{\mathrm{F}}}\right)^{1/3}$$

(e) Show the solution to this equation gives the Thomas-Fermi result for the induced screening of a point charge in an electron gas (here $r = |\mathbf{x}|$)

$$\delta\rho_{\mathrm{TF}}(r) = -Ze_p q_{\mathrm{TF}}^2 \frac{e^{-q_{\mathrm{TF}}\, r}}{4\pi r} \qquad ; \; \delta\rho = -e_p \delta n$$

(f) Show that the integrated induced density completely screens the point charge.

[20]Problems 7.24–7.25 are long, but invaluable, and the steps are clearly laid out.

7.25 The Thomas Fermi theory of the structure of an isolated atom follows from the expressions in Prob. 7.24(b,c), written with $n_0 = 0$ and away from the origin, which will be included through a boundary condition

$$\frac{2}{3}\frac{\hbar^2}{2m}(3\pi^2)^{2/3}\frac{1}{n^{1/3}}\boldsymbol{\nabla}n = e_p\boldsymbol{\nabla}\Phi$$
$$\boldsymbol{\nabla}^2\Phi = 4\pi e_p n$$

(a) Show from the first equation that for a neutral atom the electron density and electrostatic potential are related by

$$e_p\Phi(r) = \frac{\hbar^2}{2m}[3\pi^2 n(r)]^{2/3}$$

(b) Substitute this in the second relation to arrive at

$$\frac{1}{r}\frac{\partial^2}{\partial r^2}(r\Phi) = \kappa\Phi^{3/2}$$

where (recall that a_0 is the Bohr radius)

$$\kappa \equiv \frac{4e_p}{3\pi}\left(\frac{2me_p}{\hbar^2}\right)^{3/2}$$
$$= \frac{8\sqrt{2}}{3\pi}\frac{1}{a_0^2}\left(\frac{1}{e_p/a_0}\right)^{1/2} \qquad ; \; a_0 \equiv \frac{\hbar^2}{m_e e^2}$$

(c) Go to dimensionless variables

$$\phi \equiv \frac{\Phi}{e_p/a_0} \qquad ; \; \rho \equiv \frac{r}{a_0}$$

and show that

$$\frac{1}{\rho}\frac{\partial^2}{\partial\rho^2}(\rho\phi) = \left(\frac{\phi}{b}\right)^{3/2} \qquad ; \; b \equiv \frac{1}{2}\left(\frac{3\pi}{4}\right)^{2/3}$$

(d) Introduce

$$\phi \equiv \frac{Z}{\rho}\chi(x) \qquad ; \; \rho \equiv \frac{b}{Z^{1/3}}x$$

to arrive at the following non-linear Thomas-Fermi differential equation for the shielded electrostatic potential in the atom

$$\sqrt{x}\frac{d^2}{dx^2}\chi(x) = [\chi(x)]^{3/2}$$

(e) Show that the boundary conditions $\Phi \to Ze_p/r$ as $r \to 0$, and $r\Phi \to 0$ as $r \to 0$ for the neutral atom, become

$$\chi(0) = 1 \qquad ; \ \chi(\infty) = 0$$

7.26 Integrate the non-linear differential equation in Prob. 7.25(d,e) numerically, starting at (or very near) the origin and changing the intitial slope $\chi'(0)$ until the solution vanishes for large x.[21]

This calculation represents one of the first published applications of modern computers to physics [Feynman, Metropolis, and Teller (1949)].

7.27 Assume the nuclear interactions are equivalent to a slowly varying potential $-U(r)$. Within any small volume element, assume the nucleons form a non-interacting Fermi gas filled up to an energy $-\mathcal{E}$.

(a) Show that in equilibrium $\mathcal{E}$ must be constant throughout the nucleus;

(b) Derive the Thomas-Fermi result for the baryon number density[22]

$$n(r) = \frac{2}{3\pi^2}\left(\frac{2m}{\hbar^2}\right)^{3/2} [U(r) - \mathcal{E}]^{3/2}$$

(c) Derive this result by balancing the hydrostatic force $-\boldsymbol{\nabla}P$ and the force from the potential $n\boldsymbol{\nabla}U$;

(d) Show the total baryon number and energy are given by

$$B = \int_0^R d^3r\, n(r) \qquad ; \ U(R) = \mathcal{E}$$

$$E = \int_0^R d^3r \left[\frac{3}{5}\frac{\hbar^2}{2m}\left(\frac{3\pi^2}{2}\right)^{2/3} n(r)^{5/3} - U(r)n(r)\right]$$

(e) Construct the zero-temperature thermodynamic potential $\Phi = E - \mu B$, where μ is the chemical potential. Show that if Φ is made stationary under arbitrary variations of the density $\delta n(r)$, one recovers the result in part (b). Identify μ.

This is the simplest example of *density functional theory*, a now widely-used tool for calculating the structure of many-body assemblies [Kohn (1999)].

7.28 The quanta of the sound-wave excitations in a solid can be considered to form a Bose gas of *phonons*, with a phonon energy of $\varepsilon = h\nu$, frequency cut-off of ν_m, and spectral density given by Eq. (5.29).

---

[21] See Fig. 5.11 in [Walecka (2008)].

[22] Assume equal numbers of neutrons and protons (recall Prob. 7.11).

(a) Show that the phonon chemical potential must vanish [recall Eq. (7.30)];

(b) Combine the resulting Bose-Einstein distribution function in the second of Eqs. (7.18) with the phonon energy to re-derive the Debye expressions for the energy and heat capacity of the phonon gas in Eqs. (5.33).

8.1 Show that as long as the volume V is held fixed, one can just as well use the full Helmholtz free energy $A_c \to A_c + A_g$ in computing the free energy difference in a perfect solution at a given (M_1, M_2) in Eq. (8.17).

8.2 Show the employment of Eq. (8.26) in Eq. (8.24) gives Raoult's law.

8.3 Make a good numerical plot of $\Delta\tilde{A} = \Delta A/k_{\rm B}T(N_A + N_B)$ in Fig. 8.5 for various values of the parameter $\gamma \equiv \omega Z/k_{\rm B}T$ in the first of Eqs. (8.44). Verify that the phase transition occurs at $\gamma = 2$.

8.4 Suppose one starts with (N_A, N_B) systems in the condensed phase below T_C, with a mole fraction x_0 lying between the two crosses in Fig. 8.5. The mole fractions of the two new phases are $x = 1/2 \pm \delta$.

(a) How much of each new phase will be formed?

(b) Show that the new phases indeed have a lower free energy than the original solution.

8.5 Consider a two-dimensional square lattice of N_s sites. Suppose N_p particles are placed on those sites. Assume there is a constant energy shift of $-\epsilon$ when two particles sit next to each other on the lattice (see Fig. 9.4).

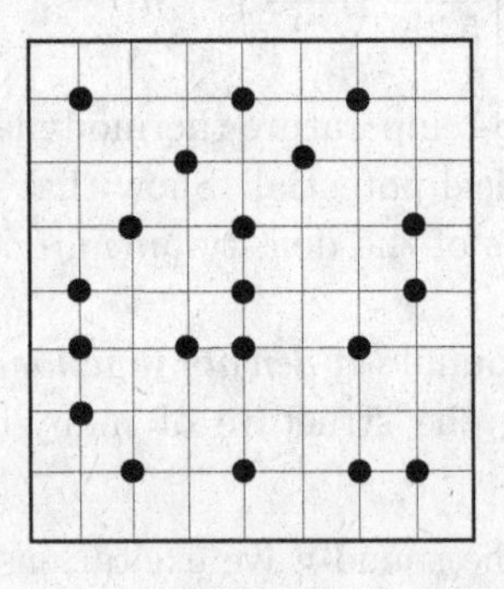

Fig. 9.4 Lattice gas. There are N_p particles on N_s sites, with an energy shift of $-\epsilon$ for each nearest-neighbor pair.

(a) Periodic boundary conditions will be assumed in both directions so that all sites are equivalent.[23] Show that a physical realization of the periodic boundary conditions is achieved by putting the lattice on the surface of a torus. Explain the relevance to Fig. 9.4;

(b) Work in the canonical ensemble. Show the configuration free energy and configuration partition function for this *lattice gas* are given by

$$A_C = -k_B T \ln (\text{P.F.})_C$$
$$(\text{P.F.})_C = \sum_{N_{pp}} g(N_s, N_p, N_{pp}) e^{\epsilon N_{pp}/k_B T}$$

Define all quantities.

8.6 (a) The Bragg-Williams approximation for the lattice gas problem formulated in Prob. 8.5 replaces N_{pp} in the sum in the configuration partition function by the value for a random distribution of particles on the lattice. Show this number is given by

$$\bar{N}_{pp} = \frac{2N_p^2}{N_s}$$

(b) Show that if $(N_s, N_p) \gg 1$, the configuration free energy is then given by

$$\frac{A_C}{k_B T N_s} = -\frac{2\epsilon}{k_B T} x^2 + [x \ln x + (1-x) \ln (1-x)]$$
$$x \equiv \frac{N_p}{N_s}$$

8.7 (a) At high temperature, the free energy in Prob. 8.6(b) is concave up as a function of the filling fraction x. Show that as soon as such a free energy develops a region that is concave down, then it possible to lower the free energy of the assembly by separating it into two phases with fractions (x_1, x_2) determined from the points of tangency of a straight line tangent to the curve at two points.

(b) How much of each phase will be formed?

(c) At what temperature will this phase separation occur in this simple lattice gas model?

8.8 Show that for temperatures $T < T_C$, the solution with the non-zero pair $\pm s$ in Fig. 8.9 yields a lower Helmholtz free energy than the solution at the origin.

---

[23] One can alternatively assume a very large lattice, so that boundary effects are unimportant.

8.9 (a) Solve Eq. (8.64) numerically for $s(T/T_C)$, and make a good plot of the molar entropy S_C/R in Fig. 8.10;

(b) Use the results from (a), compute the heat capacity from Eq. (8.69), and make a good plot of the Bragg-Williams curve in Fig. 8.11.

8.10 Solve Eq. (8.110) numerically for the exact value of k_BT_C/J in the two-dimensional Ising model, and compare with the number quoted in Table 8.1.[24]

8.11 (a) Solve Eq. (8.125) numerically, and verify the plot in Fig. 8.20 of the magnetization per particle m as a function of T/T_C for the Ising model in MFT in any number of dimensions $d \geq 2$;

(b) What is the slope of this result at $T = T_C$?

(c) Locate the exact result for m in the two-dimensional Ising model, and compare with the result in (a).

8.12 (a) Solve the one-dimensional Ising model in the presence of an external magnetic field, which produces an additional interaction

$$H' = -\mu_0 B \sum_{\kappa=1}^{N} S_\kappa$$

(b) Show there is no phase transition to a ferromagnetic state.

8.13 Demonstrate that one can achieve a physical realization of the configuration in Fig. 8.18 by placing the lattice on the surface of a *torus*.

8.14 (a) Show the following real orthogonal matrix satisfies $\underline{U}^T = \underline{U}^{-1}$

$$\underline{U} = \frac{1}{\sqrt{2}} \begin{pmatrix} 1 & 1 \\ 1 & -1 \end{pmatrix}$$

(b) Show $\underline{U}\,\underline{P}\,\underline{U}^{-1} = \underline{P}_D$, where $\underline{P}$ is the real symmetric matrix in Eq. (8.89), and $\underline{P}_D$ is the diagonal matrix in Eq. (8.93) with the real eigenvalues in Eq. (8.99).

8.15 (a) Consider the one-dimensional Ising model with periodic boundary conditions. Show

$$\sum_{\kappa=1}^{N} S_\kappa S_{\kappa+1} = N - 2N_{AB}$$

where N_{AB} is the number of unlike nearest-neighbor pairs;

---

[24]Note $J \equiv \omega/2$.

(b) Show the mean energy of the assembly of spins is[25]

$$E = -\frac{\omega}{2}N + \omega \langle N_{AB} \rangle$$

where the mean number of unlike pairs at a temperature T is given by

$$\langle N_{AB} \rangle = \frac{\sum_{\{S\}} N_{AB}\, e^{-\omega N_{AB}/k_B T}}{\sum_{\{S\}} e^{-\omega N_{AB}/k_B T}}$$

Here the sum is over all spin configurations $\{S\} = (S_1, S_2, \cdots, S_N)$;

(c) Show the Bragg-Williams approximation for $\langle N_{AB} \rangle$ is

$$\overset{o}{N}_{AB} = \frac{N}{2}$$

(d) Use the result in Eq. (8.103) to show that for large N, the exact answer for $\langle N_{AB} \rangle$ is

$$\langle N_{AB} \rangle = \frac{N}{2}\left[1 - \tanh\left(\frac{1}{x}\right)\right] \quad ; \; x \equiv \frac{2k_B T}{\omega}$$

8.16 (a) Carry out a *numerical Monte Carlo* calculation of the quantity $\langle N_{AB} \rangle$ in Prob. 8.15(b) through the following series of steps:

- Generate a random spin configuration;
- Compute N_{AB} for this configuration;
- Add the appropriate contributions in the numerator and denominator;
- Repeat

(b) Start with $N = 20$. (What is the total number of spin configurations in this case?) Compute $\langle N_{AB} \rangle / N$ for $0 < x < 40$, where $x = 2k_B T/\omega$. Be sure and include some points in the interval $0 < x < 2$. Keep as many configurations as you can in the Monte Carlo calculation;

(c) Make a plot comparing with Bragg-Williams and with the exact answer. Discuss the convergence of your numerical calculation with respect to the number of configurations employed, and with respect to N.[26]

8.17 (a) Show that the heat capacity in the one-dimensional Ising model is given for large N by

$$\frac{C_V}{Nk_B} = 2\frac{\partial}{\partial x}\frac{\langle N_{AB} \rangle}{N} = \left(\frac{1}{x^2}\right)\text{sech}^2\left(\frac{1}{x}\right)$$

[25] As usual, $\langle E \rangle \equiv E$.

[26] Problems 8.16 and 8.18 are as long as you want to make them, but they are two problems with which the students (and the author!) have had the most fun.

Include this quantity in your plot in Prob. 8.16(c);

(b) Explain why it is much more difficult to obtain the heat capacity with your Monte Carlo calculation.

8.18 An improvement of the Monte Carlo calculation is achieved with the use of the *Metropolis algorithm*, which provides increased convergence [Metropolis, Rosenbluth, Rosenbluth, Teller, and Teller (1953)]. Although the derivation goes beyond the scope of this work,[27] it is easy to implement. The summand in Prob. 8.15(b) involves a probability distribution

$$P(N_{AB}) = \frac{e^{-\omega N_{AB}/k_{\rm B}T}}{\sum_{\{S\}} e^{-\omega N_{AB}/k_{\rm B}T}}$$

The idea is to replace

$$\sum_{\{S\}} P(N_{AB}) \to \sum_{\{C\}}$$

where the sum now goes over all members of a *new ensemble of spin configurations*, with the number of members of the new ensemble in a region around N_{AB} reflecting the probability distribution $P(N_{AB})$. Then $\langle N_{AB}\rangle$, for example, is simply given by

$$\langle N_{AB}\rangle = \frac{\sum_{\{C\}} N_{AB}}{\sum_{\{C\}}}$$

The Metropolis algorithm provides a method for finding the new ensemble. After letting the assembly *thermalize* for several rounds:

(1) Start with a spin configuration $\{S\}$ with its N_{AB};
(2) Generate a new spin configuration by changing one or more spins;
(3) Compute the new $\tilde{N}_{AB}$ and the ratio

$$r = \frac{P(\tilde{N}_{AB})}{P(N_{AB})} = \frac{e^{-\omega \tilde{N}_{AB}/k_{\rm B}T}}{e^{-\omega N_{AB}/k_{\rm B}T}}$$

(4) If $r > 1$, retain the configuration, and go back to step (2);
(5) If $r < 1$, generate a random number R between $[0, 1]$. If $R \leq r$, retain the configuration; if not, discard it. Now go back to step (2).

The accumulated spin configurations then yield the desired ensemble. The numerical accuracy, of course, depends on the number of members of the ensemble.

[27] It is derived, for example, in [Walecka (2004)].

Carry out the numerical calculation in Prob. 8.16 using the Metropolis algorithm.

8.19 (a) Show that for imaginary time, the lagrangian density for the scalar field in Eq. (8.130) gives $\mathcal{L}(\mathbf{x},\tau) \equiv [\mathcal{L}(x_\mu)]_{t=-i\tau} = -\mathcal{H}(\mathbf{x},\tau)$, where $\mathcal{H}$ is the hamiltonian density;

(b) Since the total hamiltonian $H = \int d^3x\, \mathcal{H}$ is a constant of the motion, show that the exponential in Eqs. (8.131)–(8.133) is $e^{-\beta H}$.[28]

8.20 Use the two-dimensional version of Stokes theorem

$$\oint_{\mathcal{C}} \mathbf{v}\cdot d\mathbf{l} = \int_A (\boldsymbol{\nabla}\times\mathbf{v})\cdot d\mathbf{S}$$

to verify the continuum limit in Eq. (8.144) in $1+1$ dimensions ($d=2$).[29] Show that in this case σ must be chosen so that $2\sigma = 1/e_0^2a^2$.

8.21 Extend Prob. 8.20 to $3+1$ dimensions ($d=4$). Show that in this case $2\sigma = 1/e_0^2$. Note that here σ is independent of a. (This problem requires more effort.)

8.22 (a) Expand the exponential in Eq. (8.143), and derive the following result for $U(1)$ lattice gauge theory in the strong-coupling limit

$$\langle S_\Box\rangle = 2\sigma \qquad ; \sigma\to 0$$

(b) Calculate the next term and show

$$\langle S_\Box\rangle = 2\sigma - \frac{1}{2}(2\sigma)^2 \qquad ; \sigma\to 0$$

8.23 (a) Solve Eq. (8.150) numerically, and verify Eq. (8.151);

(b) Reproduce the MFT result in Fig. 8.25.

8.24 Use the results in Fig. 8.26 to add the following row to Table 8.2: [6, 0.5576, 0.33 ⋯].

8.25 Verify the thermodynamic relations in Eqs. (8.160)–(8.161) for $\beta = 1/k_{\rm B}T$.

8.26 (a) Show it is no loss of generality to assume $b_0 = 1$ in Eq. (8.157);

(b) Construct the [1,1] Padé approximant for $c_0 + c_1\beta + c_2\beta^2$.

---

[28] We remind the reader that for this final set of LGT problems, we use units where $\hbar = c = 1$, and now $\beta \equiv +1/k_{\rm B}T$.

[29] The continuum limit here is obtained by keeping the leading term in the expansion of the first of Eqs. (8.137), and assuming the rest of the series is well behaved in the limit $a\to 0$.

8.27 (a) Use the results in Prob. 8.22 to show that in the strong-coupling limit

$$E_{\Box} = 1 - \frac{1}{2}(2\sigma) + \cdots \qquad ; \text{ strong-coupling}$$

Compare with the results in Fig. 8.26 for small $\beta_{\text{eff}} = 2\sigma$;

(b) Solve Eq. (8.150) numerically for m in six dimensions ($d = 6$, $z = 5$), and calculate the MFT value of

$$E_{\Box} = (1 - m^4) \qquad ; \text{ MFT}$$

Compare with the results in Fig. 8.26 for large $\beta_{\text{eff}} = 2\sigma$, and extend the figure in β_{eff}.

8.28 Since the link variable in Eq. (8.137) is periodic in the phase ϕ_l, show the measure in Eq. (8.142) is gauge-invariant.

8.29 At high temperature and pressure, nuclear matter dissolves into a *quark-gluon plasma*. Model this as a collection of free, massless, quarks, antiquarks, and gluons (bosons). Assume degeneracies of (γ_Q, γ_G) respectively. Use $H - \mu B$ in defining the (G.P.F.), where B is the baryon number, and recall that quarks and antiquarks carry baryon number $(1/3, -1/3)$.

(a) Show the gluon chemical potential must vanish;[30]

(b) Show the parametric equation of state is given by (here $\omega_k \equiv |\mathbf{k}|c$)

$$\frac{E}{V} = \frac{\gamma_Q}{(2\pi)^3}\int d^3k\; \hbar\omega_k \left[\frac{1}{e^{(\hbar\omega_k - \mu/3)/k_{\text{B}}T} + 1} + \frac{1}{e^{(\hbar\omega_k + \mu/3)/k_{\text{B}}T} + 1}\right]$$
$$+ \frac{\gamma_G}{(2\pi)^3}\int d^3k \frac{\hbar\omega_k}{e^{\hbar\omega_k/k_{\text{B}}T} - 1}$$
$$\frac{B}{V} = \frac{1}{3}\frac{\gamma_Q}{(2\pi)^3}\int d^3k \left[\frac{1}{e^{(\hbar\omega_k - \mu/3)/k_{\text{B}}T} + 1} - \frac{1}{e^{(\hbar\omega_k + \mu/3)/k_{\text{B}}T} + 1}\right]$$

(c) What is the pressure?

8.30 In QED, the sum of leading logarithms gives a dependence of fine structure constant on momentum transfer of[31]

$$\alpha(q^2) = \frac{\alpha}{1 - (\alpha/3\pi)\ln(q^2/M_e^2)} \qquad ; \; q^2 \gg M_e^2$$

where M_e is the inverse Compton wavelength of the electron. Suppose the lattice spacing in $U(1)$ lattice gauge theory is related to the momentum

---

[30] Compare Prob. 7.28(a).

[31] See, for example, [Walecka (2010)].

transfer by $a^2 = 1/q^2$. Show that with the above expression, a singularity is reached in the coupling constant *before* one gets to the continuum limit.

In asymptotically-free theories, the sign in the denominator of the corresponding expression is reversed.

A.1 (a) Show explicitly that if the one-body distribution function has the form $f(H) = e^{-H/k_{\rm B}T}$, with $H = E$, the Boltzmann collision term in Eq. (A.18) vanishes;

(b) Then show explicitly that if $f(H) = e^{-H/k_{\rm B}T}$, with a one-body hamiltonian of the form in Eq. (A.19), the one-body distribution function in Eq. (A.20) satisfies

$$\frac{\partial f}{\partial t} = \boldsymbol{\nabla}_q U \cdot \boldsymbol{\nabla}_p f - \mathbf{v} \cdot \boldsymbol{\nabla}_q f = 0$$

Hence conclude that at a given point in phase space, f is independent of time.

A.2 It was shown in the text that the one-body Fermi distribution in Eq. (A.37) leads to a vanishing of the collision term in the Nordheim-Uehling-Uhlenbeck equation. Show that with the one-body hamiltonian of Eq. (A.19), it is then still true that

$$\frac{\partial f}{\partial t} = \boldsymbol{\nabla}_q U \cdot \boldsymbol{\nabla}_p f - \mathbf{v} \cdot \boldsymbol{\nabla}_q f = 0$$

A.3 Consider two non-relativistic, equal-mass, non-identical particles interacting through a zero-range potential $V = \lambda\delta^{(3)}(\mathbf{x})$, where the relative and center-of mass (C-M) coordinates are given by $\mathbf{x} = \mathbf{x}_1 - \mathbf{x}_2$ and $\mathbf{R} = (\mathbf{x}_1 + \mathbf{x}_2)/2$.[32]

(a) Use Fermi's Golden Rule, and show the cross section for scattering into a region $d^3p'_1$ about $\mathbf{p}'_1$ in the scattering process $\mathbf{p}_1 + \mathbf{p}_2 \to \mathbf{p}'_1 + \mathbf{p}'_2$ is[33]

$$d\sigma_{fi} = \frac{\lambda^2}{v_{12}}\frac{2\pi}{\hbar}\delta(E_1 + E_2 - E'_1 - E'_2)\frac{d^3p'_1}{(2\pi\hbar)^3}$$

(b) Verify that overall momentum is conserved, and that the quantization volume drops out of this expression.

---

[32] This problem is longer, but central to the arguments in Appendix A.
[33] See, for example, [Walecka (2008)].

(c) Write the result in (a) as $\sigma d^3p_1'/(2\pi\hbar)^3$, and hence identify σv_{12}

$$d\sigma_{fi} \equiv \sigma \frac{d^3p_1'}{(2\pi\hbar)^3}$$

$$\sigma v_{12} = \frac{2\pi\lambda^2}{\hbar}\delta(E_1 + E_2 - E_1' - E_2')$$

(d) Use the result in (a) to derive the following expression for the differential cross section in the C-M system

$$\left(\frac{d\sigma}{d\Omega}\right)_{\text{CM}} = |f|^2 \qquad ; \; f = \frac{2\mu\lambda}{4\pi\hbar^2}$$

where μ is the reduced mass. Verify that this expression has the correct dimensions.

Note that here σv_{12} depends only on $E_1 + E_2 = E_1' + E_2'$, and the differential cross section in the C-M system is independent of scattering angle.

A.4 Consider the thermodynamics of a uniform medium at temperature $T = 0$. Start from the Gibbs free energy.

(a) Show

$$\mathcal{H} = E + PV = N\mu$$

$$\mu = \left(\frac{\partial \mathcal{H}}{\partial N}\right)_P$$

where $\mathcal{H}$ is the *enthalpy* (recall Prob. 1.8), and μ is the chemical potential;

(b) Suppose the assembly is self-bound and in equilibrium with $P = 0$. Show the energy per system can be written in terms of the energy change upon insertion of a system as

$$e \equiv \frac{E}{N} = \left(\frac{\partial E}{\partial N}\right)_{P=0}$$

This is the Hugenholtz-Van Hove theorem.

A.5 (a) Assume the uniform assembly in Prob. A.4 has an energy per system of the form $E/N = e(n)$ with $n = N/V$. Suppose it is under a non-zero pressure P. Verify the equation of state is given by

$$P(n) = n^2\frac{de(n)}{dn} \qquad ; \; n = \frac{N}{V}$$

(b) Suppose one has a Fermi gas with a repulsive two-body interaction, held together under pressure. In the Hartree approximation, the single-particle potential $U(n)$, and energy per system in the assembly $e_{\rm Hartree}(n)$, are given by[34]

$$U(n) = \tilde{V}(0)n$$
$$e_{\rm Hartree}(n) = \frac{3}{5}\varepsilon_F(n) + \frac{1}{2}\tilde{V}(0)n$$

Here $\tilde{V}(0)$ is the volume integral of the two-body potential, and $\varepsilon_F(n)$ is the Fermi energy in the second of Eqs. (7.122). Show that the corresponding pressure is given by

$$P(n) = P_F(n) + \frac{1}{2}nU(n)$$

where $P_F(n)$ is the Fermi gas pressure in the third of Eqs. (7.122). In this case, a measurement of the single-particle potential $U(n)$ yields the equation of state $P(n)$.

[34]See, for example, [Fetter and Walecka (2003)]. This is another illustration of density functional theory.

Appendix A

Non-Equilibrium Statistical Mechanics

The text has focused on *equilibrium* statistical mechanics and its wide variety of applications. Little has been said about *mechanisms* and the *approach* to equilibrium. This appendix presents some simple considerations concerning non-equilbrium statistical mechanics within the framework of the Boltzmann equation and some of its immediate extensions.[1]

A.1 Boltzmann Equation

A.1.1 *One-Body Dynamics*

Consider a collection of identical, non-localized, randomly prepared systems, which in statistical equilibrium becomes the *microcanonical ensemble.* We now just do classical mechanics and assume to start with the systems are independent. Place them in their appropriate position in six dimensional phase space $\{\mathbf{p}, \mathbf{q}\}$.[2] The *distribution function* $f(\mathbf{p}, \mathbf{q}, t)$ is defined in the following manner

$$
\begin{aligned}
dN &= \text{number of systems in } d^3p\, d^3q \\
&\equiv f(\mathbf{p}, \mathbf{q}, t)\frac{d^3p\, d^3q}{(2\pi\hbar)^3}
\end{aligned}
\tag{A.1}
$$

The quantity $d^3p\, d^3q/(2\pi\hbar)^3 = d^3p\, d^3q/h^3$ counts the number of cells in this small volume in phase space. A value $f = 1$ would then imply that every cell is occupied by one particle.

The *probability* of finding a system in this region of phase space is the probability of picking a member of the ensemble at random, which is dN/N.

[1] See [Boltzmann (2011)], and a good reference here is [Stöcker and Greiner (1986)].
[2] Here $\mathbf{q} = (x, y, z)$ and $\mathbf{p} = (p_x, p_y, p_z)$.

This quantity is used to compute expectation values.

The goal now is to follow the *time evolution of the distribution function.* As a function of time, a particle at $\{\mathbf{p}_0, \mathbf{q}_0\}$ at time t_0 moves to the point $\{\mathbf{p}, \mathbf{q}\}$ at time t (see Fig. A.1). Let $d\lambda_0 = d^3p\, d^3q$ be the phase space

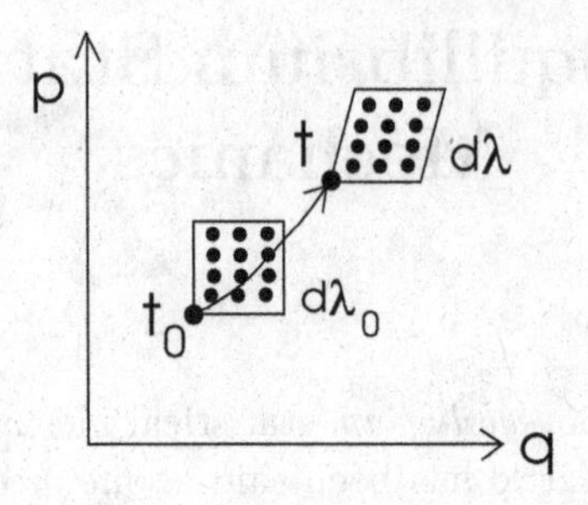

Fig. A.1 An ensemble of systems with $dN = f(\mathbf{p}, \mathbf{q}, t)\, d^3p\, d^3q/(2\pi\hbar)^3$ members in a six-dimensional phase space volume $d\lambda_0 = d^3p\, d^3q$, at a position $\{\mathbf{p}_0, \mathbf{q}_0\}$ at a time t_0, is followed along a phase trajectory to a time t.

volume at the time t_0

$$d\lambda_0 = d^3p\, d^3q \qquad ; \text{ phase-space volume at } t_0 \tag{A.2}$$

Then with hamiltonian dynamics, one has *Liouville's theorem*, which states that the phase space volume is *unchanged* along a phase trajectory[3]

$$d\lambda = d\lambda_0 \qquad ; \text{ Liouville's theorem} \tag{A.3}$$

Since the number of systems is conserved, the number of systems within this phase-space volume does not change

$$dN = dN_0 \qquad ; \text{ number conserved} \tag{A.4}$$

One concludes that *the distribution function is unchanged along a phase trajectory*

$$f[\mathbf{p}(t), \mathbf{q}(t), t] = f(\mathbf{p}_0, \mathbf{q}_0, t_0) \quad ; \text{ unchanged along phase trajectory} \tag{A.5}$$

Now write out the total differential of Eq. (A.5), and divide by dt

$$\frac{df}{dt} = \frac{\partial f}{\partial t} + \boldsymbol{\nabla}_p f \cdot \frac{d\mathbf{p}}{dt} + \boldsymbol{\nabla}_q f \cdot \frac{d\mathbf{q}}{dt} = 0 \tag{A.6}$$

[3] A proof of Liouville's theorem can be found in [Walecka (2000)], or [Fetter and Walecka (2006)].

Hamilton's equations of motion for a particle state that

$$\frac{d\mathbf{q}}{dt} = \mathbf{v} = \nabla_p H \qquad ; \text{Hamilton's equations}$$
$$\frac{d\mathbf{p}}{dt} = \mathbf{F} = -\nabla_q H \tag{A.7}$$

Equation (A.6) can then be re-written as

$$\frac{\partial f}{\partial t} = \nabla_q H \cdot \nabla_p f - \nabla_p H \cdot \nabla_q f = \{H, f\}_{\text{P.B.}} \tag{A.8}$$

where the last equality identifies the *Poisson bracket* of classical mechanics. In *equilbrium*, the time derivative of the distribution function at a given point $\{\mathbf{p}, \mathbf{q}\}$ vanishes

$$\frac{\partial f}{\partial t} = 0 \qquad ; \text{equilbrium} \tag{A.9}$$

A solution to Eqs. (A.8) and (A.9) is then provided by

$$f = f(H) \qquad ; \text{equilbrium} \tag{A.10}$$

where $H = E$ is a constant of the motion for the particle.

A.1.2 *Boltzmann Collision Term*

Consider an extension to what in equilbrium is the canonical ensemble, and include zero-range two-particle collisions. The *goal* here is to project the exact dynamics of the many-body distribution function $f(p_1, \cdots, p_{3N}, q_1, \cdots q_{3N}; t)$ down to an approximate equation for the one-body distribution function $f(\mathbf{p}, \mathbf{q}, t)$. As the ensemble evolves with time, particles are now scattered in and out of the phase-space volume $d\lambda$ (see Fig. A.2). Assume the r.h.s. of Eq. (A.8) is augmented by a collision term, so that Eq. (A.6) becomes

$$\frac{df}{dt} = \left(\frac{\partial f}{\partial t}\right)_{\text{collisions}} \tag{A.11}$$

Momentum is conserved in the collisions so that

$$\mathbf{p}_1 + \mathbf{p}_2 = \mathbf{p}_1' + \mathbf{p}_2' \tag{A.12}$$

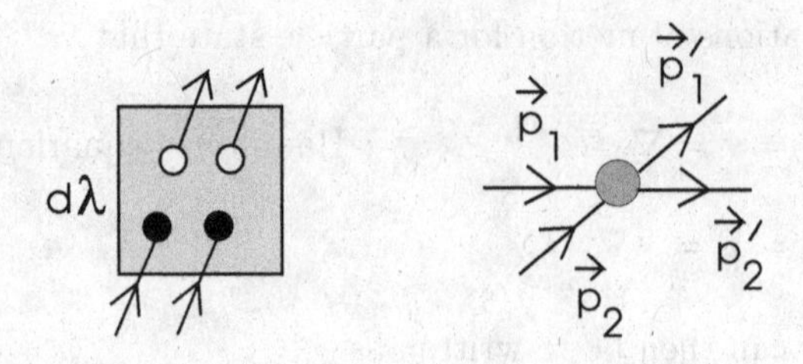

Fig. A.2 Collisions take particles in and out of the phase space volume $d\lambda$, and momentum is conserved in these collisions.

Detailed balance in a collision states that[4]

$$\text{Rate}_{i\to f} = \text{Rate}_{f\to i}$$

$$\text{or ;} \qquad \sigma\, v_{12} = \sigma'\, v_{1'2'} \tag{A.13}$$

We assume zero-range collisions where σv_{12}, while depending on $E_1 + E_2 = E_1' + E_2'$, is otherwise independent of the kinematics.[5]

The number of transitions per unit time in the direction $i \to f$ is given by

$$\left(\frac{\#\text{ of transitions}}{\text{time}}\right)_{i\to f} = (\text{incident flux}) \times \sigma \times (\#\text{ of target particle})$$

$$= (n_1 v_{12})\, \sigma\, (n_2 d^3q) \tag{A.14}$$

---

[4]It helps here to think of the quantum mechanical expressions ("Golden Rule") for the rate and cross section

$$R_{fi} = \frac{2\pi}{\hbar}\delta(E_i - E_f)|\langle f|H'|i\rangle|^2$$

$$\sigma_{fi} = \frac{R_{fi}}{\text{Flux}}$$

In these expressions:

- Energy conservation is built in;
- This is for a transition to a given final state. One still needs the number of states $d^3p'/(2\pi\hbar)^3$ in a large volume V;
- All factors of V have already been removed from these expressions;
- The hamiltonian is hermitian so that $|\langle f|H'|i\rangle|^2 = |\langle i|H'|f\rangle|^2$

However, the calculation is still classical until quantum mechanics later explicitly appears.

[5]See Prob. A.3.

where the particle density n is defined by

$$n \equiv \frac{dN}{d^3q} \qquad ; \text{ particle density} \tag{A.15}$$

Since the interaction is of zero range, everything is evaluated at the same spatial point $\mathbf{q}$, and in the same spatial volume d^3q.

The rate of change of the number of systems in the phase space volume $d\lambda$ is then given by the difference of the number of systems scattered in and the number scattered out per unit time. A combination of Eqs. (A.11)–(A.15) gives

$$\begin{aligned}
\left(\frac{\partial f}{\partial t}\right)_{\text{coll}} \frac{d^3p\,d^3q}{(2\pi\hbar)^3} = \int \cdots \int (\sigma v_{12}) \\
\times \Bigg\{\left[f(\mathbf{p}_1', \mathbf{q}; t)\frac{d^3p_1'}{(2\pi\hbar)^3}\right]\left[f(\mathbf{p}_2', \mathbf{q}; t)\frac{d^3p_2'\,d^3q}{(2\pi\hbar)^3}\right]\frac{d^3p}{(2\pi\hbar)^3}\left[\delta^{(3)}(\Delta\mathbf{p})d^3p_2\right] \\
- \left[f(\mathbf{p}, \mathbf{q}; t)\frac{d^3p}{(2\pi\hbar)^3}\right]\left[f(\mathbf{p}_2, \mathbf{q}; t)\frac{d^3p_2\,d^3q}{(2\pi\hbar)^3}\right]\frac{d^3p_1'}{(2\pi\hbar)^3}\left[\delta^{(3)}(\Delta\mathbf{p})d^3p_2'\right]\Bigg\}
\end{aligned} \tag{A.16}$$

The final factor in each line is just unity. A cancellation of common factors then gives

$$\begin{aligned}
\left(\frac{\partial f}{\partial t}\right)_{\text{collision}} &= \int \cdots \int d^3p_2 \frac{d^3p_1' d^3p_2'}{(2\pi\hbar)^6}\delta^{(3)}(\mathbf{p}+\mathbf{p}_2-\mathbf{p}_1'-\mathbf{p}_2')(\sigma v_{12}) \\
&\quad \times [f(\mathbf{p}_1', \mathbf{q}; t)f(\mathbf{p}_2', \mathbf{q}; t) - f(\mathbf{p}, \mathbf{q}; t)f(\mathbf{p}_2, \mathbf{q}; t)]
\end{aligned} \tag{A.17}$$

This can be written in an obvious shorthand notation as

$$\begin{aligned}
\left(\frac{\partial f}{\partial t}\right)_{\text{collision}} &= \int \cdots \int d^3p_2 \frac{d^3p_1' d^3p_2'}{(2\pi\hbar)^6}\delta^{(3)}(\mathbf{p}+\mathbf{p}_2-\mathbf{p}_1'-\mathbf{p}_2') \\
&\quad \times (\sigma v_{12})\,[f_1'f_2' - ff_2]
\end{aligned} \tag{A.18}$$

A few comments:

- This is the *Boltzmann collision term*;
- It is a classical result;
- It is the difference of the number of particles scattered into the phase-space volume $d\lambda$ per unit time, minus those scattered out;
- It assumes a zero-range interaction so everything occurs at the same point $\mathbf{q}$;
- The angular distribution of scattered particles is correspondingly isotropic.

- The Boltzmann collision term represents an approximate projection of the full N-body dynamics onto the space of the one-body distribution function;
- The result is a coupled, non-linear, integral contribution to the time development of $f(\mathbf{p}, \mathbf{q}, t)$.

A.1.3 *Vlasov and Boltzmann Equations*

Suppose the particles move in a one-body mean-field potential so the starting particle hamiltonian has the form

$$H = \frac{\mathbf{p}^2}{2m} + U(\mathbf{q}) \qquad \text{; one-body hamiltonian} \qquad (A.19)$$

Equations (A.8) and (A.11) then take the form

$$\frac{\partial f}{\partial t} + \mathbf{v} \cdot \boldsymbol{\nabla}_q f - \boldsymbol{\nabla}_q U \cdot \boldsymbol{\nabla}_p f = \left(\frac{\partial f}{\partial t}\right)_{\text{collision}} \qquad \text{; Vlasov eqn} \qquad (A.20)$$

This is the *Vlasov equation.* It is a non-linear, integro-differential transport equation in a mean field.

If one sets $U = 0$, the result is the *Boltzmann equation*

$$\frac{\partial f}{\partial t} + \mathbf{v} \cdot \boldsymbol{\nabla}_q f = \left(\frac{\partial f}{\partial t}\right)_{\text{collision}} \qquad \text{; Boltzmann eqn} \qquad (A.21)$$

A.1.4 *Equilibrium*

In equilibrium, as many particles must be scattered into $d\lambda$ as are scattered out, and the collision term must vanish

$$\left(\frac{\partial f}{\partial t}\right)_{\text{collision}} = 0 \qquad \text{; equilibrium} \qquad (A.22)$$

From Eq. (A.18), this will be true if

$$f_1' f_2' - f f_2 = 0 \qquad (A.23)$$

Insertion of the result in Eq. (A.10), with $H = E$, then leads to the following relation

$$f(E_1')f(E_2') = f(E)f(E_2) \qquad (A.24)$$

Energy is conserved in the collision so that

$$E + E_2 = E_1' + E_2' \qquad (A.25)$$

Since Eq. (A.24) is to hold for all (E_1', E_2'), it must be of the form

$$g(E+E_2) = f(E)f(E_2) \tag{A.26}$$

Differentiation of this relation with respect to E, and then E_2, leads to

$$f'(E)f(E_2) = f(E)f'(E_2) \tag{A.27}$$

Hence

$$\frac{f'(E)}{f(E)} = \text{constant} \tag{A.28}$$

Define the constant to be $-1/k_{\rm B}T$, so that

$$\frac{f'(E)}{f(E)} = -\frac{1}{k_{\rm B}T} \tag{A.29}$$

Then

$$f(E) = e^{-E/k_{\rm B}T} \tag{A.30}$$

With the restoration of $E = H$, this is

$$f(H) = e^{-H/k_{\rm B}T} \quad ; \text{ Boltzmann distribution} \tag{A.31}$$

We recover the Boltzmann distribution!

In *summary*:

- This is classical physics;
- If particles move with the one-body hamiltonian in Eq. (A.19), then the one-body distribution function develops in time according to Eq. (A.8)

$$\begin{aligned}\frac{\partial f}{\partial t} &= \boldsymbol{\nabla}_q U \cdot \boldsymbol{\nabla}_p f - \mathbf{v}\cdot\boldsymbol{\nabla}_q f \\ &= \{H, f\}_{\rm P.B.}\end{aligned} \tag{A.32}$$

The *equilbrium solution* to this equation, with $(\partial f/\partial t) = 0$, leads to

$$f = f(H) \tag{A.33}$$

- The Boltzmann collision term in Eq. (A.17), which includes the effect of particles scattering into and out of $d\lambda$ through zero-range, two-body collisions, augments Eq. (A.32) to

$$\frac{\partial f}{\partial t} + \mathbf{v}\cdot\boldsymbol{\nabla}_q f - \boldsymbol{\nabla}_q U \cdot \boldsymbol{\nabla}_p f = \left(\frac{\partial f}{\partial t}\right)_{\rm collision} \tag{A.34}$$

where $(\partial f/\partial t)_{\rm collision}$ is given by Eq. (A.18).

- This is the extension of the *Boltzmann* equation to the *Vlasov* equation, which includes the one-body mean field U;
- The equlibrium solution to Eq. (A.34), which now includes the additional requirement that $(\partial f/\partial t)_{\rm collision} = 0$, provides the *functional form* of $f(H)$

$$f(H) = e^{-H/k_{\rm B}T} \tag{A.35}$$

- This is quite a remarkable result. One has derived the distribution of equilibrium statistical mechanics by studying the dynamical evolution in phase space of the one-body distribution function.[6]

A.1.5 *Molecular Dynamics*

With modern computing capabilites, it is now possible to follow the full distribution function $f(p_1, \cdots, p_{3N}, q_1, \cdots, q_{3N}; t)$ in phase space with a large, finite number of systems and model interactions. These are referred to as *molecular dynamics* simulations.

A.2 Nordheim-Uehling-Uhlenbeck Equation

The extension of Nordheim, Uehling, and Uhlenbeck includes the *Pauli principle* for identical fermions. It is here that quantum mechanics first enters the dynamics. It is assumed that

> *The scattering process cannot lead to states already filled with one identical fermion per unit cell in phase space, implied by* $f = 1$ *in Eq. (A.1).*

For these particles, one makes the following replacement in the Boltzmann collision term of Eq. (A.18)

$$f_1' f_2' - f f_2 \longrightarrow f_1' f_2'(1-f)(1-f_2) - f f_2(1-f_1')(1-f_2') \tag{A.36}$$

As justification, we show that in equilibrium, where the collision term vanishes, one recovers the Fermi distribution.

Let

$$f(E) = \frac{1}{e^{\beta(\mu - E)} + 1} \equiv \frac{1}{D(E)} \tag{A.37}$$

---

[6] Boltzmann was also interested in why entropy increases and just how equilibirium is established [Boltzmann (2011)].

Then the goal is to show that

$$f(E_1')f(E_2')[1-f(E)][1-f(E_2)] = f(E)f(E_2)[1-f(E_1')][1-f(E_2')]$$
$$\text{or ;} \qquad f(E_1')f(E_2')[1-f(E)-f(E_2)] = f(E)f(E_2)[1-f(E_1')-f(E_2')] \tag{A.38}$$

With the substitution of Eq. (A.37), this becomes

$$\frac{1}{D(E_1')}\frac{1}{D(E_2')}\frac{1}{D(E)}\frac{1}{D(E_2)}[D(E)D(E_2)-D(E_2)-D(E)]$$
$$= \frac{1}{D(E_1')}\frac{1}{D(E_2')}\frac{1}{D(E)}\frac{1}{D(E_2)}[D(E_1')D(E_2')-D(E_2')-D(E_1')] \tag{A.39}$$

A cancellation of common factors leads to

$$D(E)D(E_2)-D(E_2)-D(E) = D(E_1')D(E_2')-D(E_2')-D(E_1') \tag{A.40}$$

This equality is now established by direct substitution of Eq. (A.37) and the use of energy conservation. The l.h.s. is

$$\begin{aligned}
\text{l.h.s.} &= \left[e^{\beta(\mu-E)}+1\right]\left[e^{\beta(\mu-E_2)}+1\right]-\left[e^{\beta(\mu-E_2)}+1\right]-\left[e^{\beta(\mu-E)}+1\right] \\
&= e^{\beta(2\mu-E-E_2)}-1 \\
&= e^{\beta(2\mu-E_1'-E_2')}-1 \qquad ; \; E+E_2 = E_1'+E_2' \\
&= \text{r.h.s.}
\end{aligned} \tag{A.41}$$

Thus the one-body Fermi distribution in Eq. (A.37) makes the Nordheim-Uehling-Uhlenbeck expression in Eq. (A.36) vanish, which leads to a vanishing of the new collision term. Furthermore, with the one-body hamiltonian of Eq. (A.19), it is then still true that[7]

$$\frac{\partial f}{\partial t} = \boldsymbol{\nabla}_q U \cdot \boldsymbol{\nabla}_p f - \mathbf{v}\cdot\boldsymbol{\nabla}_q f = 0 \tag{A.42}$$

A.3 Example—Heavy-Ion Reactions

As an application of Boltzmann transport theory, consider a reaction between two heavy nuclei [Stöcker and Greiner (1986)]. Here a program

---

[7]See Prob. A.2

applying the Vlasov-Uehling-Uhlenbeck model exists in the literature and is available to all [Hartnack, Kruse, and Stöcker (1993)].

In phase space, the initial configuration is as sketched in Fig. A.3.

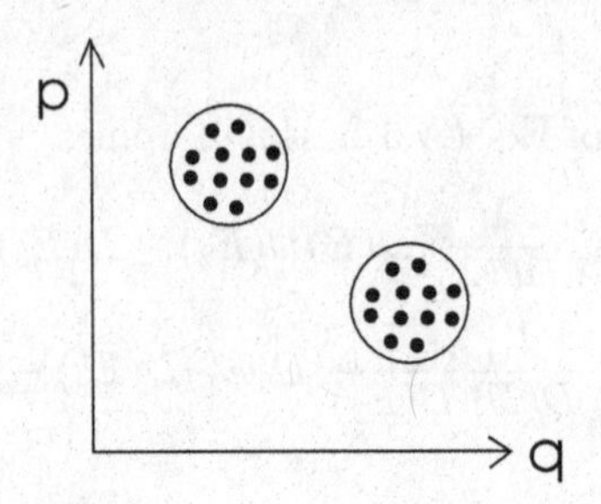

Fig. A.3 Sketch of initial phase-space configuration for two colliding heavy nuclei.

The program proceeds through the following series of steps:

(1) Choose random positions and momenta from the initial Fermi gases in the two nuclei;
(2) Follow the particles with classical dynamics and zero (short)-range two-body collisions leading to random final states, while conserving energy and momentum;
(3) Take a *statistical average* over many "events" run in parallel to determine the one-body distribution function[8]

$$f(\mathbf{p}, \mathbf{q}; t) \equiv \frac{1}{N_{\text{events}}} \sum_{\text{events}} f^{\text{event}}(\mathbf{p}, \mathbf{q}; t)$$

(4) Compute the nuclear density n at time t from f;
(5) Use a density-dependendent mean-field potential $U(n)$ in determining the one-body motion;
(6) Take an average of "runs" to get the best f;
(7) Use this f to compute particle distributions, mean values, *etc.*

The authors include inelastic N-N processes producing (π, Δ) in their program, where $\Delta(1236\,\text{MeV})$ with $(J^\pi, T) = (3/2^+, 3/2)$ is the first nucleon resonance. Locate the program and learn to run it. It's fun.

The experimental results are typically used to study the nuclear equation of state through $U(n)$.[9]

[8] As in an actual experiment.
[9] Compare Probs. A.4–A.5.

Bibliography

Amit, D. J., (2005). *Field Theory: The Renormalization Group and Critical Phenomena, 3rd ed.*, World Scientific Publishing Company, Singapore

Barmore, B., (1999). *Acta. Phys. Polon.* **B30**, 1055

Bohr, A., and Mottelson, B. R., (1975). *Nuclear Structure Vol. II, Nuclear Deformations*, W. A. Benjamin, Reading, MA

Boltzmann, L., (2011). *Lectures on Gas Theory*, Dover Publications, Mineola, NY; originally published as *Vorlesungen über Gastheorie*, Leipzig, GR (1896)

Bose, S. N., (1924). *Z. für Phys.* **26**, 178

Chandler, D., (1987). *Introduction to Modern Statistical Mechanics*, Oxford University Press, New York, NY

Colorado, (2011). *The Bose-Einstein Condensate*, http://www.colorado.edu/physics/2000/bec

Darwin, C. G., and Fowler, R. H., (1922). *Phil. Mag.* **44**, 450, 823

Davidson, N., (2003). *Statistical Mechanics*, Dover Publications, Mineola, NY; originally published by McGraw-Hill, New York, NY (1962)

Debye, P. J. W., (1912). *Ann. der Phys.* **39**, 789

Debye, P. J. W., (1988). *The Collected Papers of P. J. W. Debye*, Ox Bow Press, Woodbridge, CT

Dennison, D. M., (1927). *Proc. Roy. Soc. (London)*, **A115**, 483

Dirac, P. A. M., (1926). *Proc. Roy. Soc. (London)*, **A112**, 664

Dubach, J., (2004). Private communication, quoted in [Walecka (2004)]

Edmonds, A. R., (1974). *Angular Momentum in Quantum Mechanics*, 3rd printing, Princeton University Press, Princeton, NJ

Einstein, A., (1907). *Ann. der Phys.* **22**, 180

Einstein, A., (1924). *Sitz. der Preuss. Akad. der Wiss., Phys.-Math. Klass*, 261

Fermi, E., (1926). *Rend. Accad. Naz. Lincei* **3**, 145

Fermi, E., (1927). *Rend. Accad. Naz. Lincei* **6**, 602

Fetter, A. L., and Walecka, J. D., (2003). *Quantum Theory of Many-Particle Sys-*

tems, Dover Publications, Mineola, NY; originally published by McGraw-Hill, New York, NY (1971)

Fetter, A. L., and Walecka, J. D., (2003a). *Theoretical Mechanics of Particles and Continua*, Dover Publications, Mineola, NY; originally published by McGraw-Hill, New York, NY (1980)

Fetter, A. L., and Walecka, J. D., (2006). *Nonlinear Mechanics: A Supplement to Theoretical Mechanics of Particles and Continua*, Dover Publications, Mineola, NY

Feynman, R. P., Metropolis, N., and Teller, E., (1949). *Phys. Rev.* **75**, 1561

Fowler, R. H., and Guggenheim, E. A., (1949). *Statistical Thermodynamics, rev. ed.*, Cambridge University Press, Cambridge, UK

Gibbs, J. W., (1960). *Elementary Principles in Statistical Mechanics*, Dover Publications, Mineola, NY; originally published by Yale University Press, New Haven, CT (1902)

Gibbs, J. W., (1993). *The Scientific Papers of J. Willard Gibbs, Vol. 1: Thermodynamics*, Ox Bow Press, Woodbridge, CT

Gutiérrez, G., and Yáñez, J. M., (1997). *Am. J. Phys.* **65**, 739

Hartnack, C., Kruse, N., and Stöcker, H., (1993). *The Vlasov-Uehling-Uhlenbeck Model*, in *Computational Nuclear Physics Vol. 2*, eds. K. Lananke, J. A. Maruhn, and S. E. Koonin. Springer-Verlag, New York, NY p.128

Herzberg, G., (2008). *Molecular Spectra and Molecular Structure, Vol. I. Spectra of Diatomic Molecules*, Reitell Press, Paris, FR

Huang, K., (1987). *Statistical Mechanics, 2nd ed.*, John Wiley and Sons, New York, NY

Ising, E., (1925). *Z. Phys.* **31**, 253

Kadanoff, L. P., (2000). *Statistical Physics*, World Scientific Publishing Company, Singapore

Kittel, C., and Kroemer, H., (1980). *Thermal Physics, 2nd ed.*, W. H. Freeman and Co., New York, NY

Kittel, C., (2004). *Introduction to Solid State Physics, 8th ed.*, John Wiley and Sons, New York, NY

Kohn, W., (1999). *Rev. Mod. Phys.* **71**, 1253

Kramers, H. A., and Wannier, G. H., (1941). *Phys. Rev.* **60**, 252

Kubo, R., (1988). *Statistical Physics: An Advanced Course with Problems and Solutions*, North-Holland, Amsterdam, NL

Landau, L. D., and Lifshitz, E. M., (1980). *Statistical Physics, 3rd ed.*, Pergamon Press, London, UK

Landau, L. D., and Lifshitz, E. M., (1980a). *Quantum Mechanics Non-Relativistic Theory, 3rd ed.*, Butterworth-Heinemann, Burlington, MA

Langmuir, I., (1916). *J. Am. Chem. Soc.* **38**, 2221

Lattice 2002, eds. Edwards, R., Negele, J., and Richards, D., (2003). *Nuclear Physics* **B119** (Proc. Suppl.)

Ma, S. K., (1985). *Statistical Mechanics*, World Scientific Publishing Company, Singapore

Mayer, J. E., and Mayer, M. G., (1977). *Statistical Mechanics, 2nd ed.*, John Wiley and Sons, New York, NY

Metropolis, N., Rosenbluth, A., Rosenbluth, M., Teller, A., and Teller, E., (1953). *J. Chem. Phys.* **21**, 1087

Negele, J. W., and Ormond, H., (1988). *Quantum Many-Particle Systems*, Addison-Wesley, Reading, MA

Ohanian, H. C., (1995). *Modern Physics, 2nd ed.*, Prentice-Hall, Upper Saddle River, NJ

Onsager, L., (1944). *Phys. Rev.* **65**, 117

Pauli, W., (2000). *Statistical Mechanics: Vol. 4 of Pauli Lectures on Physics*, Dover Publications, Mineola, NY

Pauling, L., (1935). *J. Am. Chem. Soc.*, **57**, 2680

Plischke, M., and Bergersen, B., (2006). *Equilibrium Statistical Mechanics, 3rd ed.*, World Scientific Publishing Company, Singapore

Reif, F., (1965). *Fundamentals of Statistical and Thermal Physics*, McGraw-Hill, New York, NY

Rushbrooke, G. S., (1949). *Introduction to Statistical Mechanics*, Oxford University Press, Oxford, UK

Schiff, L. I., (1968). *Quantum Mechanics, 3rd ed.*, McGraw-Hill, New York, NY

Stöcker, H., and Greiner, W., (1986). *Phys. Rep.* **137**, 277

Ter Haar, D., (1966). *Elements of Thermostatics, 2nd ed.*, Holt Reinhart and Winston, New York, NY

Thomas, L. H., (1927). *Proc. Cam. Phil. Soc.* **23**, 542

Tolman, R. C., (1979). *The Principles of Statistical Mechanics*, Dover Publications, Mineola, NY; originally published by Oxford University Press, Oxford, UK (1938)

Van Vleck, J. H., (1965). *The Theory of Electric and Magnetic Susceptibilities*, Oxford University Press, Oxford, UK

Walecka, J. D., (2000). *Fundamentals of Statistical Mechanics: Manuscript and Notes of Felix Bloch, prepared by J. D. Walecka*, World Scientific Publishing Company, Singapore; originally published by Stanford University Press, Stanford, CA (1989)

Walecka, J. D., (2004). *Theoretical Nuclear and Subnuclear Physics, 2nd ed.*, World Scientific Publishing Company, Singapore; originally published by Oxford University Press, New York, NY (1995)

Walecka, J. D., (2008). *Introduction to Modern Physics: Theoretical Foundations*, World Scientific Publishing Company, Singapore

Walecka, J. D., (2010). *Advanced Modern Physics: Theoretical Foundations*, World Scientific Publishing Company, Singapore

Wannier, G. H., (1987). *Statistical Physics*, Dover Publications, Mineola, NY; originally published by John Wiley and Sons, New York, NY (1966)

Wiki (2010). *The Wikipedia*, http://en.wikipedia.org/wiki/(topic)

Wilson, E. B., Decius, J. C., and Cross, P. C., (1980). *Molecular Vibrations: The Theory of Infrared and Raman Vibrational Spectra*, Dover Publications, Mineola, NY; originally published by McGraw-Hill, New York, NY (1955)

Wilson, K., (1971). *Phys. Rev.* **B4**, 3174

Wilson, K., (1974). *Phys. Rev.* **D10**, 2445

Zemansky, M. W., (1968). *Heat and Thermodynamics: an Intermediate Textbook, 5th ed.*, McGraw-Hill, New York, NY

Index

absolute activity, 119, 123, 171, 183, 202
classical statistics, 196
quantum statistics, 196
angular momentum, 68, 77, 86
eigenvalues, 70, 87
magnetic moment, 101
principal axes, 86
angular velocity, 86
assembly, 12, 56, 96, 99, 112, 114, 127, 137, 181, 182, 184, 207, 249, 250, 277, 289, 315
asymmetric top, 89
asymptotic freedom, 283, 288, 331
atomic structure
Thomas-Fermi theory, 322
Avogadro's number, 29, 47

binomial theorem, 19, 30
black-body spectrum, 206
Bohr magneton, 101, 319
Bohr radius, 322
Bohr-Van Leeuwen theorem, 233, 319
Boltzmann distribution, 21, 25, 127, 129, 189, 191, 341
Boltzmann equation, 335, 340
Boltzmann distribution, 341
collision term, 331, 337, 339
detailed balance, 338
energy conservation, 338, 340
momentum conservation, 337
zero range, 331, 339
distribution function, 335, 336
equilibrium, 337, 340
Liouville's theorem, 336
one-body dynamics, 335
phase-space volume, 336
Boltzmann statistics, 21, 49, 195, 199, 210, 221
classical statistics, 49, 196
criterion, 49, 195, 199
distribution numbers, 27, 198, 199
grand partition function, 196
localized systems, 17
magnetic susceptibility, 110, 232
non-localized systems, 39, 48, 199
validity, 48, 49, 199
Boltzmann's constant, 16, 43, 47
Born-Oppenheimer approximation, 66, 304
Bose condensation, 207
Bose-Einstein, 217
condensates, 218
two dimensions, 315
Bose gas, 207
Bose condensation, 207
chemical potential, 209, 210, 215
energy, 208, 213
equation of state, 207, 209
gluons, 330
heat capacity, 213, 315
slope discontinuity, 214, 217
liquid ^{4}He, 207, 217
non-relativistic, 207, 209

particle number, 209, 213
phonons, 323
photons, 199
pressure, 208, 209, 213
spin degeneracy, 208
transition temperature, 209, 210, 212
two dimensions, 315
Bose-Einstein statistics, 198
bosons, 49, 74, 196, 323
Bose gas, 207
gluons, 330
liquid ^{4}He, 217
phonons, 323
photons, 201
boundary conditions, 42, 147, 310
clamped boundaries, 145
periodic, 147, 153, 155, 200, 207, 219, 272, 285, 320, 325, 326
Bragg-Williams approximation, 258, 264, 325

canonical ensemble, 127, 189
and microcanonical ensemble, 129, 130, 132, 137
applications, 141, 158, 249, 263
assembly, 128
degeneracies, 128
energies, 128, 189
Helmholtz free energy, 129
mean energy, 129
mean entropy, 132
Boltzmann distribution, 129
canonical partition function, 127, 129, 141, 159
classical limit, 132
configuration integral, 159
configuration partition function, 257
normal modes, 143
solutions, 250
classical limit, 132
configuration integral, 159, 160
configuration partition function, 257
constant-T partition function, 127
energy distribution, 135, 137, 189, 308
hamiltonian, 130, 158
Helmholtz free energy, 129, 141, 159
imperfect gases, 158
independent localized systems, 130
independent non-localized systems, 132
Ising model, 272
large N, 132, 134, 137, 163
largest term in sum, 130, 137, 250
mean energy, 132, 190
mean-square deviation, 190
mean entropy, 132
normal modes, 142
order-disorder transitions, 263
perfect gas, 133
effective Ω, 133, 135
energy distribution, 137
partition function, 134
solids, 141
canonical partition function, 127, 129
classical limit, 132, 159
configuration integral, 159
configuration partition function, 257, 263
hamiltonian, 130
imperfect gases, 159
Ising model, 272
lattice gauge theory [$U(1)$], 288
normal modes, 143
perfect solution, 250
solutions, 249
Carnot cycle, 3, 297
catalysis, 123
center-of-mass (C-M), 65, 332
chemical equilibria, 112
absolute activity, 119, 123
chemical potential, 118, 123, 229
components, 119
species, 118
chemical reaction, 114, 116, 306, 307
complexions, 115, 121
components, 8, 119

concentration, 307
entropy, 116
equilibrium constant, 116, 307
Helmholtz free energy, 117, 122
law of mass action, 114, 116, 307
mean values, 116, 122
number constraints, 115, 121
open sample, 117, 120
partition functions, 116
solid-vapor equilibrium, 120
chemical potential, 123
component, 120
Helmholtz free energy, 122
localized systems, 121
non-localized systems, 121
partition functions, 122
species, 120
vapor pressure, 122, 308
species, 8, 112, 117
surface adsorption, 123
complexions, 124
component, 123
fractional site occupation, 125
Langmuir isotherm, 125, 308
localized sites, 124
number constraints, 124
partition functions, 125
perfect gas, 125
species, 123
temperature, 117
total energy, 115
chemical potential, 8, 112, 118, 183, 185, 229, 332
Bose gas, 209
chemical reaction, 118
components, 8, 119, 120
Fermi gas, 220–222, 227
Gibbs free energy, 9, 171, 311
gluon gas, 330
Helmholtz free energy, 9
Landau diamagnetism, 240
open sample, 8
Pauli spin paramagnetism, 229
perfect gas, 47, 210
phase equilibrium, 11, 123, 301
phonon gas, 324
photon gas, 202
quark-gluon plasma, 330
solid-vapor, 123
species, 8, 118, 120
thermodynamic potential, 184
classical limit, 50, 53
entropy, 60, 62
number of complexions, 55, 56, 59
partition function, 53, 78, 132, 159, 303
weighting, 50
classical mechanics, 50
canonical momenta, 50, 68
generalized coordinates, 50
Hamilton's equations, 50, 236, 337
hamiltonian, 50
lagrangian, 50
Liouville's theorem, 50
Newton's second law, 152, 154, 234
normal coordinates, 142
normal modes, 63, 142
phase space, 50
Poisson bracket, 337
rigid-body motion, 85
classical statistics, 49, 53, 176, 196, 300
Bohr-Van Leeuwen theorem, 319
Boltzmann statistics, 49, 196
criterion, 49, 195, 300
equipartition theorem, 55, 302
validity, 48, 199
cluster, 165, 171
commutation relations, 236
complex variables, 32
analytic function, 32, 299
Cauchy-Riemann equations, 33, 299
contour integral, 32
harmonic functions, 34
Laurent series, 33
modulus, 34, 299
power series, 32
residue theorem, 33
Taylor series, 35, 316
complexion, 13, 17, 40, 56, 74, 95, 115, 124

component, 8, 119, 249
composites, 207
configuration partition function, 257, 264
  Bragg-Williams approximation, 258
  Ising solution (Z=2), 268
  largest term in sum, 264, 269
  lattice gas model, 324
  mean number of pairs, 258
  nearest neighbors, 257
  order-disorder transitions, 263
  properties, 257
coordinate representation, 237
coordination number, 255, 279
  effective, 279, 280, 290
corresponding states, 175
critical opalescence, 193
critical point, 174, 178, 180

De Haas-Van Alphen effect, 247
Debye T^3-law, 149
Debye equation, 109
Debye model, 145, 148, 324
Debye temperature, 148
deformed nuclei, 90
degeneracy, 94, 128, 130
  spin, 208
density functional theory, 323, 333
diamagnetism, 111
  Bohr-Van Leeuwen theorem, 319
  Landau diamagnetism, 233
diamond, 149, 309, 310
dielectric medium, 104
  Clausius-Mosotti relation, 108
  Debye equation, 109
  dielectric constant, 106
  electric susceptibility, 104
  polarization, 106
Dirichlet integral, 58, 301
Dulong and Petit law, 29, 148

Einstein model, 27, 141, 150, 249
Einstein's theory of specific heat, 27, 308
  energy, 28
  heat capacity, 28
  Helmholtz free energy, 28
  limiting cases, 29
  partition function, 28
elasticity
  keratin molecules, 314
  one-dimensional chain, 157, 314
electric dipole moment, 95, 97, 304
electromagnetic radiation, 200
  absolute activity, 202
  black-body spectrum, 206
  boundary conditions, 200
  chemical potential, 202
  Coulomb gauge, 200
  energy in cavity, 203, 204
    energy density, 203
    Rayleigh-Jeans law, 205
    Stefan-Boltzmann law, 207
    Wien's law, 205
  equation of state, 204
  field energy, 201
  fields, 200
  heat capacity, 204
  Maxwell's equations, 200
  normal modes, 200, 201, 314
  photon gas, 201
  photons, 201
  Planck distribution, 206
  pressure, 315
  spectral weight, 203
  ultraviolet catastrophe, 205
  vacuum, 201
  vector potential, 200
energy, 2, 21, 57, 115, 127, 128, 135, 190, 208, 310
  conservation, 2, 338, 343
  electromagnetic field, 201
  external field, 97, 229
  first law, 2, 7, 298
  internal, 2, 21, 56, 310
  normal modes, 142
  of assembly, 25, 28, 129, 143, 186, 199, 257, 275
  of mixing, 259
ensemble, 12, 27, 127, 137, 180–182, 184, 328
enthalpy, 298, 304, 332

entropy, 5, 6, 24, 39, 64, 184, 297, 311
  and number of complexions, 15, 16, 55, 60
  calorimetric, 92, 93
  canonical ensemble, 132–134, 137
  chemical equilibria, 116
  classical limit, 60
  configuration entropy, 266
  extensive, 15
  Fermi gas, 227
  grand canonical ensemble, 184
  Helmholtz free energy, 7
  microcanonical ensemble, 25, 39, 132, 134, 137
  of mixing, 259
  perfect gas, 46, 47, 137
  properties, 15
  randomness, 15
  second law, 5, 7, 298
  spectroscopic, 92, 93
  spin entropy, 277, 298
  thermodynamic potential, 184
  third law, 12, 16
  variation, 9, 299
equation of state, 175
  Bose gas, 207, 209
  Fermi gas, 219
  imperfect gases, 170, 177, 178, 188
  isotherms, 173, 175
  Landau diamagnetism, 241
  nuclear, 344
  perfect gas, 47, 122, 159, 186, 297
  photon gas, 204, 209
  quark-gluon plasma, 330
  relativistic Fermi gas, 318
  Van der Waal's, 160, 173–175, 295, 311
equilibrium, 1, 10, 299
  Gibbs criteria, 9, 10
equipartition theorem, 46, 55, 149, 205
  derivation, 302
ergodic hypothesis, 14
euclidian metric, 285
Euler's theorem, 311
extensive, 178, 192
Fermi energy, 221–223, 227
  metals, 227, 318
  nuclear matter, 317
  white dwarf, 318
Fermi gas, 219
  chemical potential, 220–222
  distribution numbers, 221, 343
  energy, 219, 333
  equation of state, 219, 220, 318
  Fermi energy, 221–223, 227, 319
  heat capacity, 224
  Landau diamagnetism, 233, 319
    Bohr-Van Leeuwen theorem, 233, 319
    chemical potential, 240, 244
    counting of states, 238, 239
    De Haas-Van Alphen effect, 247
    eigenvalue spectrum, 238
    grand partition function, 240, 243
    high-temperature limit, 240
    low-temperature limit, 242
    magnetization, 240, 243, 244
    orbits, 235
    particle in magnetic field, 234, 235, 238
    particle number, 240, 243
    susceptibility, 242
  low temperature, 224
    basic integral, 226
    chemical potential, 226
    entropy, 227
    heat capacity, 224, 227
    particle number, 226
    thermodynamic potential, 226
  metal, 315, 318
  non-relativistic, 219, 220
  nuclear matter, 315–317
    symmetry energy, 317
  particle number, 220
  Pauli spin paramagnetism, 228, 316
    Boltzmann result, 232
    chemical potential, 229
    grand partition function, 229
    hamiltonian, 229

magnetic susceptibility, 231, 232
magnetization, 230, 319
particle number, 230
thermodynamics, 232
zero temperature, 316
pressure, 219, 220, 317, 333
spin degeneracy, 219
sum to integral, 222, 320
symmetry energy, 317
thermionic current, 319
two dimensions, 247, 320
ultra-relativistic, 317
energy, 317
equation of state, 318
pressure, 317
white-dwarf star, 318
zero temperature, 222
energy, 223
Fermi energy, 224
particle density, 223, 224
pressure, 223, 224
ultra-relativistic, 317
Fermi wavenumber, 222
Fermi-Dirac statistics, 198
fermions, 49, 74, 77, 197, 218, 219, 342
electrons, 218, 239, 318, 321, 322
Landau diamagnetism, 233
low temperature, 224
nucleons, 77, 218, 316
Pauli spin paramagnetism, 228
quarks, 218
ultra-relativistic, 317
zero temperature, 222
ferromagnetism, 270
Heisenberg hamiltonian, 270
Ising model, 277, 278, 280, 326
linear chain, 277
fluctuations, 189
fundamental constants, 43, 47, 48

Gamma function, 58, 135, 211
gas constant, 29, 47
Gibbs free energy, 7, 10, 183, 311, 332
equilibrium, 10
Euler's theorem, 311
grand partition function, 313
one-dimensional chain, 313
gluons, 283, 330
Golden Rule, 331, 338
grand canonical ensemble, 181, 189, 207
and canonical ensemble, 181
and microcanonical ensemble, 181
applications, 186, 195
assembly, 184
Boltzmann statistics, 195
bosons, 199
chemical potential, 207
distribution numbers, 198, 199
Boltzmann, 198, 199
bosons, 198, 199
fermions, 198, 199
energy, 199
fermions, 218
fluctuations, 189
grand partition function, 182, 184
Landau diamagnetism, 233
largest term in sum, 183
mean particle number, 191
mean-square deviation, 191
particle distribution, 190
Pauli spin paramagnetism, 229
quantum statistics, 196
bosons, 196
fermions, 197
thermodynamic potential, 184
grand partition function, 169, 182
absolute activity, 183, 185, 186
Boltzmann statistics, 196
Bose gas, 208
chemical potential, 183
energy, 185, 186
entropy, 184
Fermi gas, 219
Gibbs free energy, 313
imperfect gases, 188
independent localized systems, 187
independent non-localized systems, 186
Landau diamagnetism, 240

one-dimensional chain, 313
particle number, 184
Pauli spin paramagnetism, 229
photon gas, 204
pressure, 184
quantum statistics, 197, 313
bosons, 197
fermions, 198
thermodynamic potential, 184
volume, 312

hadron, 283
Halley's formula, 300
Hamilton's equations, 50, 236, 337
hamiltonian, 50
canonical ensemble, 130
diatomic molecule, 65
Heisenberg, 271
Ising model, 271
many-body, 158
mean field, 340
normal modes, 142
one-body, 340
one-dimensional rotor, 91
particle in magnetic field, 236
perfect gas, 133
rigid rotor in external field, 97
scalar field, 285, 329
spin-1/2 Fermi gas, 229
stretched spring, 310
symmetric top, 86
hard-sphere gas, 312
second virial coefficient, 312
third virial coefficient, 312
harmonic oscillator, 27, 142, 201, 238
partition function, 28, 70, 143, 241
spectrum, 27, 238
wave functions, 238, 320
Hartree approximation, 212, 333
heat capacity, 28, 92, 190, 300
Bose gas, 217
Debye model, 148, 310
diatomic molecule, 72, 302
Einstein model, 27, 310
energy distribution, 190
Fermi gas, 227
Ising model, 275, 278
one-dimensional rotor, 92
order-disorder transitions, 267
perfect gas, 300, 304
phase transition
λ-point, 262
first order, 262
second order, 262
phonon gas, 324
photon gas, 207, 315
specific heat, 28
heavy-ion reactions, 343
Heisenberg hamiltonian, 270, 271
Heisenberg uncertainty principle, 51
Helmholtz equation, 42
Helmholtz free energy, 7, 39, 41, 64, 129, 141, 187
Bose gas, 207
canonical ensemble, 129, 141, 159
chemical equilibria, 117, 122
diatomic molecules, 72
Einstein model of solid, 28
equilibrium, 10
extensive, 192
imperfect gases, 159
in external field, 111
independent localized systems, 39, 187
independent non-localized systems, 41, 187
Ising model, 274
microcanonical ensemble, 39, 41, 138
order-disorder transitions, 266
perfect gas, 46, 62
photon gas, 202
solutions, 251
stretched springs, 311
hindered rotation, 90
classical partition function, 91
heat capacity, 92
one-dimensional rotor, 90
vibration, 91
Hugenholtz-Van Hove theorem, 332

ice, 95

imperfect gases, 158
absolute activity, 171
and perfect gas, 159, 176
classical theory, 158
configuration integral, 159, 160, 162, 163, 165, 177
equation of state, 177, 178
linked-clusters, 165, 168
pair-interaction, 165
summation of series, 169
critical isotherm, 174
critical point, 174, 178, 180
density, 159, 170
density fluctuations, 193
equation of state, 170, 178, 188
general, 177
interpretation, 171
grand partition function, 188
hard-sphere gas
virial coefficients, 312
Helmholtz free energy, 159
law of corresponding states, 175, 178
critical point, 178
derivation, 176
equation of state, 177, 178
experimental results, 179, 180
reduced variables, 178
linked-cluster analysis, 162, 188
phase separation, 174
pressure, 159, 170
reduced quantities, 175
two-body potential
Lennard-Jones, 176, 179, 311
Van der Waal's, 176, 311
Van der Waal's gas, 160, 173, 312
virial expansion, 160, 171
second coefficient, 160, 163, 164, 172, 179, 311, 312
third coefficient, 312
intensive, 178
Ising model, 270, 275
ferromagnetism, 277, 278, 280
hamiltonian, 271
mean field theory, 278, 281
critical temperature, 281
dimension d, 278, 281
effective coordination number, 279, 280
magnetization m, 279, 280, 282, 326
total spin, 279
Metropolis algorithm, 328
Monte Carlo calculation, 327
nearest neighbors, 271, 277, 278, 280
one-dimension, 275, 282
canonical partition function, 272, 274
energy, 275
hamiltonian, 271
heat capacity, 275, 328
Helmholtz free energy, 274
magnetic field, 326
matrix solution, 272–274, 326
Metropolis algorithm, 328
Monte Carlo calculation, 327
no phase transition, 277
number of unlike pairs, 326
periodic boundary conditions, 272, 278, 279, 326
phase transition, 277, 278, 280
two-dimensions, 277
energy, 278
heat capacity, 278
mean field theory, 281, 326
Onsager solution, 277, 281
phase transition, 278
transition temperature, 278, 326
isothermal compressibility, 191
number distribution, 191

Kronecker delta, 314

Lagrange's method of undetermined multipliers, 23, 48, 115, 121, 124, 181, 250
Landau diamagnetism, 233, 319
Bohr-Van Leeuwen theorem, 233, 319
chemical potential, 240, 244

counting of states, 238, 239
De Haas-Van Alphen effect, 247
eigenvalue spectrum, 238
gauge, 235
grand partition function, 240, 243
harmonic oscillator, 238
high-temperature limit, 240
low-temperture limit, 242
magnetization, 240, 243, 244
orbits, 235, 320
particle in magnetic field, 234
particle number, 240, 243
susceptibility, 242
Langevin function, 99, 103, 111
Langevin response, 103
Langmuir adsorption isotherm, 125, 308
Laplace's equation, 34
lattice gas model, 324, 325
Bragg-Williams approximation, 325
configuration partition function, 325
filling fraction, 325
Helmholtz fee energy, 325
periodic boundary conditions, 325
phase transition, 325
lattice gauge theory, 283
$U(1)$ lattice gauge theory, 285
continuum limit, 288, 329
coupling, 288
coupling constant, 330
dimension, 291, 293
effective energy, 293
effective temperature, 288, 293
euclidian metric, 285
four dimensions, 291, 329
gauge invariance, 287, 330
imaginary time, 285
improved approximations, 292
mean field theory, 289, 290, 293, 330
measure, 287, 330
numerical results, 291–293, 330
partition function, 287, 288
phase transition, 291, 293
six dimensions, 293, 330
strong-coupling limit, 293, 329, 330
three dimensions, 290
two dimensions, 289, 329
$U(1)$ mean field theory
critical coupling, 290
dimension, 291
effective coordination number, 290
effective magnetization, 289
gauge invariance, 289
numerical results, 291–293, 329, 330
phase transition, 291
plaquette action, 289
unit cell, 290
continuum limit, 288, 329
fields, 286
gauge invariance, 289
improved approximations, 292
lattice spacing, 286
link, 286
mean field theory, 289, 293
non-abelian theory $SU(n)$, 295
periodic boundary conditions, 286
plaquette, 286
quantum chromodynamics (QCD), 295
quantum electrodynamics (QED), 285
site, 286
strong-coupling limit, 291, 293, 329
law of corresponding states, 173, 178
law of mass action, 116, 119, 307
Legendre transformation, 298
Lennard-Jones potential, 176, 311
Liouville's theorem, 50, 336
liquid ^{4}He, 207, 217, 300
heat capacity, 315
mass, 315
mass density, 217
phase transition, 218
λ-point, 218
two fluids, 218
superfluid, 218

liquid hydrogen, 318
liquids, 301
  solutions, 249
long-range order, 264
Lorentz force, 234

magnetic dipole moment, 97, 101, 110, 315
  experimental value, 101, 305
  in quantum mechanics, 101
Maxwell's equations, 200
Maxwell-Boltzmann distribution, 300
mean field theory, 278, 280, 282, 289
mean values, 25, 116, 122, 129, 189, 253, 292, 336
mean-square deviation, 189
metastable equilibrium, 79, 94
method of steepest descent, 29, 48, 112, 131, 169
  contour integral, 32, 33
  distribution numbers, 300
  entropy, 39
  generating function, 30, 169
  Helmholtz free energy, 39, 171
  large N, 34, 37, 300
  linked-cluster analysis, 136, 169
  number of complexions, 32, 33, 37
  partition function, 39
  saddle point, 34, 35, 113, 170
  several species, 112, 114
Metropolis algorithm, 282, 328
microcanonical ensemble, 17, 27
  and canonical ensemble, 130
  applications, 41, 63, 64, 97, 112, 300
  Boltzmann distribution, 21, 25, 300
  chemical equilibria, 112
  classical limit, 50, 53
  classical statistics, 49
  distribution numbers, 21, 27, 29, 64, 98, 300
  entropy, 24, 39, 41, 64
  Helmholtz free energy, 26, 39, 41, 64, 117, 122, 129
  high-temperature limit, 53
  identical systems, 40
  independent localized systems, 17, 29, 55, 113, 124
  independent non-localized systems, 39, 41, 48, 55, 113
    validity, 48, 49, 199
  internal partition function, 63
  mean displacement, 311
  mean energy, 27, 311
  mean number, 27, 116, 122
  method of steepest descent, 29, 112
  molecular spectroscopy, 64
  most probable distribution, 19
  number of complexions, 22, 29, 40
  paramagnetic and dielectric assemblies, 97
  partition function, 26, 39, 41, 98
    classical limit, 53
    degeneracy, 26
    harmonic oscillator, 28
    internal, 63
    perfect gas, 45, 62
    permanent dipole, 98, 101
    rigid rotor, 71, 72, 90, 98
    stretched spring, 310
    surface site, 125, 308
    symmetric top, 87
  quantum statistics, 49
  summary, 137, 138
  temperature, 26, 39
  two levels, 18
molecular dynamics, 342
molecular spectroscopy, 64
  and deformed nuclei, 90
  Born-Oppenheimer approximation, 66, 304
  C-M system, 65
  conversion factors, 80
  diatomic molecules, 64, 89
    angular momentum, 68
    carbon monoxide CO, 95
    electric dipole moment, 81, 84, 95, 303
    electronic state, 72, 75
    hamiltonian, 65, 66, 70
    heat capacity, 72
    high-T limit, 78, 303

homonuclear, 76, 77, 84
hydrogen H_2, 79
moment of inertia, 69, 83
nuclear spin degeneracy, 77, 84, 303
nuclear statistics, 74, 76–78
oxygen O_2, 78
partition function, 70, 77, 307
rotation, 70
rotation-vibration coupling, 304
rotational bands, 82
selection rules, 81
small oscillations, 67
symmetry factor, 73
thermal population, 81, 83
translation, 70
typical energies, 80
vibration, 70
wave function, 74
ortho- and para-H_2, 79
heat capacity, 80, 303
metastable assembly, 79, 303
partition function, 79
statistical equilibrium, 303
statistical mixture, 79
polyatomic molecules, 85
asymmetric top, 89
eigenfunctions, 87
high-T limit, 87, 89
hindered rotation, 90–92
internal symmetry, 304
partition function, 87
spectrum, 87
spherical top, 89
symmetric top, 85
vibrations, 90
spectroscopic and calorimetric entropies, 92–94
sources of $k_B \ln \Omega_0$, 93
table of, 94
symmetry factor, 74, 91, 304
molecular structure, 75
LCAO method, 75
moment of inertia, 69, 83, 85
momentum, 13, 50, 51, 235, 337
canonical, 50
conservation, 331, 337
Monte Carlo calculation, 282, 327
multinomial theorem, 30, 131, 169

neutron stars, 315
Newton's second law, 152, 154, 234
non-equilibrium statistical mechanics, 335
Boltzmann equation, 335
heavy-ion reactions, 343
molecular dynamics, 342
Nordheim-Uehling-Uhlenbeck eqn, 342
Vlasov equation, 340
Nordheim-Uehling-Uhlenbeck eqn, 331, 342
collision term, 342
energy conservation, 343
equilibrium, 342
Fermi distribution, 343
normal modes, 63, 90, 142, 201
sum to integral, 146, 147, 308, 309
nuclear matter, 316
density, 317
equation of state, 332, 344
Fermi energy, 317
Fermi wavenumber, 316
symmetry energy, 317
nuclear structure
Thomas-Fermi theory, 323
numerical methods, 282
Metropolis algorithm, 282, 328
Monte Carlo calculation, 282, 327

order-disorder transitions, 261
λ-point, 261
Bragg-Williams approximation, 264
configuration entropy, 266
configuration partition function, 263
Helmholtz free energy, 266
Ising solution (Z=2), 268
long-range order, 264
one dimension, 270

quasi-chemical approximation, 267, 269
specific heat, 267
transition temperature, 266, 267, 269

Padé approximant, 293, 329
paramagnetic and dielectric
assemblies, 97
diamagnetism, 111
Bohr-Van Leeuwen theorem, 319
Landau diamagnetism, 233
dielectric medium, 104
capacitor, 104
Clausius-Mosotti relation, 108
Debye equation, 109
dielectric constant, 106, 109
effective field, 108
electric susceptibility, 104, 107
ferroelectrics, 111
Lorentz-Lorenz effect, 108
polarizability, 305
polarization, 104, 105, 109
ferromagnetism, 270
Ising model, 270, 277, 278, 280
field analogies, 110, 305
paramagnetic medium, 109
Curie temperature, 111
Curie's constant, 110
Curie-Weiss law, 111
effective field, 110, 306
ferromagnet, 111
field analogies, 110, 305
Langevin function, 111
magnetic charge, 306
magnetic dipole moment, 110
magnetic susceptibility, 109
solenoid, 109, 305
Pauli spin paramagnetism, 228, 316
permanent dipoles
classical gas, 97
distribution numbers, 98
electric dipole moment, 97
external field, 98
induced moment, 99, 102
Langevin function, 99, 103
Langevin response, 103
magnetic dipole moment, 97, 101
partition function, 98, 101
polarizability, 106, 305
thermodynamics, 111, 232
Helmholtz free energy, 111
induced dipole moment, 111
partition function, 26, 39, 129
canonical ensemble, 129
canonical partition function, 127, 129, 250
classical limit, 159
configuration integral, 159
configuration partition function, 257
normal modes, 143
chemical equilbria, 307
classical limit, 53, 132, 138, 139
diatomic molecule, 72
external field, 98
harmonic oscillator, 54
imperfect gases, 158, 159
one-dimensional rotor, 91
perfect gas, 54, 133
stretched spring, 310
configuration integral, 159
configuration partition function, 257, 264
constant-T, 127
degeneracy, 26, 128, 138, 308
diatomic molecule, 70
electronic, 72
homonuclear, 77, 303
rotation, 70, 303
translation, 70
vibration, 70
Einstein model of solid, 28
field theory, 284
$U(1)$ lattice gauge theory, 287
action, 285
imaginary time, 284
lattice gauge theory, 285
path integral, 285

periodic boundary conditions, 285
scalar field, 284
grand partition function, 182, 313
imperfect gases, 188
independent localized systems, 187
independent non-localized systems, 186
quantum statistics, 197
harmonic oscillator, 28, 91
high-temperature limit, 53, 71, 78, 87, 302, 303
hindered rotation, 91
imperfect gases, 159, 188
internal partition function, 63
Ising model, 272
microcanonical ensemble, 26
normal modes, 143
perfect gas, 45, 62, 134
permanent dipole, 98, 101
quantum statistics, 197
solutions, 250
stretched spring, 310
sum replaced by integral, 43, 302
symmetric top, 87
system on surface, 125, 308
zero of energy, 117
Pauli matrices, 295
Pauli principle, 76, 197, 232, 271, 342
Pauli spin paramagnetism, 228, 316
perfect gas, 3, 41, 47, 159, 176, 190
Carnot cycle, 3
chemical potential, 47, 210
energy, 46, 56
enthalpy, 304
entropy, 46, 47, 60–62
equation of state, 47, 61, 125, 133, 186, 241, 297
heat capacity, 46, 304
Helmholtz free energy, 46, 62
isothermal compressibility, 192
number of complexions, 55, 56, 59
particle number, 192
partition function, 43, 45, 134, 186
phase space volume, 57, 59, 133
solutions, 249
phase equilibria, 11, 262, 295, 301, 325
phase space, 50
and number of complexions, 60
area, 52, 302
canonical transformation, 50
cells, 51, 52, 56
classical limit, 60
dimension, 50
Dirichlet integral, 59
distribution function, 335
energy distribution, 59, 133, 137
equipartition theorem, 302
Hamilton's equations, 50
harmonic oscillator, 301
Heisenberg principle, 51
Liouville's theorem, 50, 336
of assembly, 56
particle in box, 51, 52
perfect gas, 60
phase orbit, 301
phase trajectory, 50, 336
Sommerfeld-Wilson, 52
volume, 50, 133, 336
weighting, 51–53
phase transition, 207, 260, 261
λ-point, 261
Bose gas, 217, 315
critical point, 174
first-order, 261
Ising model, 277, 278, 280
lattice gas model, 325
lattice gauge theory, 291, 294, 295
linear chain, 277
liquid ^{4}He, 217
one dimension, 270
order-disorder, 261
regular solutions, 260
amount formed, 324
transition temperature, 260, 261, 324
second-order, 261
phonon, 323
energy, 323
phonon gas, 323

chemical potential, 324
energy, 324
heat capacity, 324
photon, 201
energy, 206
polarization, 201
photon gas, 199
absolute activity, 202
chemical potential, 202
energy, 204
energy density, 203
equation of state, 204, 209
heat capacity, 204
number, 201
pressure, 204, 315
Planck distribution, 29, 206
Planck's constant, 43, 48, 206
Poisson bracket, 337
polar-spherical coordinates, 58, 301
potential
diatomic molecule, 67
interatomic, 158, 161, 163, 164, 311
Lennard-Jones, 176, 179, 311
many-body, 158, 160
one-body, 333, 340
thermodynamic, 184
two-body, 158, 161, 163–165
general form, 176
Van der Waal's, 176
zero-range, 331
pressure, 7, 10, 46, 122, 125, 159, 170, 184, 298, 318
Bose gas, 208, 209
Fermi gas, 219, 220, 317
Hartree approximation, 333
Helmholtz free energy, 7
imperfect gases, 159, 177, 178
partial, 252
perfect gas, 46
photon gas, 204, 315
pressure-volume work, 7, 298
Raoult's law, 252
thermodynamic potential, 184
Thomas-Fermi theory, 224, 318, 321, 323
two dimensions, 320
quantized
circulation, 212
flux, 212
normal modes, 142
oscillators, 142, 201
particle in box, 42
rotor, 70
quantum chromodynamics (QCD), 283, 295
asymptotic freedom, 283
color, 283
confinement, 283
gluons, 283, 330
lattice gauge theory, 283, 295
quarks, 283, 330
quantum electrodynamics (QED), 284
action, 284
coupling constant, 330
gauge invariance, 284
lagrangian density, 284
quantum statistics, 49, 76, 196, 210, 300
Boltzmann statistics, 199
Bose gas, 207
bosons, 196, 207
distribution numbers, 198, 199
electromagnetic radiation, 202
energy, 199
Fermi gas, 218
fermions, 197, 219
grand partition function, 197, 313
bosons, 197
fermions, 198
interactions, 195
occupation numbers, 196
photon gas, 199
quark-gluon plasma, 218, 330
quarks, 218, 283, 330
quasi-chemical approximation, 260, 267
Bethe-Peierls, 260
Guggenheim, 260

Raoult's law, 252, 255
Rayleigh-Jeans law, 205
reduced mass, 66

Riemann zeta function, 211

Sackur-Tetrode equation, 47
Schrödinger equation, 41, 237
screening, 321
  Thomas-Fermi theory, 321
semi-empirical mass formula, 317
solids, 141
  Debye T^3-law, 149
  Debye model, 145, 148
  Debye temperature, 148, 149, 309
  Dulong and Petit law, 148
  Einstein model, 141, 150
  lattice model, 154
    frequency cut-off, 156, 157
    spectral weight, 155, 157, 310
  longitudinal waves in rod, 151
    spectral weight, 154, 157
    wave equation, 152
  Nernst-Lindemann approximation, 158
  normal modes, 151, 153, 155
    clamped boundaries, 145
    Debye model, 145, 148, 150, 154
    Einstein model, 144
    frequency cut-off, 144, 149
    lattice model, 154
    minimum wavelength, 309
    number of, 144, 148, 154
    periodic boundary conditions, 147
    quantization, 142
    sound waves, 145, 150, 151
    spectral weight, 143, 144, 147, 151, 153, 157, 310
    sum to integral, 146, 147, 153, 308, 309
  sound waves, 150
    longitudinal, 150, 151
    transverse, 150
    velocity, 150, 309
  thermodynamics, 148
solutions, 249
  canonical partition function, 254
  localized sites
    Einstein model, 250
  mixing
    energy, 253, 256, 259
    entropy, 253, 259
    free energy, 253, 259
  mole fractions, 252
  nearest-neighbor energy, 255
  number of rearrangements, 255
  partial pressures, 252
  perfect gas, 249
  perfect solution, 249
    canonical partition function, 250
    free energy of mixing, 253
    Helmholtz free energy, 251, 324
    mixing energy, 253
    mixing entropy, 253
    Raoult's law, 252, 255, 324
  regular solutions, 254, 257
    Bragg-Williams approximation, 258
    configuration energy, 256
    configuration partition function, 257
    coordination number, 255
    Einstein model, 254
    energy of mixing, 257, 259
    entropy of mixing, 259
    free energy of mixing, 257, 259, 324
    mean number of pairs, 258
    number of nearest neighbors, 256
    phase transition, 260, 324
    quasi-chemical approximation, 260
    transition temperature, 260, 261
Sommerfeld-Wilson quantization, 52
sound waves, 145, 152
  in solids, 150
  phonons, 323
  velocity, 145, 150, 152
  wave equation, 152
species, 8, 112, 114, 117, 118

specific heat, 28, 300
heat capacity, 28
spectral weight, 143, 144, 203
spherical top, 89
spin-spin interaction, 271
spontaneous symmetry breaking, 283
standard model, 283
statistical hypotheses
assumption I, 14
assumption II, 16
statistical mechanics, 14, 16, 17, 26
assembly, 12, 128
basic problem, 2
Boltzmann statistics, 195
canonical ensemble, 127, 129, 141, 180
canonical partition function, 129, 141
classical limit, 50, 53, 132
classical statistics, 48, 49, 199
complexion, 13
ensemble, 12, 128, 180
equipartition theorem, 46, 302
ergodic hypothesis, 14
external field, 111
fluctuations, 189
Gibbs free energy, 7
grand canonical ensemble, 180, 181
grand partition function, 182
Helmholtz free energy, 7, 26, 129
high-temperature limit, 53
internal partition function, 63
mean values, 25, 116, 129, 137, 183, 189
mean-square deviation, 189
metastable assembly, 79
microcanonical ensemble, 17
non-equilibrium, 335
partition function, 26, 129
quantum statistics, 49, 196
second fundamental relation, 129
state average, 14
statistical equilibrium, 79, 80
statistical hypotheses
assumption I, 14, 302
assumption II, 16
system, 12, 63, 128
thermodynamic potential, 184
third fundamental relation, 184
time average, 13
Stefan-Boltzmann law, 207
Stirling's formula, 20, 21, 23, 30, 46, 115, 121, 124, 131, 134, 137, 187, 250, 264, 308
derivation, 137, 308
symmetric top, 85
diatomic molecule, 89
eigenfunctions, 87
high-T limit, 87, 89
internal symmetry, 304
partition function, 87
spectrum, 87
system, 12, 63, 92, 112, 114, 127, 255

thermionic current, 319
Richardson-Dushman, 319
thermodynamic potential, 184
entropy, 184
equilibrium, 185
external field, 233
Fermi gas, 226
grand partition function, 184
particle number, 184
pressure, 184
thermodynamics, 1, 2
Carnot cycle, 3, 297
chemical potential, 8, 11, 112, 119, 171, 311
closed sample, 6, 8
enthalpy, 298
entropy, 5, 6, 25, 259, 297
equilibrium, 9, 10, 298, 299
Gibbs criteria, 9, 10, 185, 298, 299
thermodynamic potential, 185
external field, 112, 232
first and second laws, 7, 25, 61, 117, 138, 139, 298
first law, 2, 297–299
free energy
Gibbs, 7, 171, 311
Helmholtz, 7, 26, 129

heat, 2, 7, 92, 298
heat capacity, 28, 92
isolated sample, 6
isothermal compressibility, 191
magnetic work, 112
Nernst heat theorem, 12
open sample, 8, 117, 120
phase equilibria, 11, 301
pressure-volume work, 7, 298
quasistatic, 4, 7, 298
reversible processes, 4, 7, 298
sample, 2, 12
second law, 3, 298, 299
state function, 3
stretched springs, 311
thermodynamic potential, 184
thermodynamic properties, 1
third law, 12, 16, 92
two dimensions, 320
work, 2, 298, 311

Thomas-Fermi theory, 224
atomic structure, 322
nuclear structure, 323
screening, 321
white dwarfs, 318

ultra-relativistic Fermi gas, 317
ultracentrifuge, 300
units, 80, 101, 200, 284, 329

vacuum, 201
Van der Waal's
attraction, 158
equation of state, 160, 173, 295
critical isotherm, 174
critical point, 174
reduced quantities, 175, 312
potential, 164, 176, 311
second virial coefficient, 160, 164
third virial coefficient, 312
vapor pressure, 122, 308
Raoult's law, 252
virial expansion, 160
virial coefficients, 160, 163, 164, 173, 311, 312
Vlasov equation, 340

white-dwarf star, 318
work function, 319

Yang-Mills theory, 283, 284